AF577431

Phytoremediation and Innovative Strategies for Specialized Remedial Applications

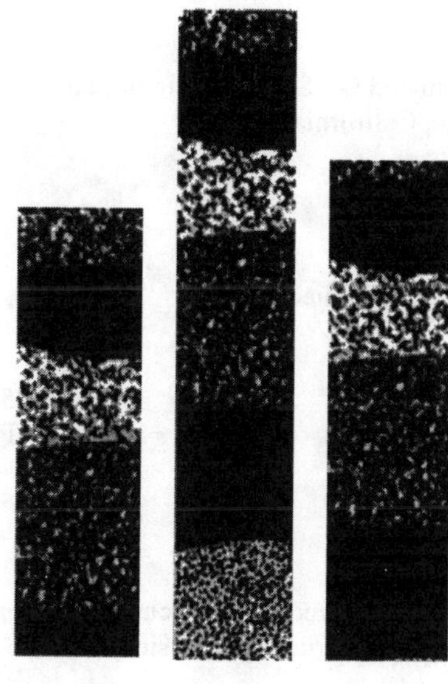

Editors

Andrea Leeson
and Bruce C. Alleman
Battelle

The Fifth International In Situ and On-Site Bioremediation Symposium

San Diego, California, April 19–22, 1999

Library of Congress Cataloging-in-Publication Data

Phytoremediation and innovative strategies for specialized remedial applications / editors,
Andrea Leeson and Bruce C. Alleman
p. cm.
Proceedings from the Fifth International In Situ and On-Site Bioremediation
Symposium, held April 19–22, 1999, in San Diego, California.
Includes bibliographical references and index.
ISBN 1-57477-079-9 (hardcover : alk. paper)
1. Phytoremediation Congresses.
I. Leeson, Andrea, 1962– . II. Alleman, Bruce C., 1957– .
III. International Symposium on In Situ and On-Site Bioremediation
(5th : 1999 : San Diego, Calif.)
TD192.75.P473 1999
628.5--dc21 99-23328
CIP

Printed in the United States of America

Battelle Press
505 King Avenue
Columbus, Ohio 43201, USA
614-424-6393 or 1-800-451-3543
Fax: 1-614-424-3819
Internet: press@battelle.org
Website: www.battelle.org/bookstore

For information on future environmental conferences, write to:

Battelle
Environmental Restoration Department, Room 10-123B
505 King Avenue
Columbus, Ohio 43201-2693
Phone: 614-424-7604
Fax: 614-424-3667
Website: www.battelle.org/conferences

CONTENTS

FOREWORD

The Fifth International In Situ and On-Site Bioremediation Symposium was held in San Diego, California, April 19–22, 1999. The program included approximately 600 platform and poster presentations, encompassing laboratory, bench-scale, and full-scale field studies being conducted worldwide on a variety of bioremediation and supporting technologies used for a wide range of contaminants.

The author of each presentation accepted for the program was invited to prepare a six-page paper, formatted according to specifications provided by the Symposium Organizing Committee. Approximately 400 such technical notes were received. The editors conducted a review of all papers. Ultimately, 389 papers were accepted for publication and assembled into the following eight volumes:

Natural Attenuation of Chlorinated Solvents, Petroleum Hydrocarbons, and Other Organic Compounds – Volume 5(1)
Engineered Approaches for In Situ Bioremediation of Chlorinated Solvent Contamination – Volume 5(2)
In Situ Bioremediation of Petroleum Hydrocarbon and Other Organic Compounds – Volume 5(3)
Bioremediation of Metals and Inorganic Compounds – Volume 5(4)
Bioreactor and Ex Situ Biological Treatment Technologies – Volume 5(5)
Phytoremediation and Innovative Strategies for Specialized Remedial Applications – Volume 5(6)
Bioremediation of Nitroaromatic and Haloaromatic Compounds – Volume 5(7)
Bioremediation Technologies for Polycyclic Aromatic Hydrocarbon Compounds – Volume 5(8)

Each volume contains comprehensive keyword and author indices to the entire set.

Many of the articles in this volume deal with phytoremediation, a site remediation strategy that has found increasing application in the past few years, with field implementations taking place at a wide variety of sites. From laboratory studies on plant uptake to full-scale phytoremediation treatment strategies, this volume covers the use of plants to treat contaminants such as hydrocarbons, metals, pesticides, perchlorate, and chlorinated solvents. In addition to the phytoremediation studies, this volume also covers specialized remediation approaches such as bioremediation of acid mine drainage, sequential anaerobic/aerobic in situ treatment, membrane bioreactors, and Fenton's reagent oxidation.

We would like to thank the Battelle staff who assembled the eight volumes and prepared them for printing. Carol Young, Lori Helsel, Loretta Bahn, Gina Melaragno, Timothy Lundgren, Tom Wilk, and Lynn Copley-Graves spent many hours on production tasks—developing the detailed format specifications sent to each author; examining each technical note to ensure that it met basic page layout requirements and making adjustments when necessary; assembling the volumes; applying headers and page numbers; compiling the tables of contents and author and keyword indices, and performing a final check of the pages before submitting them to the publisher. Joseph Sheldrick, manager of Battelle Press, provided valuable production-planning advice and coordinated with the printer; he and Gar Dingess designed the covers.

The Bioremediation Symposium is sponsored and organized by Battelle Memorial Institute, with the assistance of a number of environmental remediation organizations. In 1999, the following co-sponsors made financial contributions toward the Symposium:

Celtic Technologies
Gas Research Institute (GRI)
IT Group, Inc.
Parsons Engineering Science, Inc.
U.S. Microbics, Inc.
U.S. Naval Facilities Engineering Command
Waste Management, Inc.

Additional participating organizations assisted with distribution of information about the Symposium:

Ajou University, College of Engineering
American Petroleum Institute
Asian Institute of Technology
Conor Pacific Environmental Technologies, Inc.
Mitsubishi Corporation
National Center for Integrated Bioremediation Research & Development (University of Michigan)
U.S. Air Force Center for Environmental Excellence
U.S. Air Force Research Laboratory Air Base and Environmental Technology Division
U.S. Environmental Protection Agency
Western Region Hazardous Substance Research Center (Stanford University and Oregon State University)

The materials in these volumes represent the authors' results and interpretations. The support of the Symposium provided by Battelle, the co-sponsors, and the participating organizations should not be construed as their endorsement of the content of these volumes.

Andrea Leeson and Bruce Alleman, Battelle
1999 Bioremediation Symposium Co-Chairs

DEMONSTRATION OF A SEQUENTIAL ANAEROBIC/AEROBIC IN-SITU TREATMENT SYSTEM AT A SUPERFUND SITE

Craig C. Lizotte, P.E. (Envirogen, Canton, MA)
Michael C. Marley (XDD,LLC, Portsmouth, NH)
Scott C. Crawford (Envirogen, Canton, MA)
Annette M. Lee (XDD, LLC, Portsmouth, NH)
Robert J. Steffan Ph.D. (Envirogen, Lawrenceville, NJ)

ABSTRACT: A large-scale enhanced bioremediation demonstration has been operating at a Superfund landfill site in New England since January 1998. The demonstration uses sequential anaerobic and aerobic treatment zones to treat a wide array of chemicals present in site groundwater. The anaerobic zone is designed primarily to facilitate reductive dehalogenation of chlorinated VOCs, whereas the aerobic treatment zone facilitates the biological destruction of non-chlorinated VOCs and the immobilization of arsenic. The objective of the project is to demonstrate enhanced biodegradation activity through the treatment zone, and to calculate compound specific biodegradation rates for use in the design of a full-scale treatment system that will prevent off-site migration of contaminants of concern (COCs) from the landfill. This treatment zone approach will provide a cost-effective alternative to the existing record of decision (ROD)-specified remedy of a landfill cap and pump and treat currently planned for the site. Data to date demonstrate that biostimulation additives are being well distributed in both the anaerobic and aerobic treatment areas. Field data indicates a 6 to 8 month acclimation period for enhanced biological activity, and is consistent with predictions derived from laboratory testing and site characterizations. The data further suggests that enhanced degradation of *cis*-DCE to ethene and ethane is occurring in the monitored portion of the anaerobic zone.

INTRODUCTION

This paper describes work conducted to date at a Superfund landfill site (site) in New England, related to the demonstration of sequential in situ anaerobic (primarily for chlorinated VOC degradation)/aerobic (primarily for non-chlorinated VOC degradation and arsenic immobilization) bioremediation of COCs in ground water migrating from the site. The objective of the demonstration is to show enhanced biodegradation through the treatment zones (Bouwer, E.J., 1992). In addition, the data will be used to calculate compound specific biodegradation rates in order to expand the system design across the entire toe of the landfill, and to obtain an evaluation of the comparability of laboratory and field-measured enhanced biodegradation rates.

To collect sufficient information to design a full-scale treatment zone for the site a treatment zone demonstration (TZD) was designed in the summer of 1996. The design utilized a 61 by 91 meter area of the site at the toe of the landfill slope. The design was composed of two amendment injection systems (one for electron donor and one for oxygen) and an aquifer monitoring network (piezometers and small diameter wells). Construction of the TZD began in early October 1996 and was completed in mid-January 1997. Pre-start-up samples were collected from the TZD in September 1997. TZD operations (the injection of amendments) began in January 1998. The TZD has operated since January 1998 and it is estimated that TZD operations will continue through the year 2000. A description of the TZD design and construction, as well as operational data collected to date, is presented in the following sections.

MATERIALS AND METHODS

Anaerobic and Aerobic Treatment Area Target COCs. To achieve an adequate reduction in the levels of the COCs found at the site, the treatment zone was designed as a sequential anaerobic and aerobic treatment system. The anaerobic area was designed to add electron donor compounds to enhance the biodegradation of chlorinated COCs. Laboratory treatability studies evaluated the use of sodium benzoate and ethanol as potential electron donors, and both substrates stimulated the reductive dehalogenation of chlorinated VOCs in the site samples (data not presented). Based on these results, safety and cost considerations, and a previous successful application of sodium benzoate in a field-scale treatment system (Beeman et al, 1994) sodium benzoate was chosen as the electron donor for use in the anaerobic treatment zone. The anaerobic area is located hydraulically upgradient from the aerobic area of the treatment zone. The aerobic area was designed to enhance the biodegradation of non-chlorinated COCs and reduce soluble arsenic levels through immobilization by precipitation with soluble iron present in the ground water. The aerobic area uses the injection of gaseous oxygen to maintain aerobic conditions.

Treatment Zone Demonstration Design. The TZD location, orientation and design were selected based on the results of ground water quality measurements, hydrogeologic investigations and treatability testing performed at the site. The TZD was located in the area of the site that demonstrated the highest levels of the combined COCs. The dimensions of the TZD area were determined based on the results of the treatability testing that provided an estimate of degradation rates. The TZD width (dimension perpendicular to the ground water flow direction) was 61 meters. This width was considered sufficient to minimize the potential of "edge" effects on the ground water flow patterns. The treatment zone width is also considered sufficient to evaluate the impact of the addition of fluids (both electron donor and oxygen) on the ground water flow patterns.

The length of the TZD (dimension parallel to the ground water flow direction) was 91 meters. The anaerobic treatment area extends for the first 46 to 61 meters and the aerobic treatment area is the final 30 to 45 meters of the TZD. The lengths were determined based on the results of the laboratory treatability studies.

The depth of the treatment zone was designed to include the three aquifers present at the site. The three aquifers are identified as shallow, intermediate and deep. The shallow aquifer extends from grade to approximately 7.6 meters below grade. The shallow aquifer soils are primarily fine to medium sands. The intermediate aquifer extends from approximately 7.6 meters below grade to approximately 12.2 meters below grade. The intermediate aquifer soils are primarily fine to medium sands with interbedded silt and clay lenses. The deep aquifer extends from approximately 12.2 meters below grade to approximately 15.2 meters below grade. The deep aquifer soils are primarily fine to medium sands with significant lenses of silt and clay.

The TZD system occupies approximately one half hectare of land at the down gradient toe of the landfill site and is comprised of:

- 30 sodium benzoate injection points (SIPs) and 30 oxygen injection points (OIPs);
- transfer piping to deliver sodium benzoate and oxygen to the SIPs and OIPs respectively;
- a sodium benzoate mixing and storage system (housed in a prefabricated metal building);
- a 5,700 liter oxygen storage tank and controls for the delivery systems;
- approximately 125 small diameter monitoring points; and
- 27 ground water piezometers to provide ground water elevation data throughout the TZD.

RESULTS AND DISCUSSIONS

Distribution of Amendments. The injection of amendments (sodium benzoate and oxygen) began in January 1998. Target levels of dissolved amendments, 3.7 mg/L sodium benzoate and 2 mg/L dissolved oxygen, for the site aquifer were established based on the results of the laboratory treatability study, an evaluation of contaminant levels, and the limited air and oxygen injection study. For example, based on previous studies (Sewell and Gibson, 1991) it was assumed that approximately 1% of the electrons liberated during oxidation of sodium benzoate would be used to reduce the chlorinated solvents in the groundwater. For this study, a 10 fold molar excess of electron donor was added to account for variability in VOC concentrations and dehalogenation efficiency. Similar levels had successfully stimulated dehalogenation in the laboratory treatability study.

Although the TZD has only operated for a relatively short period, the most recent site data indicates that the existing system is able to distribute amendments sufficiently to meet the above referenced target additive concentrations. Sodium benzoate concentrations in the hundreds of mg/l have been measured at monitoring points 10 to 15 meters from the injection gallery across a large portion of the anaerobic zone. The oxygen injection system has also resulted in elevated dissolved oxygen concentrations throughout a large portion of the aerobic treatment zone. Although the injection of oxygen has proved somewhat difficult to control resulting in occasional incursions of gaseous oxygen up-gradient into the anaerobic zone

Anaerobic Biodegradation of COCs. The rate of biodegradation of the target COCs can be described by first order biodegradation kinetics. The net decrease in *cis*-DCE concentration was assumed to be instantaneous and stoichiometrically converted to VC on a molar basis (i.e., 100% conversion efficiency; 1 mole of *cis*-DCE degraded is converted to 1 mole of VC). Therefore, the rate of accumulation of VC is proportional to the rate of *cis*-DCE degradation. While VC was being produced, it was also assumed to be simultaneously degrading according to first order kinetics. The total VC degraded over time was then assumed to be completely converted to ethene and ethane. The set of equations were solved using time step iterations to estimate the degradation/production of the COCs. The biodegradation rates for *cis*-DCE and VC were input variables.

A family of concentration curves, presented in Figure 1, was generated for the degradation of *cis*-DCE, accumulation/degradation of the daughter product of VC, and the accumulation of ethene and ethane. The curves were fit to the actual field data trends by adjusting the first order biodegradation rates until a reasonably unique solution was found. The estimated biodegradation rates for *cis*-DCE and VC were 4 x 10^{-3} day^{-1} and 2 x 10^{-2} day^{-1}, respectively. These rates are approximately 1.3 and 5 times greater, respectively, than the rates used to design the TZD. It should be noted, however, that in other studies biodegradation rates typically become slower as reductive dechlorination proceeds from *cis*-DCE to VC (i.e., biodegradation rate of *cis*-DCE > rate for VC). Our data suggests that other processes, perhaps desorption of *cis*-DCE from soil, may be affecting our measurements of actual degradation.

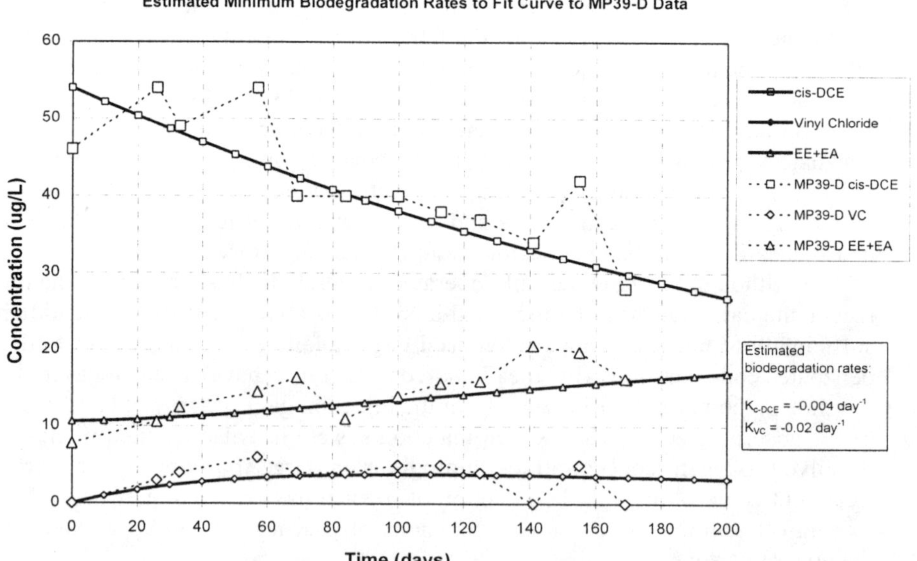

Figure 1. Curve Fit for Estimated Biodegradation of *cis*-DCE and Production of Daughter Products to Field Data (Monitoring Point MP39-D).

Aerobic Biodegradation of COCs. Two methods were used to estimate the biodegradation rate of aerobically degradable COCs at the site: 1) the BIOSCREEN computer model (USEPA 1996), and 2) mass balance calculations. For the purposes of the estimates, it was assumed that the influent concentration of COCs entering the aerobic treatment zone was stable and consistent and that steady-state conditions were established prior to implementation of the aerobic treatment zone. Therefore, at the start of the oxygen injection, the ground water concentrations entering and exiting the aerobic zone were essentially constant; the net difference in concentration was due primarily to the background rates of biodegradation (and possibly other process such as retardation and dispersion of COCs). These are reasonable assumptions based upon review of the data from monitoring points. After the start-up of aerobic zone oxygen injection (and allowing time for microbial acclimation) the enhanced biodegradation rates could be estimated.

BIOSCREEN modeling was performed to match the model predicted concentration profile to the actual field data points using the biodegradation rate as the fitting parameter. The model took into account effects of dispersion and retardation of the COCs. The fit was largely qualitative, because there were only two rows of data points along the flow path of the plume and the curve is first order. An example curve fit for benzene is presented in Figure 2.

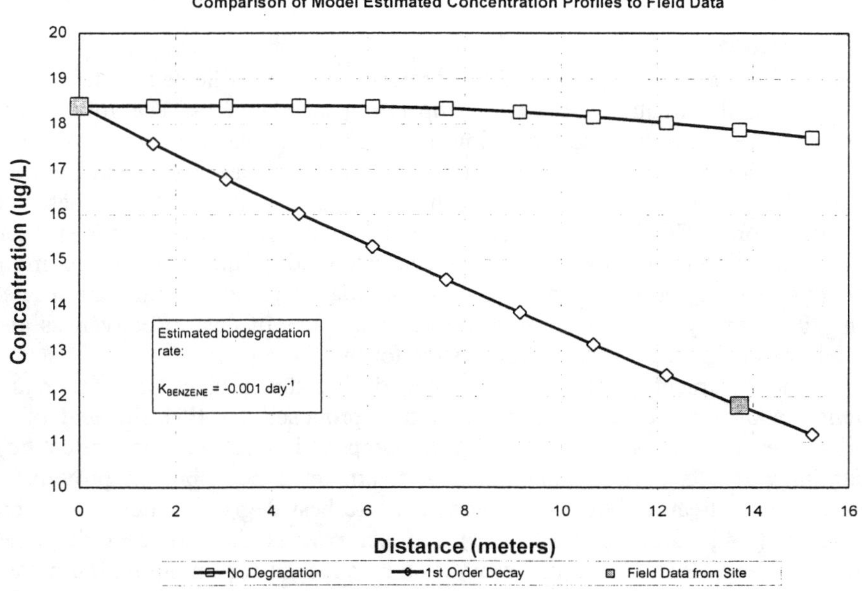

Figure 2. BIOSECREEN model simulation output for estimation of the first order biodegradation rate for benzene. The model curve was fit to actual field data prior to aerobic zone startup.

The mass balance approach involved estimating the mass flux into and out of the aerobic zone while ignoring the effects of dispersion and COC retardation. The net difference between the mass in and mass out was assumed to be due to biodegradation. The first order biodegradation rate was estimated based on the net change in mass and the time interval considered.

Preliminary background biodegradation rates were estimated for benzene that ranged between approximately 1×10^{-4} to 4×10^{-3} day^{-1}. BIOSCREEN estimates were usually more conservative. Sensitivity analyses determined that dispersion did not have a significant impact on results of BIOSCREEN modeling. However the effects of retardation caused the estimated biodegradation rates to decrease by up to five times over the assumed range of fraction of organic carbon content (0.12% to 0.94%).

The distance between monitoring points along the path of the plume is such that the ground water travel time across the aerobic zone could be greater than one year. Therefore, the ground water migrating into the aerobic zone at the start of system implementation may only be just reaching the downgradient monitoring points. In addition, if a 3 to 6 month acclimation period is assumed, the tail end of the ground water slug that was not impacted by the treatment zone may still be passing the downgradient monitoring points, especially if COC retardation is significant. Steady state conditions are not likely to have yet been established, so these analyses can not yet be used to determine the degree to which the biodegradation rate has been enhanced, at this time.

CONCLUSIONS

Although it is early in the life of the project, the data collected to this point suggests that the treatment system is operating as planned. The initial data collected to evaluate the biodegradation of COCs in the anaerobic zone indicates that the addition of an electron donor (sodium benzoate) will enhance the biodegradation rates of target COCs. The initial data collected to evaluate the biodegradation of COCs in the aerobic zone, however, indicates that it is still too early in the project life reliably calculate accurate biodegradation rates. As more data are collected, we will be able to more accurately calculate degradation rates at the site, thereby allowing more accurate predictions of cost effectiveness and efficiency of a full scale remediation system for the site.

The treatment system demonstrated during this project provides many advantages over remedial measures currently proposed for this site and other similar sites (i.e., impermeable cap with pump and treat leachate collection). Preliminary results indicate that the dual treatment zone approach provides a mechanism for treating both compounds that are best degraded under anaerobic conditions (e.g., chlorinated solvents) and compounds that are best degraded aerobically (e.g., BTEX). It also provides a mechanism for immobilizing metals like arsenic (via oxidation in the aerobic zone) or others. The ability to treat the contaminants in-situ eliminates the need to pump the contaminated groundwater to the ground surface, thereby limiting the need to handle or dispose of the contaminants. It also allows large volumes of groundwater to be treated simultaneously, without the flow limitations imposed by engineered aboveground systems like strippers, carbon canisters, and/or bioreactors. Another advantage of the system is that it allows flushing of the landfill to remove existing contaminants that will ultimately be degraded as they pass through the treatment system. The use of a permeable cap on the landfill also will allow some contaminants to be degraded within the landfill footprint, while still providing a capped landfill thus allowing beneficial re-use of the site.

REFERENCES

Beeman, R.E., J.E. Howell, S.H. Shoemaker, E.A. Salazar, and J.R. Buttram. 1994. A field evaluation of in situ microbial dehalogenation by the biotransformation of chlorinated ethenes. In Hinchee, R.E., A. Leeson, S. Semprini, and S.K. Ong (Eds.), *Bioremediation of Chlorinated and Polyaromatic Hydrocarbon Compounds*, pp. 14-27. Lewis Publishers, Boca Raton, FL.

Bouwer, E.J. 1992. Bioremediation of chlorinated solvents using alternate electron acceptors.
In. H*andbook of Bioremediation*, pp. 149 –174.

Sewell, G.W. and S.A. Gibson. 1991. Stimulation of the reductive dechlorination of tetrachloroethene in anaerobic aquifer microcosms by the addition of toluene. Environ. Sci. Technol. 25:982-984.

U.S. Environmental Protection Agency, 1996, BIOSCREEN Natural Attenuation Decision Support System, Version 1.3, User's Manual. EPA/600/R-96/087, August, 1996.

PHYTOREMEDIATION AND MODELING OF LAND CONTAMINATED BY HYDROCARBONS

M. Yavuz Corapcioglu[1], Clyde Munster[1], Malcolm Drew[1], Robert Rhykerd[1], Kijune Sung[1] and Yoon-Young Chang[2]
[1]Texas A&M University, College Station, TX, USA
[2] Korea Inst. Of Science and Technology, Seoul, Korea

Abstract: The use of deep-rooted plants for phytoremediation of contaminated soils has been offered as an alternative to excavation and incineration or landfilling for waste management. Plant root exudates may stimulate microbial activity and thereby enhance biodegradation of recalcitrant compounds by co-metabolism. A predictive model has recently been developed for predicting the fate of recalcitrant hydrocarbons in soil. The model requires extensive testing with field data for validation and calibration. The objective of this paper is to test the model using field data from plants grown in contaminated soil. The soil was contaminated with approximately 10 $\mu g\ kg^{-1}$ of 2,4,6-trinitrotoluene (TNT), chrysene and 2,2',5,5'-tetrabromobiphenyl (PBB). Treatments were either vegetated (Johnsongrass) or fallow (control). Experimental units consisted of polyvinyl chloride (PVC) columns, 0.1 m in diameter x 1.5 m in length packed with the contaminated soil to a bulk density of 1.4 $g\ cm^{-3}$. Soils were watered to near field capacity twice weekly. Time domain reflectometry was used to monitor soil moisture at depths of 10, 75, and 140 cm in the columns. On days 30 and 90 soil microbial biomass was measured using the chloroform fumigation method, root density was determined by image analysis, and chemical concentrations in the soil and herbage were performed by immunoassay techniques. After 90 days there was no significant difference in microbial biomass between vegetated and fallow treatments. Soil moisture was significantly lower in the vegetated treatment than in the fallow treatment. No significant differences in chemical loss were observed between the treatments. Approximately 99% of the TNT had disappeared from the soil while decreases of chrysene and PBB were 60-80%. Trace amounts of the contaminants were detected in the herbage. It appears that the decreased water potential in the soil offset the stimulatory effects of plant roots on microbial activity. The computer model has successfully simulated the disappearance of these recalcitrant compounds through 90 days.

INTRODUCTION

Spills from military and industrial activities have resulted in soils contaminated with PAHs, PCB's and TNT. These soils are at risk of contaminating surface and ground waters by runoff or leaching and, thus posing an environmental health risk to vegetation, wildlife, and humans. Contaminated soils are often excavated and placed into landfills or incinerated to remove the contaminants from the environment. Recent public concern and rising costs regarding these procedures point to the need of a less expensive and publicly accepted remediation option. Phytoremediation, the use of plants grown in contaminated soils to enhance remediation, is one such option.

Phytoremediation is still a relatively new technology, but shows great promise to enhance bioremediation. Plants may sequester, absorb, and translocate organic contaminants, which removes them from the soil environment (Cunningham et al., 1995). Perhaps the greatest contribution of phytoremediation is the stimulation of soil microorganisms by plant root activity. The carbon exudates from plant roots act as microbial substrates to increase soil microbial populations and activity, which in turn increases metabolic and cometabolic transformations of contaminants to less toxic products (Aprill and Sims, 1990; Nair, 1993; McFarlane et al., 1990; and Sims and Overcash, 1983). Root exudates increase soil microbial populations and activity and build up organic matter in the root zone due to necromass (Keister and Cregan, 1990; Anderson and Coates, 1994; and Foster et al., 1983). Roots also improve soil aeration by directly giving off oxygen to the root zone as well as allowing improved entry of oxygen into the soil by diffusion along old root channels (Nye and Tinker, 1977; Schnoor, 1993). Roots may also reduce movement of contaminants in soil by extracting excess water, thus reversing the vertical hydraulic gradient (Topp et al., 1986; Nair et al., 1993; Vlamis et al., 1985; Schnoor, 1993).

Computer models assist in predicting the fate of contaminants in soil using various remediation processes. Our model examines the root water relations of crops by a simple predictive technique that is usable in a wide range of applications, such as *in situ* plant remediation for contaminated soil near the surface zone (Chang and Corapcioglu, 1997; Chang and Corapcioglu, 1998). Incorporating a time-specific root distribution model into a vertical unsaturated soil water flow equation developed this simulation model. Our computer model will help save users time and money by providing valuable remediation information. This model will guide the user to collect only the relevant data while minimizing the acquisition of irrelevant data. The validated phytoremediation model will be used as a predictive tool to assess the effectiveness of various plants on remediation of soil contaminated by recalcitrant compounds. The model's flexibility allows simulations to show degradation of several classes of contaminants using varying plant species. Once it is established that phytoremediation is effective for a specific site, the model will aid in plant selection and predict the time required for phytoremediation to achieve cleanup goals. Because many simulations can be examined, a phytoremediation strategy may be planned that will save the user time and money. Output from this computer model may prevent users from opting for an unnecessarily expensive remediation technique, such as excavation and incineration, save users money by minimizing the number of samples required for analysis and save time by utilizing plants that will optimize remediation at a specific site. This model will be available to industries that will directly benefit from its application, such as consulting firms, the military, and the agricultural extension service.

The overall purpose of this research is to obtain field data comparing phytoremediation to bioremediation of soil contaminated with chrysene, 2,2',5,5'-tetrabromobiphenyl, and TNT. This information will also be used to validate and further refine a newly developed phytoremediation computer model. Data will be collected to determine partitioning of the contaminants between soil, the soil solution, and vegetation. Important modeling features to be developed are the temporal and spatial changes in root quantity and root interaction with soil moisture and the contaminants.

MATERIALS AND METHODS

Lysimeters were constructed from cylinders, 9.8 cm i.d. and 1.5 m length (Figure 1). Each lysimeter was sectioned longitudinally to give two separable halves to simplify destructive sampling of roots and soil. The two halves were cemented together with adhesive, and clamped in position with hose-clamps fitted at three positions along the column. The bottoms of the lysimeters were fitted with porous caps to allow drainage of excess moisture into a collection jar. This construction was found to be robust, permitting the lysimeters to be filled with moist soil, which was compacted to the required bulk density (1.4 g cm^{-3}).

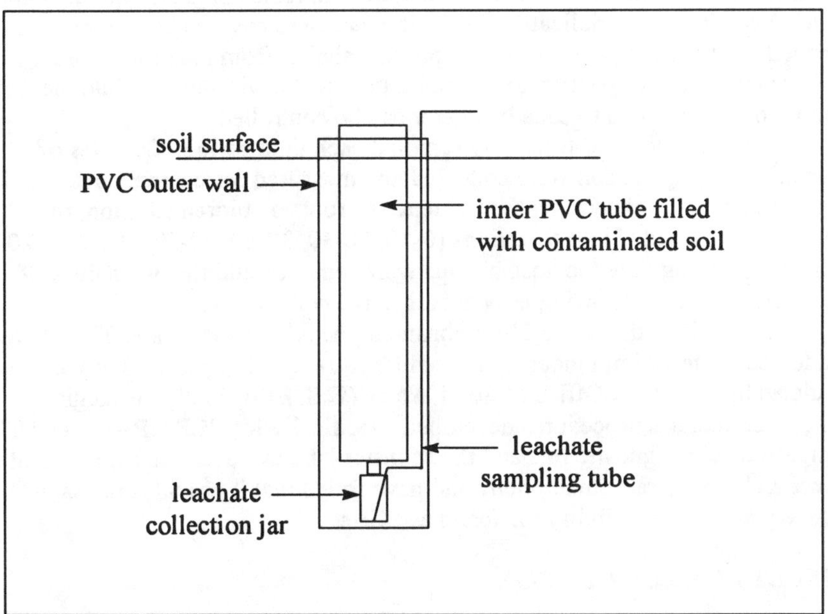

Figure 1. Diagram of lysimeter and supporting structure for use in phytoremediation of contaminated soil.

The lysimeters were placed in the ground at the Texas A&M University Research Farm near College Station, TX in a cylindrical PVC tube with a 15 cm i.d. and 1.75 m in length. A leachate collection system was placed in the bottom of the sleeve just below the porous cover (Figure 1). Leachate was collected in glass mason jars placed in the bottom of the PVC cylinder and was extracted via a vacuum tube running from the bottom of the glass jar to the surface where it will connect to a brown collection jar fitted to a vacuum source.

The lysimeters were filled with a Weswood loam soil. This soil has a pH of 7.9, organic carbon matter content of 1.0%, and a sand, silt and clay content of 23, 47, and 30% respectively. This soil type is representative of soils at many contaminated sites, but at the site we are using, has not been previously exposed to spills of PAHs, PBBs, or TNT. Contaminants were mixed into the soil to a concentration of 10 mg kg^{-1} soil each and then packed in the lysimeters to a bulk density of 1.4 g cm^{-3}.

Soil moisture in the lysimeters was maintained near field capacity gravimetrically. Lysimeters were lifted out of the ground with hoist block and tackle assistance and weighed using a load cell. The load cell was calibrated in the weight range of the lysimeters and connected to a data logger and computer which was programmed to convert load cell output to kilograms. Measured amounts of water were added to the soil surface to restore that lost by drainage and evapotranspiration.

The soil in all lysimeters was contaminated with 10 mg of each contaminant kg^{-1} soil. Treatments were vegetated (Johnsongrass) and unvegetated (fallow) conducted in triplicate. The lysimeters were covered by a translucent, corrugated fiberglass panel system to provide shelter from the rain. The rain shelter permits the vegetation to be grown in a field environment while the amount of water added to each lysimeter can be controlled.

Columns were destructively sampled each time period. Samples of vegetation, roots, and soil were collected and measured for contaminant concentrations. To determine the influence of roots on bioremediation, the columns were divided into 6 sections (0-10, 10-30, 30-60, 60-90, 90-120, 120-150 cm). At sampling time the hose clamps were removed and the two halves of lysimeter separated, allowing easy access to the soil and roots.

Analysis of the 2,2',5,5'-tetrabromobiphenyl, chrysene, and TNT were conducted using an immunoassay procedure (Strategic Diagnostic Inc.) as validated by U.S. EPA Office of Solid Waste (U.S. EPA, 1995). Strategic Diagnostic Inc. has procedures developed specifically for PCBs, PAHs, and TNT. Both Strategic Diagnostic Inc. and the Munster lab have used the PCB test kit to detect 2,2',5,5'-tetrabromobipheny and have found that it provides the same accuracy and reproducibility for PBBs as PCBs.

RESULTS AND DISCUSSION

The concentration of all three chemicals decreased in both vegetated and unvegetated treatments but differences after 90 days were not statistically significant between vegetated and fallow treatments or with depth. Of the three contaminants, TNT showed the largest loss with more than 99% undetectable. The concentration of PBB decreased by 75-85% while the chrysene concentration fell by 60-90%. Enumeration of microbial biomass indicated the presence of a large population, approximately 1.5 g kg^{-1}, but again no significant differences were observed between vegetated and fallow treatments. Major differences were observed in the water use between the treatments. Losses from the vegetated treatment were nearly 10 times larger than from the fallow treatment. Much of the loss from the vegetated treatment is attributed to transpiration. Comparing the water content of the two soils at three depths reveals the vegetated soils tended to show much more extensive water losses throughout the column compared to the fallow treatments.

It appears that water may have been limiting bioremediation of the recalcitrant hydrocarbons in the vegetated treatment. We did not observe an increase in microbial populations in the vegetated treatment as expected. While roots are known to stimulate microbial activity and increase microbial biomass, particularly in the rhizosphere, this was not observed. Transpiration by the plants removed significant amounts of water from the soils at all depths, which may have limited microbial activity.

Preliminary results of computer model simulations fit well with the observed data. Important parameters to accurately simulate the fate of recalcitrant compounds appear to be soil organic carbon, soil moisture, and chemical K_{OC} and water solubility.

CONCLUSIONS

The use of plants to enhance bioremediation of recalcitrant hydrocarbons was not observed. However, this may have been due to the rapid depletion of soil moisture by rapidly growing plants in a hot and dry climate where the only additions of water was through biweekly irrigation. The use of plants to stimulate bioremediation of recalcitrant hydrocarbons may require a more intense water management program than would be necessary for fallow bioremediation systems. The incorporation of rapidly growing plants in saturated soils may be beneficial to reduce downward movement of the chemicals because of the increased removal of water from the soil in the rooting zone by plant transpiration.

Computer simulations may be useful for predicting plant-assisted remediation of recalcitrant compounds in soil. Further validation of the model with additional plant species is necessary to predict the type of vegetation and rotations required to optimize bioremediation for a given site.

ACKNOWLEDGMENTS

This research was funded through U.S. EPA grant R 8254-01-0.

REFERENCES

Anderson, T.A., and J.R. Coates. 1994. Bioremediation through rhizosphere technology. American Chemical Society. Washington, DC. pp. 10-26, 123-57.

Aprill, W., and R.C. Sims. 1990. Evaluation of the use of prairie grasses for simulating poly-cyclic aromatic hydrocarbon treatment in soil. Chemosphere. 20:253-265.

Chang, Y., and M.Y. Corapcioglu. 1997. Effect of roots on water flow in unsaturated soils. J. Irrigation and Drainage Engineering. 123:202-209.

Chang, Y., and M.Y. Corapcioglu. 1998. Plant-enhanced subsurface bioremediation of nonvolatile hydrocarbons. *J. Env. Engr*. 124:162-169.

Cunningham, S.D., W.R. Berti, and J.W. Huang. 1995. Remediation of contaminated soils and sludges by green plants. *In* Hinchee et al. (eds.) Bioremediation of Inorganics. pp. 33-54. Third International In Situ and on site Bioreclamation Symposium, San Diego, CA.

Foster, R.C., A.D. Rovira, and T.W Cock. 1983. Ultrastructure of the root-soil interface. The American Phytopathological Society. MN. pp. 5-11.

Keister, D.L. and P.B. Cregan. 1990. The rhizosphere and plant growth. Kluwer Academic Publishers. Boston, MA pp. 1-11

McFarlane, C., T. Pfleeger, and J. Fletcher. 1990. Effect of uptake and disposition of nitro-benzene in several terrestrial plants. Environ. Toxicol. Chem. 9:513-520.

Nye, P.H., and P.B. Tinker. 1977. Solute Movement in the soil-root system. 1st ed., Blackwell Scientific Publications. Oxford. pp. 100-150.

Nair, D.R., J.G. Burken, L.A. Licht, and J.L. Schnoor. 1993. Mineralization and uptake of triazine pesticide in soil-plant systems. J. Env.Eng., 119:842-854.

Schnoor, J.L. 1993. Vegetative remediation of hazardous waste sites. AEEP/NSF Research Opportunities Conference. Ann Arbor, MI.

Sims, R.C. and M.R. Overcash. 1983. Fate of polynuclear aromatic compounds in soil-plant systems. Residue Rev. 88:1-68.

Topp, E., I. Schneunert, A. Attar, and F. Korte. 1986. Factor affecting the uptake of ^{14}C-labeled organic chemicals by plants from soil. Ecotoxicol. Environ. Saf. 11:219-228.

Vlamis, J., D.E. Williams, J.E. Corey, A.L. Page, and T.J. Ganje. 1985. Zinc and cadmium uptake by barley in field plots fertilized seven years with urban and suburban sludge. Soil Sci., 139:81-87.

DEGRADATION OF ORGANIC CONTAMINANTS IN SLUDGE-AMENDED AGRICULTURAL SOIL

Frank Laturnus, Christian Grøn, Gerda K. Mortensen, Per Ambus (Risø National Laboratory, Roskilde, Denmark)
Susan Bennetzen (VKI, Hørsholm, Denmark)
Erik S. Jensen (Royal Agricultural University, Taastrup, Denmark)

ABSTRACT: The impact of plant growth and organic amendment upon biodegradation of organic contaminants was investigated. Soil was mixed in different ratios with sewage sludge, and carrots were planted and grown for 84 days. Several compounds like nonylphenols, LAS, trimethylnaphthalenes were removed up to 95% in soil highly loaded with organic compounds, whereas DEHP remained more recalcitrant. An increase in degradation was found when lower sludge dosages were applied to the soil. Lacking degradation was due to reduced bioavailability, as compounds directly added to the soil were almost completely removed. The growth of plants increased the degradation of organic compounds in soil loaded with high sludge dosages (90 t dry weight / ha). The degradation of organic compounds in soil may be due to uptake and metabolization in the plants, or it may be due to enhanced microbial activity induced by the root growth.

INTRODUCTION

The composition of organic wastes reflects the patterns of consumption in modern society, especially with respect to remains of industrial chemicals. The use of chemicals in industrialized countries leads not only to a widespread distribution in terrestrial and aquatic ecosystems, but also to a high residue of anthropogenic organic compounds in sewage sludge, manure, organic industrial waste and organic household waste (Kjølholt et al., 1995; Smith, 1996). Recently, high concentrations of household tensides and their degradation products (*e.g.* nonylphenols and nonylphenol ethoxylates, linear alkylbenzene sulfonates), plasticizers (*e.g.* phthalates), combustion by-products (*e.g.* polyaromatic hydrocarbons) and PCBs were found in Danish sewage sludges (Kristensen et al., 1996). As some of these anthropogenic compounds are of a high toxicity, their presence in agro-ecosystems is of particular concern because they can be transferred to humans *via* agricultural products or leaching to groundwater.

Generally, the introduction of organic contaminants to agricultural fields occur by pesticide application, spills and deposition of chemicals, atmospheric depositions and application of organic wastes such as sewage sludge, manure, organic industrial waste and organic household waste. Biodegradation is an important mechanism to reduce the concentration of these compounds in the environment. As part of studying the fate of organic contaminants in agro-ecosystems, we have investigated the impact of plant growth and organic amendment on biodegradation of a number of organic contaminants.

MATERIALS AND METHODS

Soil samples for biodegradation experiment were prepared by mixing a sandy soil with sewage sludge of mixed industrial and domestic origin derived from a waste water treatment plant close to Copenhagen. The mixing ratios corresponded to sludge amendments of 0 to 90 t dry weight / ha. The growth and degradation experiments were accomplished in a greenhouse under constant photon fluxes, photoperiods and temperature. Carrots (*Daucus carota* L.) were sown in steel cylinders filled with soil samples, and placed together with plant-free controls under glass dust covers (Figure 1). To avoid desiccation during the incubation, 60 to 75 % of the water holding capacity of the soil was maintained by watering. After 84 days, the plants were removed and the remaining soil was thoroughly mixed, sampled and stored at –20 °C until analysis.

FIGURE 1. Greenhouse set-up for the degradation experiment with carrots grown in sludge-amended soil.

For analysis, approximately 50 g soil sample were extracted with dichloromethane after adjusting to a pH below 2. The extract was concentrated by evaporation, treated with activated copper, followed by derivatization with diazomethane and cleaning by SPE on a Florisil column. Identification and quantification was achieved by gas chromatography with mass spectrometric detection in full-scan mode, and external calibration standards. Detection limits ranged from 0.01 to 0.04 mg/kg dry weight, and recovery from 41 to 125 %. The method precision was 10 to 30 %.

The analytical method for LAS was Soxhlet extraction with methanol, SPE clean up, concentration of extracts and finally determination of total LAS using HPLC with fluorescens detection.

RESULTS AND DISCUSSION

In the greenhouse experiment with carrots grown in sludge-amended soil, nonylphenols (including mono- and diethoxylates), LAS, mono-, di-, and trimethylnaphthalenes, and phenanthrene (+ anthracene) were removed up to 95% even at high sludge dosages (Figure 2). Conversely, DEHP appeared to be more recalcitrant with degradation of 0 to 20%. DEHP is considered to be rapidly degradable in sludge amended soil (Wild and Jones 1991). However, the present data did not support this, probably because the high organic amendment caused partially anaerobic conditions in the soil and, therefore, reduced biodegradability of DEHP (Pedersen et al., 1996). Whereas 41% of DEHP was removed in soil amended with 6 t dry weight sludge / ha, this compound became more persistent with higher sludge dosages added to the soil (Table 1). Thus, with the use of sewage sludge as agricultural fertilizer there is a risk of accumulating DEHP in the soil. Similarly, biodegradation of DEHP in soils with a high organic contamination could be slow.

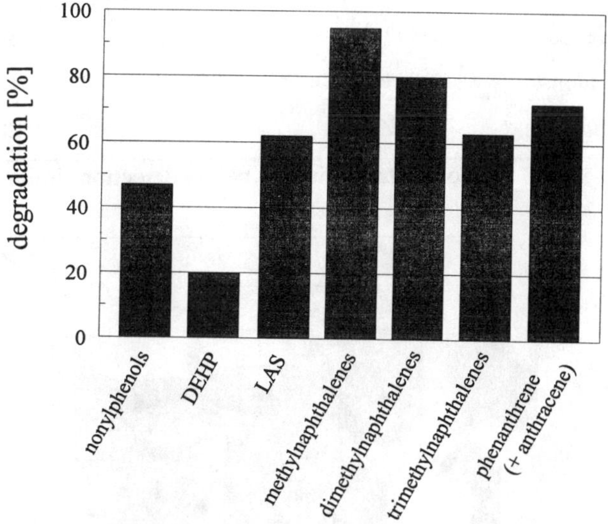

FIGURE 2. Degradation of organic compounds in soil amended with high sludge dosage (90 t dry weight/ha) after plant growth.

The growth of plants obviously increased the degradation of the organic compounds in the soil (Table 1). Whereas for methylnaphthalenes, LAS and DEHP only a minor increase was found, plant growth had a quite high effect on the removal of nonylphenols, di- and trimethylnaphthalenes. Also, a lower organic load with reduced sludge dosages increased the degradation of most of the compounds in the soil (Table 1). At the lower sludge dosages, anaerobic conditions did not develop. As all of the compounds in question are degradable under aerobic conditions, it is suggested that the growth of plants in soils with high loads of organic compounds increased the degradation due to enhanced soil aeration by the roots. Therefore, phytoremediation by plant enhanced bio-degradation may be more effective with heavily polluted soils. Also, application of sludge to agricultural fields may reduce the biodegradation of organic compounds with increasing sludge dosages.

TABLE 1. Degradation of organic compounds in soil amended with different sludge dosages planted with carrots and in plant-free experiments.

compounds	sludge dosage added		
	6 t	90 t	90 t plant-free
nonylphenols	64%	47%	2%
DEHP	41%	20%	7%
LAS	99%	62%	56%
methylnaphthalenes	>54%[a]	80%	66%
dimethylnaphthalens	>83%[a]	95%	11%
trimethylnaphthalenes	77%	63%	20%

[a]concentation in soil below detection limits

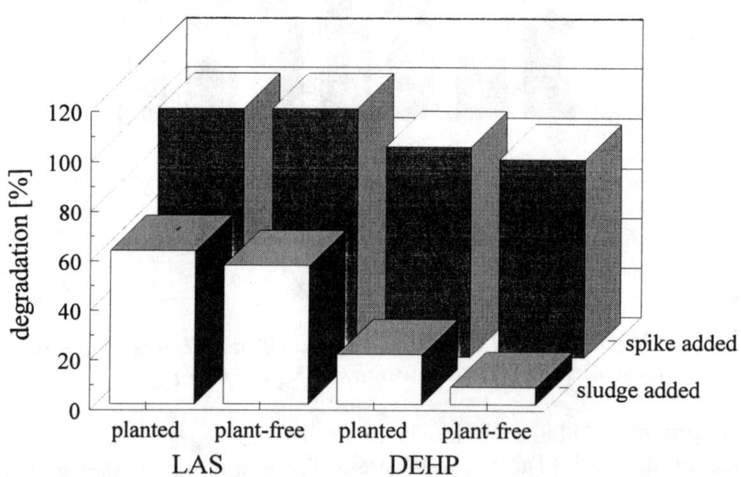

FIGURE 3. Degradation of LAS and DEHP in spiked and sludge-amended soil. The concentration of the added spike corresponded to a sludge dosage of 90 t dry weight/ha.

The removal of LAS and DEHP appeared to be almost complete, and, thus, much more efficient when added as free compounds to the soil (Figure 3). The less efficient degradation of the compounds added with sludge is probably due to a reduced bioavailability in the soil. As contaminated soils, organic waste products consists of a complex mixture of natural organic substances and of more hydrophobic organic chemicals. Interactions between organic chemicals and organic matter in the sludge may significantly decrease the bioavailability of the chemicals to the microorganism in the soil (Alexander, 1995).

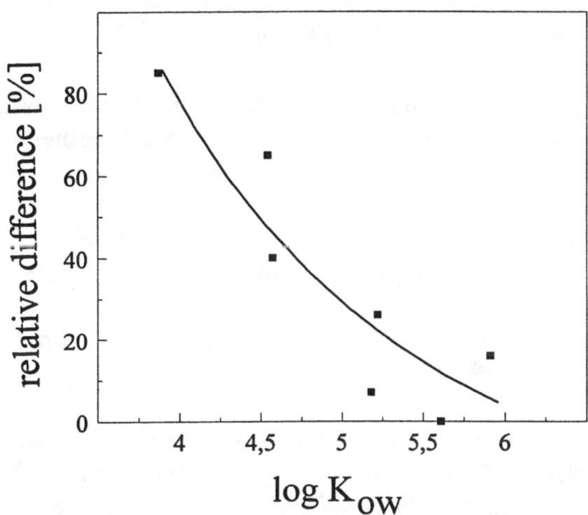

FIGURE 4. Degradation enhancement (as relative differences between the concentrations of PAH compounds found in soil with and without plant growth)* vs. *the partition coefficient (octanol – water) for the single compounds. Concentrations detected in soil amended with 90 t sludge/ha and planted with carrots.

However, the degradation enhancement effect of plant growth depended upon the properties of the single substances. In Figure 4, the degradation enhancement for a number of polyaromatic hydrocarbons (PAHs) is given as a function of their octanol/water partition coefficients (log K_{ow}). Evidently, the degradation enhancement was larger for the compounds with the lowest log K_{ow} and, thus, for the least lipophilic compounds with the higher bioavailability in soil.

It can be concluded that sludge application to soil for the purpose of fertilization or waste removal inhibited the degradation of organic contaminants and may even lead to an accumulation of some organic contaminants in the soil. On the other hand, plant growth increased the degradation of the organic compounds, probably because of uptake and metabolization in the plants or by enhanced microbial activity. Plant enhanced degradation was most efficient for soils with a high organic load and for less lipophilic compounds. The conclusions may be extrapolated to the use of phytoremediation of contaminated soils in general.

REFERENCES

Alexander, M. 1995. "How toxic are toxic chemicals in soil?" *Environ. Sci. Technol.* 29: 2713-2717.

Kjølholt, H., H. V. Andersen, and C. Poll. 1995. *Occurrence and Effects of Organic Contaminants in Sewage Sludge.* Working Report No. 15 (in Danish), Environmental Protecting Agency, Copenhagen.

Kristensen, P., J. Tørsløv, L. Samsøe-Petersen, and J. O. Rasmussen. 1996. *The Use of Waste Products in Agriculture.* Environmental Project No. 328 (in Danish), Environmental Protecting Agency, Copenhagen.

Pedersen, F., J. Larsen. *1996 Review of Environmental Fate and Effetcs of DEHP.* Working Report No. 54, Environmental Protecting Agency, Copenhagen.

Smith, S. R. 1996. *Agricultural Recycling of Sewage Sludge in the Environment.* CAB International, Wallingford.

Wild, S.R., and K.C. Jones. 1991. "Organic contaminants in wastewaters and sewage sludges: Transfer to the environment following disposal. In K. C. Jones (Ed.), *Organic contaminants in the environment – Environmental pathways & effects*, pp. 133-158. Elsevier, London.

SALT TOLERANCE OF WOODY PHREATOPHYTES FOR PHYTOREMEDIATION APPLICATIONS

Paul R. Thomas (Thomas Consultants, Inc., Cincinnati, Ohio, USA)
Jennifer J. Krueger (Thomas Consultants, Inc., Cincinnati, Ohio, USA)

ABSTRACT: A six-week-long test program was conducted to study the short-term response of live, unrooted phreatophyte cuttings of genus *Populus* and genus *Salix* to salinity in a controlled, greenhouse environment. The program tested four candidate woody phreatophytes prior to final species selection for a phytoremediation system at a former manufactured gas plant site in Charleston, SC. Growth was evaluated by biomass, shoot length, water consumption, and transpiration ratio. Hybrid poplar, *P. charcowiensis x P. incrassata*, exhibited vigorous growth at all treatment levels relative to the other species tested and was concluded to be the most salt tolerant relative to the other species under the test conditions. Another hybrid poplar, *P. maximowiczii x P. trichocarpa*, exhibited the least salt tolerance with growth severely affected by exposure to increased salinity conditions; mortality rates of 75 and 100 percent, respectively; low biomass; and short shoot length relative to the control plants. Salt stress of this species was recorded early in the study. Documented mortality, transpiration ratio, biomass and shoot length indicated *S. matsudana x S. alba* and *P. deltoides x P. nigra* were more salt tolerant than *P. maximowiczii x P. trichocarpa.*

INTRODUCTION

Vegetation selection in phytoremediation system design is necessary to ensure compatibility between candidate species, site conditions, and the proposed application. Tolerance to salt is often an important consideration when the application involves direct or potential exposure of the vegetative species to high-ionic strength solutions. For example, if a system is installed to serve as a hydraulic barrier or to change the local water balance for plume control, the ionic strength of the plume itself may be problematic to the root systems. Vegetative covers installed over landfills with leachate recirculation systems must be tolerant of high salinity conditions common to landfill leachates. Phytoremediation systems installed in coastal settings may encounter high salinity conditions in groundwater, surface water, and soil. Exposure to highly saline conditions causes stress to plants in two ways: (1) salt stress, which is disruption of the osmotic potential resulting in water stress, and (2) ion stress, which is toxicity caused by high concentrations of specific ionic species (Hale and Orcutt, 1987).

Phreatophytes are used for many phytoremediation systems because they consume and transpire large amounts of water. Phreatophytes, such as poplars (genus *Populus)* and willows (genus *Salix*) are unmatched in their ability to rapidly produce leaf, root, and vascular tissue to extract water from the subsurface. In this study, various candidate woody phreatophyte species are evaluated for salt tolerance prior to final species selection for a phytoremediation system. Remediation is being conducted at a former manufactured gas plant site located on the Charleston peninsula in South Carolina. The goal of the system is to intercept a plume of dissolved monoaromatic hydrocarbons and polycyclic aromatic hydrocarbons (PAHs) to stop migration of contaminants to the nearby Cooper River. Shallow groundwater exhibits a tidally-affected diurnal flow pattern and water quality varies in salinity from fresh to brackish with an average total dissolved solids (TDS) concentration of 2,029 mg/L (range from 720 to 5,390 mg/L) (Campbell et al., 1996).

Objective. The objective of the study was to quantitatively evaluate the relative salt tolerance of various woody phreatophytes under greenhouse conditions. Quantitative evaluation of growth was made from measurements of total biomass produced, shoot length, and water consumed. Transpiration ratio was calculated from these data.

MATERIALS AND METHODS

A six-week greenhouse study evaluated the relative salt tolerance of four woody phreatophyte species:

1. *Salix matsudana x S. alba* (hybrid willow)
2. *Populus deltoides x P. nigra* (hybrid poplar)
3. *P. charcowiensis x P. incrassata* (hybrid poplar)
4. *P. maximowiczii x P. trichocarpa* (hybrid poplar)

The selection of three hybrid poplars was made based on past performance and availability for phytoremediation applications; the hybrid willow was included because willows are fast-growing and include salt tolerant species (Newsholme, 1992).

Preparation. Live, unrooted cuttings of each species were shipped from nurseries and refrigerated until use. Each cutting was measured for stem length and mass (wet) and tagged with a unique identification. Moisture content of each species was measured in order to calculate initial plant dry mass.

Test Procedure. The study was carried out at a greenhouse in Cincinnati, Ohio. Two cuttings of each species were planted per container in buckets filled with silica sand to a depth of 10 cm. The containers, with holes drilled into the bottoms and lined with filter fabric, were placed into treatment tubs. Salt consisting of equal parts sodium chloride, magnesium sulfate, and calcium chloride was added to pre-mixed nutrient solution to elevate salinity levels corresponding to Level 1: +1,000 mg/L added salt (electrical conductivity, EC of 3,700 $\mu S\ cm^{-1}$); Level 2: +2,500 mg/L added salt (EC = 4,500 $\mu S\ cm^{-1}$); and Level 3: +5,000 mg/L added salt (EC = 8,200 $\mu S\ cm^{-1}$). The control level consisted of pre-mixed base nutrient solution (Peter's Professional 20-10-20, Scott's Manufacturing Co.) mixed to a baseline TDS level of 350 mg/L and EC of 1,950 $\mu S\ cm^{-1}$ (see Table 1).

The treatment tubs were filled with equal volumes of the three levels of salt solutions and a control solution containing no added salt. A replicate set of all treatment tubs was also established. Thus, four samples of each species were tested at each salt level. Initially, clear plastic bags were wrapped around each treatment tub and secured around the buckets to control evaporation from the tubs. The plastic bags were periodically removed and re-secured during the test to reduce algae and mold growth in the tubs. The sand surface was open to the air during the entire test. The tubs were monitored periodically for uptake of the solution over a 6-week period. Additional base nutrient and salt solution was added to the tubs to maintain saturated conditions in the bottom of each bucket.

TABLE 1. The base nutrient solution used for dosing the tree cuttings.

Nutrient	Percentage
Total N (20)	
Ammoniacal nitrogen	7.77
Nitrate nitrogen	12.23
Available phosphate	10.00
Soluble potash	20.00
Magnesium	0.15
Boron	0.02
Copper (chelated)	0.01
Iron (chelated)	0.10
Manganese (chelated)	0.06
Molybdenum	0.01
Zinc (chelated)	0.02
Total	50.36

The study commenced on May 12, 1998. Each bucket was initially irrigated with 2 liters (L) of the base nutrient solution to hydrate the sand substrate and to stimulate growth. No other irrigation was performed during the test. At the end of six weeks (June 24, 1998) the plants were removed from the containers. Each plant was logged for root growth, stem growth, and leaf measurements. The plants were assessed for moisture content by oven drying at 70°C for 48 hours. Data are summarized in Table 2.

TABLE 2. Percent increase in biomass as measured by the dry weight of shoots and leaves; no roots included.

Level	Salinity (mg/L)	*P. charcowiensis x P. incrassata*	*P. maximowiczii x P. trichocarpa*	*P. deltoides x P. nigra*	*S. matsudana x S. alba*
Control	350	161.3	170.0	175.5	284.5
1	1,350	166.2	132.7	179.8	214.5
2	2,850	160.2	23.3	111.0	128.0
3	5,350	105.8	7.7	90.7	71.6

RESULTS AND DISCUSSION

Growth. Salt stress in plants is generally manifested in yield reduction and stunted growth. As expected, all species exhibited decreased growth with exposure to increased salinity levels although the effect between species varied significantly. Measurements of above-ground dry biomass produced during the study and initial dry mass were used to calculate the percent increase in total biomass. The average growth gain for the three hybrid poplars under the control conditions was between 150 to 200 percent (Table 2). The hybrid willow biomass increased the most – nearly 300 percent under control conditions. Average biomass at Level 1 salinity conditions declined relative to the control for the hybrid willow and *P. maximowiczii x P. trichocarpa*, and continued to decline sharply with subsequent salt levels (Figure 1).

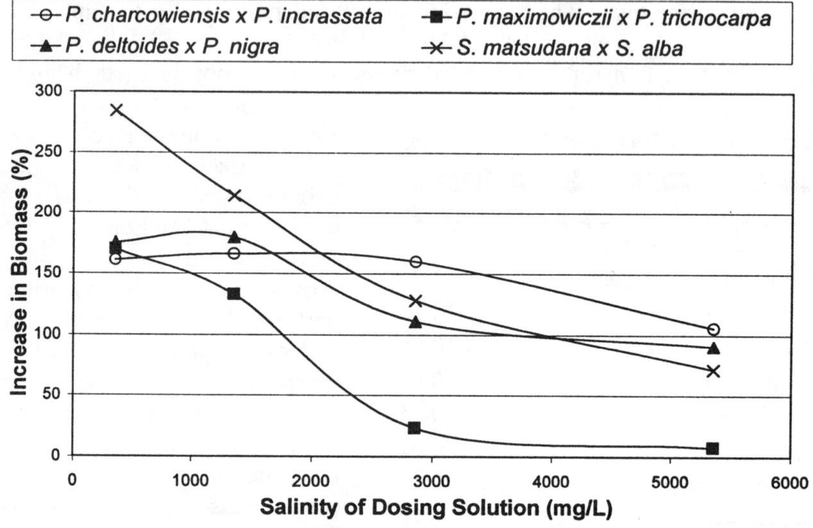

FIGURE 1. Graph of biomass versus salinity.

The growth of the other two hybrid poplars, *P. charcowiensis x P. incrassata* and *P. deltoides x P. nigra*, as measured by average biomass, was unaffected at Level 1 salt conditions (Table 2 and Figure 1). At Level 2 the biomass of *P. deltoides x P. nigra* declined although less relative to *S. matsudana x S. alba* and *P. maximowiczii x P. trichocarpa*. The growth of *P. charcowiensis x P. incrassata* as measured by average biomass was unaffected at Level 2. Declines in biomass of all species were observed under Level 3 salinity conditions. The greatest decrease was recorded for *P. maximowiczii x P. trichocarpa*.

The longest shoot length of each cutting was also recorded as a measure of relative growth. These data are similar to biomass percentages with all species showing decreased growth under the Level 2 and Level 3 salinity conditions. The growth of *P. maximowiczii x P. trichocarpa* was severely affected at the Level 2 and Level 3 salinity conditions with mortality rates of 75 percent (3 of 4 samples) and 100 percent (4 of 4 samples). Two of the *P. maximowiczii x P. trichocarpa* cuttings died within 1 week into the study. The only other species that did not survive the test was *P. deltoides x P. nigra* (25 percent mortality at both Levels 2 and 3) with death occurring between weeks 3 and 4.

Transpiration Ratio. Transpiration ratio is a parameter used to compare the relative consumption of water between plant species. Defined as the ratio of water consumed (mass) to the biomass produced over time, transpiration ratio is a normalized, unitless quantity that reflects the water requirement for plants under the conditions tested. Transpiration ratio data have principally been used by agricultural engineers to evaluate plant efficiency and to estimate water loss from crops. These data are not applicable to field conditions because of the controlled environment of the greenhouse, but are useful for comparison between species and within species while controlling a single variable such as salinity. In comparing the transpiration ratio between species, a higher ratio indicates one of two conditions: 1) either more water was consumed or 2) less biomass was produced. If a plant is salt tolerant, the transpiration ratio measured at varying salinity levels does not change significantly. The less salt tolerant a plant is, the greater the increase in transpiration ratio.

Transpiration ratio data calculated from this test are presented in Table 3. Under control conditions, the transpiration ratios for the hybrid poplars ranged between about 400 to 700; the ratio for the hybrid willow was 466. Under exposure to increased salinity levels, the transpiration ratio for all species also increased as biomass declined. The greatest increase was measured in *S. matsudana x S. alba* indicating lower salt tolerance relative to the other species. The least increase (greatest salt tolerance) was measured in *P. charcowiensis x P. incrassata.*

TABLE 3. Transpiration ratio, the ratio of water consumed to dry biomass produced during the test.

Level	Salinity (mg/L)	*P. charcowiensis x P. incrassata*	*P. maximowiczii x P. trichocarpa*	*P. deltoides x P. nigra*	*S. matsudana x S. alba*
Control	350	456.3	719.8	416.8	466.0
1	1,350	598.6	873.2	479.0	564.8
2	2,850	672.8	nc - high mortality	599.9	810.4
3	5,350	855.3	nc - high mortality	1089.0	1470.9

The data obtained from this study indicate a substantial variation in tolerance to salt within species in the same family (Salicaceae, which is comprised of poplars and willows), and even between poplars of the same genus (*Populus*). Likewise, studies of simulated leachate irrigation to hybrid poplars and willows by others have yielded mixed results (Cureton et al., 1991; Shrive et al., 1994). Salt stress has been documented due to osmoregulation disruption (Cureton et al., 1991) where vegetative stress was measured in hybrid poplar following two seasons of exposure to leachate with an EC of only 800 to 1,000 $\mu S\ cm^{-1}$. Conversely, hybrid poplar irrigated with a relatively high ionic strength leachate (EC = 4,250 $\mu S\ cm^{-1}$) exhibited increased growth and photosynthetic response (Shrive et al., 1994), with the enhanced growth attributed to increased availability of plant nutrients from the leachate. The variability in plant performance of species of similar parentage is evidence of the delicate balance of soil water availability, nutrient levels, salinity, and other factors that affect plant vigor. Because of the documented variability in tolerance of these species, pilot tests, either in the field or greenhouse, are useful to support design of full-scale phytoremediation systems.

CONCLUSIONS

A six-week-long greenhouse study was conducted to study the short-term response of live, unrooted phreatophyte cuttings of genus *Populus* and genus *Salix* to salinity. A hybrid poplar, *P. charcowiensis x P. incrassata*, exhibited vigorous growth at all treatment levels relative to the other species tested and was concluded to be the most salt tolerant relative to the other species under the test conditions. Another hybrid poplar, *P. maximowiczii x P. trichocarpa*, exhibited the least salt tolerance with growth severely affected at the Level 2 and Level 3 salinity conditions; mortality rates of 75 and 100 percent, respectively; low biomass; and short shoot length relative to the control plants. Salt stress of this species was recorded after only 1 week into the study. Documented mortality, transpiration ratio, biomass and shoot length indicated *S. matsudana x S. alba* and *P. deltoides x P. nigra* were more salt tolerant than *P. maximowiczii x P. trichocarpa.*

Tolerance to salt can be a critical factor in the performance and survival of many phytoremediation systems, including those located in arid regions, in tidally influenced areas, or leachate recirculation / wastewater treatment systems involving tree plantations or landfill covers. These results demonstrate how field or greenhouse-scale pilot tests can be used to screen candidate species prior to final selection for phytoremediation systems.

ACKNOWLEDGEMENTS

The authors thank W. Irwin of the South Carolina Electric & Gas Co. and A. Contrael and V. Tandon of IT Corporation for supporting this research.

REFERENCES

Campbell, B.G., M.D. Petkewich, J.E. Landmeyer, and F.H. Chapelle. 1996. *Geology, hydrogeology, and potential of intrinsic bioremediation at the National Park Service Dockside II Site and Adjacent Areas, Charleston, South Carolina, 1993-94.* US Geological Survey Water-Resources Investigations Report 96-4170, Columbia SC.

Cureton, P. M., P. H. Groenevelt, and R. A. McBride. 1991. "Landfill Leachate Recirculation: Effects on Vegetation Vigor and Clay Surface Cover Infiltration." *J. Environ. Qual.* 20: 17-24.

Hale, M. G. and D. M. Orcutt. 1987. *The Physiology of Plants Under Stress.* John Wiley & Sons, New York.

Newsholme, C. 1992. *Willows, The Genus Salix.* Timber Press, Inc., Portland, OR.

Shrive, S. C., R. A. McBride, and A. M. Gordon. 1994. "Photosynthetic and Growth Responses of Two Broad-Leaf Tree Species to Irrigation with Municipal Landfill Leachate." *J. Environ. Qual.* 23: 534-542.

A GREENHOUSE-BASED TEST OF PLANT-AIDED PETROLEUM HYDROCARBON DEGRADATION

Kevin R. Hosler (Conor Pacific - WTI, Burlington, Ontario, Canada)
Evelyn N. Drake (Exxon Research and Engineering, Annandale, New Jersey)

ABSTRACT:

A greenhouse-based phytoremediation experiment assessed the potential effects of three treatment factors on the degradation of weathered petroleum hydrocarbons in contaminated soils. A 2^3 factorial experimental design evaluated three factors: 1) turfgrass (a mixture of tall fescue and perennial ryegrass), 2) BuRIZE™ (a commercial mycorrhizal inoculant), and 3) fertilizer (a commercial liquid formulation). Controls consisted of untreated contaminated soil. The design allowed for testing of significant differences in contaminant degradation rates resulting from the three factors – used individually or jointly - when compared to the controls. Four sampling events were performed over a study period of 180 days.

Statistical evaluation of the experimental results revealed a consistent, significant treatment effect for the "fertilizer" factor, and periodic treatment effects for the "grass" factor and "grass / mycorrhizal inoculant" interaction.

INTRODUCTION

Methods and conditions for bioremediating organic contaminants such as petroleum hydrocarbons are well known (e.g., Sims *et al.,* 1989) and various treatments (e.g., nutrients, electron acceptors, modified soil conditions, etc.) have successfully enhanced contaminant biodegradation. Recalcitrant contaminants or unsuccessful treatments have also been identified (Bulman *et al.*, 1985, 1988; Hosler *et al.*, 1992).

Phytoremediation - or specifically phytodegradation - technologies exploit the unique abilities of plants to degrade or facilitate degradation of predominantly organic contaminants, alone or via microbial associations within the rhizosphere. Research by April and Sims (1990), Banks *et al.* (1997) and Qui *et al.* (1997) has shown that contaminant degradation may be enhanced in the rhizosphere or root-zone of various plant species. Remediation may be achieved by exploiting the unique symbiotic associations between certain fungal species and plant roots, called "mycorrhiza" (literally, *fungus-root*).

Research is needed however to show that plants used in phytoremediation applications provide distinct and significant benefits above those of general bioremediation processes.

EXPERIMENTAL METHODS

Experimental Design. A 2^3 factorial experimental design was used in a greenhouse-based experiment to assess the degradation of total petroleum hydrocarbons (TPH) in weathered, crude oil-contaminated soil. The experiment evaluated the effects of three factors:

(1) a 50:50 mixture by seed quantity of tall fescue [*Festuca arundinacea* var. "Bonsai"] and perennial ryegrass [*Lolium perenne* var. "Affinity"], applied at a rate of 5 g of seed mix /pot;

(2) BuRIZE™ (a commercial vesicular-arbuscular mycorrhizal or VAM inoculant; Buckman Laboratories of Canada Ltd.; Vaudreuil, PQ), applied at a rate of 1 propagule VAM/cm^3; and,

(3) a commercial liquid fertilizer formulation ("Schultz-Instant®" water; Schultz Co., St. Louis, MO; guaranteed analysis of 10:15:10 N:P:K), applied at a rate recommended for turfgrass.

The factors were tested at two levels each (i.e., either present or absent), and the treatments can be summarized as follows:

#1: Contaminated soil	(no factors)
#2: Contaminated soil + grass	(factor "A")
#3: Contaminated soil + BuRIZE™	(factor "B")
#4: Contaminated soil + fertilizer	(factor "C")
#5: Contaminated soil + grass + BuRIZE™	(factors "AB")
#6: Contaminated soil + grass + fertilizer	(factors "AC")
#7: Contaminated soil + BuRIZE™ + fertilizer	(factors "BC")
#8: Contaminated soil + grass + BuRIZE™ + fertilizer	(factors "ABC")

Treatment #1 represented a biotic control. There were 2 replicates of each experimental unit (pot) which contained 1833 g of soil. The pots were randomly assigned to a grid location on a greenhouse table - to avoid any confounding environmental variables - and incubated with relatively constant temperature (23°C+/- 3°C), humidity, and a 12-hour photoperiod provided by natural and supplemented light.

Soil Preparation. Contaminated soil collected from a field site hotspot was sieved to <2 mm size and then stored at 4°C until ready for use. Clean soil from the A_h horizon of a non-cultivated rural agricultural field was similarly sieved and stored. Prior to and during their use in the study, both soils were characterized for basic physico-chemical properties previously shown to influence bioremediation processes (Sims *et al.*, 1989; Hosler & Booth, 1995).

An approximate 1:4 mix ratio of contaminated to clean soil was used for this study because preliminary testing had shown that the 3.5% to 5% range of TPH in the contaminated soil was severely detrimental to plant growth. The soils were thoroughly mixed and triplicated subsamples were analyzed to verify soil homogeneity, yielding a satisfactory 7.6% coefficient of variation for TPH values.

Soil Sampling. Soil sampling was performed in a consistent manner for each of four sampling events. The timing of the sampling events was based upon TPH degradation rates reported in the literature and/or observed from the previous sampling. Appropriate and sufficient numbers of QA/QC samples (i.e., controls, spikes, duplicates, and method blanks) were included in the study. Soil sampling occurring after grass had grown involved coring the turf, separating it from the adhered soil, and mixing this soil to the portion sampled from underneath the core. Samples were well mixed prior to analysis.

Treatment Application. After the initial soil sampling, moisture was added to all pots to achieve 70% of the soil's maximum water holding capacity. Grass seeds were planted in designated treatment pots using common horticultural practices. Inoculation of designated treatment pots with BuRIZE™ VAM was performed according to the manufacturer's instructions. Liquid fertilizer was applied to designated treatments as part of the initial moisture addition.

RESULTS AND DISCUSSION

The physico-chemical characterization of both uncontaminated and contaminated soils revealed that the critical environmental factors necessary for bioremediation of petroleum hydrocarbons (e.g., temperature, moisture content, pH, cation exchange capacity, etc.) were within acceptable levels (e.g., Sims *et al.*, 1989). Characterization of the heterotrophic and TPH-degrading microbial populations revealed that both of the soils individually possessed reasonably-sized heterotrophic and TPH-degrading microbial populations (i.e., > 1.0×10^5 cfu/g).

Germination of the plants occurred after approximately seven days. After three weeks, grass growth was relatively uniform amongst all treatment pots.

The TPH replicate average concentration data for all treatments for the initial (day 0) and final (day 180) sampling periods, including the initial standard deviation values and the overall TPH percent loss values are presented in Table 1.

TABLE 1. Replicate average TPH concentration values for initial and final sampling periods, initial standard deviations, and percent loss values.

Treatment	Time (d)	Average [TPH] (mg/kg)	Standard Deviation	Time (d)	Average [TPH] (mg/kg)	% Loss
Control	0	6350	919	180	3600	43.3
A	0	7000	141	180	3750	46.4
B	0	7100	566	180	3250	54.2
C	0	6900	849	180	1900	72.5
AB	0	7200	141	180	3700	48.6
AC	0	7150	495	180	1900	73.4
BC	0	6800	707	180	1500	77.9
ABC	0	6500	283	180	2250	65.3

The experimental hypothesis for the study was that the use of turfgrass should result in greater TPH loss, presumed to be degradation, as compared to those treatments without turfgrass. All treatments, including the control, showed decreasing TPH concentrations (replicate averages) during the course of the experiment, however the control showed the lowest decrease (43.3%). The highest loss of TPH occurred in the "BC" (BuRIZE VAM inoculant plus fertilizer) treatment, followed closely by the "AC" (grass plus fertilizer) and "C" (fertilizer) treatments.

Reductions in TPH concentrations caused by physico-chemical processes not controlled by the experimental design (e.g., volatilization, leaching, hydrolysis, photodegradation) were likely negligible due to controls imposed via the experimental technique. Thus, decreasing TPH values (i.e., TPH loss) may reasonably be attributed to degradation, or more likely biodegradation.

A statistical analysis of the TPH data was performed similarly to methods outlined in Chapter 5 of Cochran & Cox (1992), using a *STATISTICA®* software package (StatSoft®, Tulsa, OK). The data was analysed for the four time periods, individually and in combination (i.e., periods #1, #2, #3, & #4, periods #1, periods #2, periods #3, and periods #1). All individual and combinations of factors were assessed to determine which effects and/or interactions were statistically significant for enhancing TPH degradation. Table 2 represents one of the analysis of variance (ANOVA) tables and shows two significant factors and one interaction (bolded); *all time periods were considered for this analysis.*

TABLE 2. ANOVA for the 2^3 factorial experiment; significant effects are bolded.

Source	Sums of Squares	Df*	Mean squares (effect)	F-value	p-value
"A" (Turfgrass)	360000	1	360000	1.669	.205
"B" (BuRIZE VAM)	15625	1	15625	.072	.789
"C" (Fertilizer)	1314000	1	1314000	**60.942**	**.000**
"T" (Time)	142655700	3	47551900	**220.530**	**.000**
"AB"	250000	1	250000	1.159	.289
"AC"	122500	1	122500	.568	.456
"BC"	105625	1	105625	.489	.489
"AT"	1046250	3	348750	1.617	.204
"BT"	153126	3	51042	.236	.870
"CT"	5213124	3	1737708	**8.058**	**.000**
"ABC"	62500	1	62500	.289	.594
"ABT"	1328751	3	442917	2.054	.125
"ACT"	116250	3	38750	.179	.909
"BCT"	655626	3	218542	1.013	.399
"ABCT"	163749	3	54583	.253	.858

* = degrees of freedom for the effect.

The statistical data analysis reveals that the "fertilizer" and "time" factors were consistently significant during the full study period (i.e., for the four time intervals tested). One would assume that the "time" factor would be significant since TPH degradation is known to occur abiotically in soils during time periods much less than the 180-days studied here. A significant effect for the "fertilizer" factor is reasonable given that bioremediation of petroleum hydrocarbons is commonly enhanced by the addition of inorganic fertilizers.

A "grass / time" and a "grass / VAM inoculant" interaction each occurred in just one instance. The "grass / time" interaction occurred between periods #2 and #3 (66 and 88 days) and it was during this time that the rate of grass growth was quite high. The "grass / VAM inoculant" interaction might possibly be attributed to a mycorrhizal growth effect that enhanced TPH degradation during the last 92 days of the study (i.e., between periods #3 and #4). Other possible reasons for these results are provided in Hosler and Drake (1998).

There are several important benefits that plants may provide to contaminated site remediation that were not tested in this study, however the simple benefit to TPH degradation provided by plants was not as great as hypothesized. Plants have an inherent ability to aerate, bind and/or stabilize soils at contaminated sites (e.g., by reducing or eliminating erosion, and regulating water infiltration) and thus their presence is undoubtedly advantageous. If contaminant degradation and/or stabilization can be achieved *in situ* by plants, remediation costs for specific sites could be considerably reduced versus the use of more intensive bioremediation technologies such as biopiles, bioventing, or physico-chemical technologies such as soil washing or thermal destruction. Further testing of the specific benefits that may be provided by plants at contaminated sites is recommended in order to facilitate optimization of phytoremediation processes.

ACKNOWLEDGEMENTS

The authors gratefully acknowledge the contributions of Dr. Terry McIntyre (Environment Canada), Dr. Ken Mullen, (University of Guelph), Ms. Laura Black (Imperial Oil), and Mr. Rob Booth (WTI / CPET). This work was funded by Environment Canada and Imperial Oil Ltd.

REFERENCES

April, W., and Sims, R. C. 1990. "Evaluation of the Use of Prairie Grasses for Stimulating Polycyclic Aromatic Hydrocarbon Treatment in Soil. *Chemosphere, 20* (1-2): 253-265.

Banks, K. M., Pekarek, S., Rathbone, K., and Schwab, A. P. 1997. "Phytoremediation of Petroleum Contaminated Soils: Field Assessment". In B. C. Alleman and A. Leeson (Symposium Chairs), *In Situ and On-Site Bioremediation: Volume 3*, pp. 305-308. Battelle Press, Columbus, OH.

Bossert, I., and Bartha, R. 1984. "The Fate of Petroleum in Soil Ecosystems". In R. M. Atlas (Ed.), *Petroleum Microbiology*, pp. 435-473. Macmillan Publishing Company, New York, NY.

Bulman, T. L., Hosler, K. R., Fowlie, P. J. A., Lesage, S., and Camilleri, S. 1988. *Fate of Polynuclear Aromatic Hydrocarbons in Refinery Waste Applied to Soil.* PACE Report 88-1. Petroleum Association for Conservation of the Canadian Environment, Ottawa, ON.

Bulman, T. L., Lesage, S., Fowlie, P. J. A., and Webber, M. D. 1985. *The Persistence of Polynuclear Aromatic Hydrocarbons in Soil.* PACE Report 85-2. Petroleum Association for Conservation of the Canadian Environment, Ottawa, ON.

Cochran, W. G., and Cox, G. M. 1992. *Experimental Designs (Second Edition).* Wiley Classics Library Edition, John Wiley & Sons Inc., New York, NY.

Hosler, K. R., and Booth, R. M. 1995. *Protocol Development for Bench-Scale Soil Bioremediation Studies (Part I - Guidance Document).* Development and Demonstration of Site Remediation Technology (DESRT) Program Report, Environment Canada, Ottawa, ON.

Hosler, K. R., and E. N. Drake. 1998. "A Test of Plant-Aided Petroleum Hydrocarbon Degradation". In *Reclamation and Restoration of Settled Landscapes; Proceedings of the 23rd Annual Meeting, Canadian Land Reclamation Association in association with the Society for Ecological Restoration Ontario Chapter*, Markham, ON.

Hosler, K. R., Bulman, T. L., and Booth, R. M. 1992. "The Persistence and Fate of Aromatic Constituents of Heavy Oil Production Waste During Landfarming". In P. T. Kostecki, E. J. Calabrese, and M. Bonazountas (Eds.) *Hydrocarbon Contaminated Soils; Volume II*, pp. 591-609. Lewis Publishers, Boca Raton, FL.

Qui, X., Leland, T. W., Shah, S. I., Sorenson, D. L., and Kendall, E. W. 1997. "Field Study: Grass Remediation for Clay Soil Contaminated with Polycyclic Aromatic Hydrocarbons". Chapter 14 in E. L. Kruger, J. R. Coats, and T. A. Anderson (Eds.) *Phytoremediation of Soil and Water Contaminants.* American Chemical Society Symposium Series 664, Washington, DC.

Sims, J. L., Sims, R. C., and Matthews, J. E. 1989. *Bioremediation of Contaminated Surface Soils*, U.S. Environmental Protection Agency Research and Development Report, EPA/600/9-89/073, R. S. Kerr Environmental Research Laboratory, Ada, OK.

USING MICROBIAL COMMUNITY STRUCTURE CHANGES TO EVALUATE PHYTOREMEDIATION

C. M. Reynolds, C. S. Pidgeon, L. B. Perry, B. A. Koenen, and D. K. Pelton. (USA-CRREL, Hanover New Hampshire), H.L. Nichols and D.C. Wolf (University of Arkansas, Fayetteville Arkansas).

ABSTRACT: Both laboratory and field research have shown that phytoremediation can be effective for treating contaminated soils, but we cannot yet confidently predict success. A trade-off inherent in using phytoremediation is that, although it is inexpensive to implement and maintain, it takes longer than traditional treatments. Acceptance of phytoremediation has been hindered because of the longer treatment times and the uncertainty of achievable rates and endpoints. If a requirement for using phytoremediation is spatially and temporally intense monitoring, much of the inherent cost savings are negated. Although we are gaining experience in using phytoremediation at a number of sites, we remain somewhat limited in our ability to effectively predict success. The operative mechanisms for phytoremediation are largely contaminant dependent. For many organic contaminants, the accepted phytoremediation mechanism is enhanced microbial activity and altered microbial community structure in the rhizosphere, which in turn enhances degradation of contaminants. We hypothesized that a potential tool for predicting long-term success and monitoring endpoints of phytoremediation could be based on changes in the soil microbiology.

At a site in Fairbanks, Alaska, we compared the effects of vegetation and nutrient additions on remediating both diesel- and crude-oil-contaminated soils in a factorial experiment with appropriate control treatments. We sampled periodically and measured changes in petroleum concentrations. Concomitantly, we characterized the microbial populations by species richness—a diversity index. We used fatty acid methyl ester (FAME) profiles to identify bacterial isolates from the soils. The most effective treatment for reducing petroleum concentrations, vegetation with nutrient additions, was also characterized by differences in the bacterial species richness compared to the control treatment. In both treatments, there was an increase in bacterial species richness as contaminant concentrations decreased, but species richness and remediation were greatest in the vegetation and nutrient treatment in the diesel-contaminated soil. Bacterial species richness in the crude-oil contaminated soil was both more constant and lower than in the diesel-contaminated soil. During the study, the TPH concentrations in the crude-oil contaminated soil did not decrease as rapidly or to the same extent as that in the diesel-contaminated soil.

An improved understanding of the time-dependent relationships between contaminant concentration changes and microbial community changes, coupled with improved techniques to readily characterize microbial communities, may provide a useful tool for monitoring the functioning of phytoremediation, evaluating desirable endpoints, or both.

INTRODUCTION

Several efforts have shown that phytoremediation can be used to clean contaminated soils inexpensively. A hindrance to more extensive use of phytoremediation is the ability to predict its success and monitor its operation. At present, phytoremediation demonstrations use extensive site characterization prior to planting followed by comprehensive monitoring to confirm that phytoremediation has reduced soil contaminant concentrations. While this approach is necessary in expanding the knowledge base for new applications of phytoremediation, greater acceptance and use of phytoremediation may depend largely on our ability to successfully predict situations where phytoremediation will work.

This issue may be addressed by several approaches. One is to develop a nationwide database of phytoremediation results from field demonstrations that span such variables as climate, soil, and petroleum concentration, age, and type. This strategy is being used by the "Total Petroleum Hydrocarbons in Soil Subgroup", a part of the U.S. Environmental Protection Agency (EPA) administered "Phytoremediation of Organics Action Team", which was established in 1997 as one of seven Action Teams under the Remediation Technologies Development Forum (RTDF). The purpose of the RTDF is to identify how government and industry can cooperatively develop and improve environmental technologies needed to address their mutual cleanup problems in the safest, most cost-effective manner.

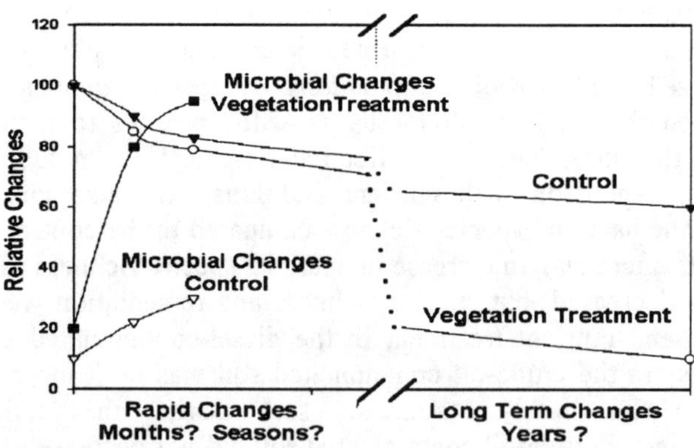

FIGURE 1. Hypothetical changes in contaminant concentrations and soil microbial characteristics during remediation.

Another approach is to use microbial changes in the contaminated soil-root system as indicators that the treatment is working (Figure 1). Measurable differences in rhizosphere microbiology have been observed in different agricultural systems (Cattelan et al. 1998), but less is known for contaminated soils. We hypothesized that, because phytoremediation of petroleum is generally agreed to result from enhanced microbial activity in the rhizosphere, changes in the soil microbiology status may be a useful predictor of decreases and endpoints in soil contaminant concentrations that may not be readily measured (Reynolds et al. 1999)

MATERIALS AND METHODS

We evaluated vegetation with nutrient additions to both diesel-contaminated and crude-oil-contaminated soils in Fairbanks, Alaska. Annual ryegrass (*Lolium multiflorum*) and Arctared red fescue (*Festuca rubra*) were chosen for their rapid growth and cold hardiness, respectively, and were planted as a mix consisting of equal seed masses of each species. Both grasses have extensive root distribution and tolerance to low-fertility soils. Nutrients were supplied by a commercially available agricultural fertilizer, granular 20-20-10, surface-applied at approximately 620 g/m^2. We used surface application because it could readily be used at remote field sites. This vegetation+nutrient treatment was compared to a control treatment—equivalent to treatment by natural attenuation—that consisted of no vegetation and no nutrient amendments.

For each treatment, triplicate soil samples were obtained by compositing numerous subsamples from either the vegetated or control plots. For microbial characterization, soil samples were serially diluted and plated on 0.1 strength TSA to determine viable numbers of bacteria (Zuberer, 1994). Numbers of aerobic, heterotrophic bacteria were determined following incubation of the plates for three days at 20 ± 1°C. Bacterial isolates were either identified to the species level or characterized by fatty acid methyl ester (FAME) techniques following the procedures outlined by Sasser (1990, 1991) using the MIDI bacteria reference library (MIDI 1995). For each soil sample characterized, we evaluated between 50 and 100 randomly chosen isolates from dilution plates having between 30 and 300 colonies. We treated isolates that were not identified by the MIDI library but having distinctively different fatty acid chromatograms as individual species. Unknowns having similar fatty acid chromatograms were characterized as multiple members of the same species. Species richness (d) was defined as (Pielou, 1975):

$$d = (S-1)/\log N \quad (1)$$

Where S = number of species
N = number of individuals

Soil TPH was extracted by sonication with CH_2Cl_2. Anhydrous Na_2SO_4 was added to the soil during extraction as a drying agent. Extracts were analyzed by GC-FID.

RESULTS

Soil TPH concentrations in both the natural attenuation and vegetation+nutrient treatments decreased relative to the initial TPH concentrations, which were approximately 8350 mg/kg and 6200 mg/kg for diesel- and crude-oil-contaminated-soils, respectively (Figures 2 and 3). The vegetation+nutrient treatment had significantly lower TPH concentrations after approximately 650 days of treatment for both the diesel and crude oil contaminated soils. We attribute the anomaly for the crude-oil concentrations in the control treatment at 300 and 400 days to high variability resulting in a low concentration at the 300 day sampling time rather than to an apparent increase in TPH concentration at 400 days.

In the diesel-contaminated soil, species richness increased after 300 days for both the control and vegetation treatments (Figure 2). Using *d* as an indicator, these data demonstrated that bacterial diversity increased concomitant with decreases in contaminant concentration in soil. This effect, as well as the decrease in contaminant concentration, was greater in the vegetation-treated soil compared to the control soil. It is difficult to conclude the functional relationship between decreases in contaminant concentrations and species richness from these data alone. We can not yet conclude if species richness changes caused or resulted from decreases in TPH concentrations. Because time series data on species richness are relatively sparse, there are likely to be changes that are occurring that we have not observed.

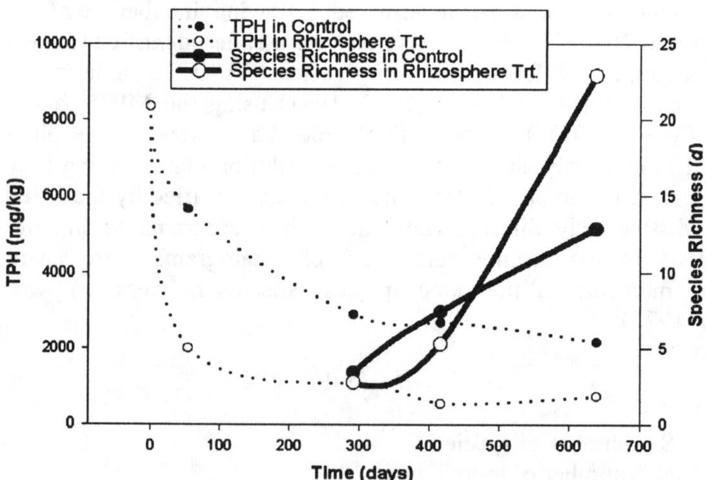

FIGURE 2. TPH concentration changes and species richness changes in diesel contaminated soil in rhizosphere-treated and control soils.

In the crude-oil-contaminated soil, species richness was lower for both the control and rhizosphere-treatments than in the diesel-contaminated soil (Figures 2 and 3). Moreover, there was little change in species richness from 300 days to 650 days, compared to the increases observed in the diesel-contaminated soils. For each treatment, TPH concentrations in the crude-oil-contaminated soil remained greater than corresponding TPH values in the diesel-contaminated soil. At 650 days, TPH values were approximately 2500 mg/kg and 1400 mg/kg for the control and vegetated treatments, respectively, in the crude-oil-contaminated soil.

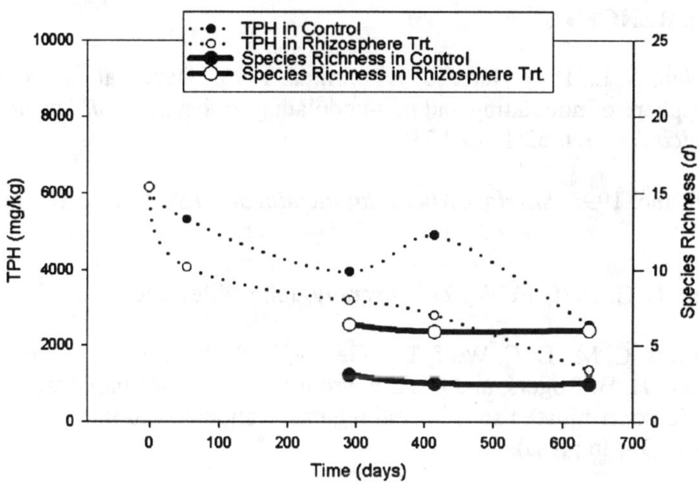

FIGURE 3. TPH concentration changes and species richness changes in crude-oil-contaminated soil in vegetation-treated and control soils.

DISCUSSION AND CONCLUSIONS

We have characterized soil bacterial communities and their changes using the species richness index. In diesel-contaminated field soils, increases in bacterial species richness were greater in the vegetation treatment than in the control. In both treatments, there was an increase in species richness as contaminant concentrations decreased. Bacterial species richness in the crude-oil-contaminated soil was both more constant and lower than in the diesel-contaminated soil. During the study, the TPH concentrations in the crude-oil-contaminated soil did not decrease as rapidly or to the same extent as that in the diesel-contaminated soil.

An improved understanding of the time-dependent relationships between contaminant concentration changes and microbial community changes, coupled with improved techniques to readily characterize microbial communities, may provide a useful tool for monitoring the functioning of phytoremediation, evaluating desirable endpoints, or both.

ACKNOWLEDGMENTS

This research was supported by the Army Environmental Quality Technology (EQT) program, Project BT25-EC-B06 "Biodegradation Processes of Explosives/Organics Using Cold Adapted Soil Systems," the Strategic Environmental Research and Development Program (SERDP), Project CU-712-

Army "Enhancing Bioremediation Processes in Cold Regions," and the Environmental Security Technology Certification Program (ESTCP), Project #1011, "Field Demonstration of Rhizosphere-Enhanced Treatment of Organics-Contaminated Soils on Native American Lands with Application to Northern FUD Sites."

REFERENCES

Cattelan, A. L., P. G. Hartel, J. J. Fuhrman. 1998. "Bacterial composition in the rhizosphere of nodulating and non-nodulating soybean." *Soil Science Society America Journal.* 62:1549-1555.

MIDI, Inc. 1995. *Sherlock Microbial Identification System*. MIDI Inc. Newark, DE.

Pielou, E. C. 1975. *Ecological Diversity*. John Wiley and Sons, New York.

Reynolds, C. M., D. C. Wolf, T. J. Gentry, L. B. Perry, C. S. Pidgeon, B. A. Koenen, H. B. Rogers, and C. A. Beyrouty. 1999. "Root-based treatment of organic-contaminated soils in cold regions: Rationale and initial results." *Polar Record* (35: In press).

Sasser, M. 1990. "Identification of bacteria through fatty acid analysis." pps. 199-204. In: *Methods in Phytobacteriology*. Z. Klement, K. Rudolf, and D. Sands, (eds) Akademiai Kiado, Budapest.

Sasser. M. and M.D. Wichman. 1991. "Identification of microorganisms through use of gas chromatography and high-performance liquid chromatography." P. 111-118, in *Manual of Clinical Microbiology.*, fifth edition, W.J. Hausler, Jr,. K.L. Herrmann, H.D. Isenberg, H.J. Shadomy, (eds.) American Society for Microbiology, Washington, D. C. 1364 pp.

Zuberer, D.A. 1994. "Recovery and enumeration of viable bacteria." In R.W. Weaver (ed.) *Methods of Soil Snalysis.* Part 2. J. Soil Sci. Soc. Am., 5, 119-144.

GROWING BIOMASS TO STIMULATE BIOREMEDIATION: TECHNICAL AND ECONOMIC PERSPECTIVE

Joop Harmsen, Antonie van den Toorn, Dethmer Boels (SC-DLO, Wageningen)
Bert Vermeulen (IMAG-DLO, Wageningen)
Wim Ma (IBN-DLO, Wageningen)
Jaap van der Waarde (Bioclear, Groningen)
Rik Duijn (De Vries & van de Wiel, Schagen)
Ruud Kampf (Uitwaterende Sluizen, Edam, The Netherlands)

ABSTRACT. Landfarming combined with biomass production is a concept developed by DLO for removal of PAH and mineral oil with natural micro-organisms from contaminated sediments. The principle is the application of wet sediments to a willow crop without special soil protection provisions. After remediation the sediments can be used for agriculture soil or earth construction material. The willow enhances the remediation process and reduces the sanitation costs through a financial yield of the biomass. The biomass is used for generating energy and contributes to the Dutch Energy Program. This project aims at the development of a simple, extensive and cost effective remediation technology with (low) acceptable risk levels.

INTRODUCTION

The Netherlands is situated in the delta of some big European rivers. Part of the country is situated below sea level. To secure accessibility of harbours, navigability and water carrying capacity of watercourses, it is necessary to dredge harbours, waterways and ditches. A significant portion of the dredged sediments is polluted with polycyclic aromatic hydrocarbons (PAH) and mineral oil. Both contaminants can be removed using biological treatment methods. Instead of expensive high-tech methods, the more extensive and cheap treatment methods should be used for cleaning large quantities of dredged sediments.

Possibilities of landfarming. Landfarming is a well-known, relatively simple and inexpensive biological treatment technique to clean up contaminated soils, even for removal of residual concentrations (Harmsen et al., 1994). First step in landfarming is an intensive treatment in which the readily available contaminants are removed. During the second step, an extensive (intrinsic) treatment, the poorly available part of the contaminant is removed.

Biodegradation of PAH and mineral oil occurs under aerobic conditions. Originally, sediments are anaerobic and all the pores are completely filled with water. Therefor landfarming of sediments includes dewatering and ripening. The sediment has to be converted into a soil with gas filled pores. Under favourable conditions this conversion requires not more than one year, while under more natural conditions this process takes several years.

Biological degradation of PAH and mineral oil already starts during the dewatering process. The biological available part is removed in one or two years.

The residual concentration is often too high for reuse of the soil and an additional extensive (intrinsic) treatment is necessary. During this treatment aerobic conditions should be maintained in the soil. In an experiment lasting several years, it has been observed that degradation continues and it seems possible to obtain a clean soil. However years to decades will be necessary (Harmsen et al., 1997).

Beneficial land use. Although intrinsic landfarming is a cheap method with respect to labour and energy, the claim on land is high. Combined with a long treatment time the costs may become high, because it is a product of treatment time in years and the annual costs of land-use. In a dense populated country as the Netherlands, these costs are high, even in rural areas. After removal of the bioavailable part of contaminants, environmental risks are low, which makes beneficial land-use possible.

To maintain aerobic circumstances for degradation, not all kinds of land use are possible. Growing certain crops on the sediment enhances soil structure development and aeration of the soil. However some limitations are encountered:

- Growing of non-consumable vegetation. Although uptake PAH or mineral oil by crops is not likely, people will not buy products grown on polluted soil.
- Growing of vegetation during several years. This offers the possibility to grow crops with a life cycle of more then one year.
- There will always be a fear for spreading of contaminants after harvesting. It is therefor necessary to process or use the crops under controlled conditions.

The following beneficial uses fulfilling these limiting conditions:

- Nature development. It takes several years before the desired nature has been developed and during this period the soil may become clean.
- Growing of biomass for production of electrical energy.

The second option has been investigated to resolve some questions:

(1) Maximum layer thickness of the sediments.
(2) Environmental risks associated with the toxic substances in the sediments.
(3) Conditions for the practical application of the concept.
(4) Remediation costs.
(5) Supply of sediments to be treated according to the proposed technology.

MATERIALS AND METHODS

Such a long-term research has to be carried out under controlled field conditions. To make this possible IMAG-DLO, SC-DLO and De Vries en van de Wiel have created an 'experimental site' (EUROJOULE) with an initial size of 10 hectares.

The concept has two possible starting points and both are investigated on the EUROJOULE-site:

- Planting of willow cuttings as soon as foot can be set on the sediment. It is necessary to do this on a regular way to make harvesting possible after 3 or 4 years. To save time, cuttings were planted in already dewatered sediment.
- Making use of an existing willow crop. Before spreading of the wet sediment, the willows were harvest and cut on such a height that the remaining length was longer then the thickness of the applied layer of sediment (figure 1)

The project has started in autumn of 1997 and includes (1) a large scale field experiment to study physical and chemical processes involved in remediation on plots with different layer thickness of the (partly dewatered) sediment and leaching risks of toxic substances, (2) a small scale field experiment to study the recovery of a willow crop after application of wet sediments, (3) laboratory research to estimate the potential degradation rate, ecological risks and physical sediment characteristics relevant for dewatering and ripening and (4) a desk study to develop and apply a method to estimate the regional distributed supply of actual and future quantities of sediments to be treated according to the concept. The research is not only focused on the biological processes. The products of the combination are clean soil and biomass. The combination has to be optimised in relation to these two products.

FIGURE 1. DLO-concept in development. Wet sediment applied in a willow crop

RESULTS AND DISCUSSION

TECHNICAL PERSPECTIVE.

Dewatering and ripening. For Dutch circumstances 1998 was a very wet year, 1200 mm of rainfall instead of the average of 800 mm. For dewatering of the sediment a period with an evaporation surplus is necessary and these periods have been very short, also in the summer period. Due to the formation of cracks drainage of the fields occurs and the surplus of rain has been drained efficiently. Despite the wet conditions, ripening continued, but at a sub-optimal rate. The oxygen supply in the dewatered sediment with cuttings, however, did not improve in 1998 (Figure 2). The wet sediment in the existing crop remained anaerobic during the whole year. Formation of new roots was restricted to 10-20 cm around the plants in both fields. The moisture content in this zone was enough for development of the willow.

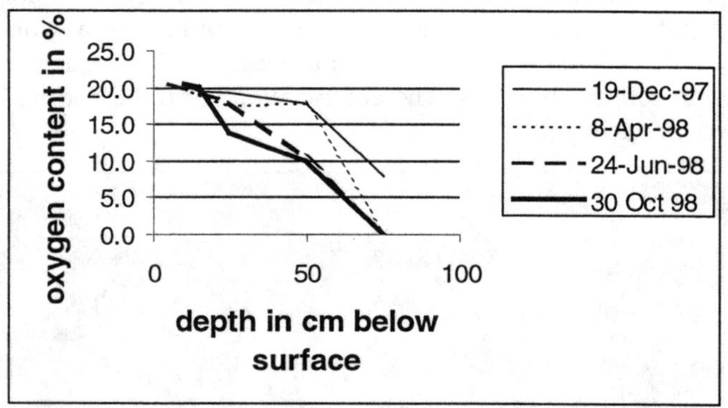

FIGURE 2. Oxygen concentration in gas filled pores in the experimental field with 75-cm dewatered sediment

In a normal, less wet year, ripening and improvement of the oxygen supply will occur. From observations and model calculations it could provisionally be concluded that full ripening is likely for thicknesses of ripened sediment as given in table 1

TABLE 1 Maximum thickness of ripened sediment with different crops

Crop	Thickness (m)
Grass	0.35 – 0.4
Normal arable	0.55 – 0.9
Willow	0.55 – 1.05

Biodegradation. Laboratory batch tests show a non-significant decay of PAH and mineral oil. Apparently, the bioavailability is very small. This shows that years will be necessary to improve the quality of the sediment. These years are available in the described bioremediation concept.

Field measurements of the soil air composition show that aerobic processes prevail; CO_2 development, CH_4 and H_2S were not present. Also in the field significant degradation did not occur. The PAH concentration remained on the same level (54 mg/kg d.m. at T=0 and 52 mg/kg d.m. at T=280 days, standard deviation were respectively 14 and 11 mg/kg d.m.). From the partition of the PAH over the different ring systems it can, however, be concluded that some biodegradation has to occur. The fraction of the more easily degradable 2+3 ring PAH decreases (Figure 3)

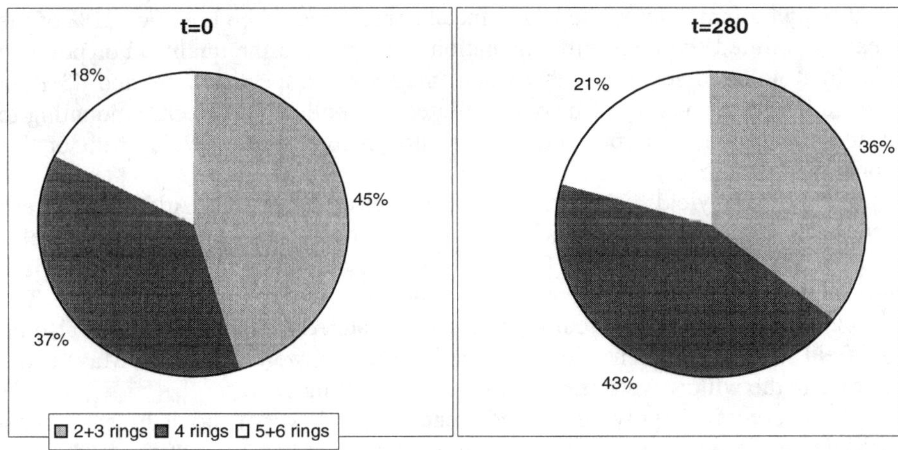

FIGURE 3. The partition of PAH over different ring systems at T = 0 and T = 280 days

Risks. PAH, mineral oil and heavy metals were measured in the water phase in the sediment, just below the sediment, in the water of the sub surface drainage system and open drains. There was no increase in concentration, except for zinc in the water just below the sediment. This increase could be attributed to changes in the soil water composition:

- Increase of salt content
- Decrease of pH
- Increase of the CO_2 pressure
- Changes of the dissolved organic matter

These changes are responsible for a shift in the partitioning of zinc over de adsorbed and dissolved fractions. The solubility of zinc already present in the layer below the sediment has increased. This is a natural effect, which always will occur when sediment is spread out on existing soil. The long-term effects need further investigation.

Results of bioassay with *Lumbricus rubellus* show no significant ecological risk (reproduction and mortality).

ECONOMIC PERSPECTIVE

The described concept will only be successful if costs are low, and if there is a social acceptance. The NIMBY effect is also for this concept very important. Especially after applying the black sediment the land looks unattractive. Demonstration projects like this are important to show that within limited time an acceptable situation is achieved.

To estimate the present and future supply of sediment for this concept data of the waterboard "Hooghoomraadschap Uitwaterende Sluizen' have been used. This waterboard is responsible for maintaining canals and ditches in the north-west part of the Netherlands. Remediation is only meaningful if quality improves. The qualifying parameters are PAH and heavy metals. To improve the quality by bioremediation the heavy metal content should be low. For 21% of the heavily polluted sediment bioremediation will improve the quality. For polluted and light polluted sediments these percentages are respectively 59 and 43%. In the study area a quantity of heavily polluted and polluted sediment amounting to 5 and 10 million m^3 is present. This is enough for further development of this concept

A high yield of biomass will decrease the costs. Adding of fresh sediments stimulates grow of willows in an existing crop. At least normal yields will be obtained. Because the first yield will be obtained after 3 or 4 years it is not clear yet if the sediment will give restrictions for harvesting. The development of the willow cuttings on the dewatered sediment was less, also on the field without sediment. The short period (cutting were planted in May) and a disease in the willow were responsible for the limiting grow.

The cost of applying the sediment will be lower if the sediment can be brought to the willows directly after dredging. Cost can be reduced with good logistics. Especially the presence of a reservoir near the willows is necessary. The costs of this concept with a remediation period of 10 years are roughly \$ 25/m^3. Traditional landfarming during a period of 10 years in a controlled containment will cost ca \$ 65/m^3.

ACKNOWLEDGEMENTS

This study was part of the Research Programme of the Ministry of Agriculture, Nature and Fishery and part of the Dutch Research Programme Bio-Technological In-Situ Bioremediation (NOBIS)

REFERENCES

Harmsen, J., H.J. Velthorst and I.P.A.M. Bennehey. 1994. "Cleaning of residual concentrations with an extensive form of landfarming". *In R.E. Hinchee, D.B. Anderson, F.B. Blaine and G.D. Sayles (Eds). Applied Biotechnology for Site Remediation*, pp 84-91. Lewis Publishers, Boca Raton.

Harmsen, J., H.J.J. Wieggers, J.J.H. van den Akker, O.M. van Dijk-Hooyer, A. van den Toorn and A.J. Zweers. 1997. "Intensive and extensive treatment of dredged sediments on landfarms". *In In Situ and On-Site Bioremediation*: Volume 2, pp 153-158. Battelle Press, Columbus.

PHYTOREMEDIATION OF CREOSOTE-CONTAMINATED SURFACE SOIL

Glendon J. Fetterolf (Black & Veatch Corporation, Charlotte, North Carolina)
John T. Novak, Scott B. Crosswell and Mark A. Widdowson (Virginia Polytechnic Institute & State University, Blacksburg, Virginia)

ABSTRACT: At the site of a former creosote treatment facility, a grass phytoremediation field study was initiated in July 1997. This study is part of a larger site remediation project. The site is contaminated with polycyclic aromatic hydrocarbons (PAHs). A grass study area and a test matrix consisting of 36 planted (fescue and rye grasses) and unplanted cells was designated. The focus of the study was to evaluate PAH remediation in fertilized unplanted, fescue and rye grass soil. Cell soil samples were collected from a depth of 15 to 21 cm. Data from four sampling periods, t=0, t=9, 12 and 17 months, is presented. The six most prevalent PAHs, acenaphthene, fluorene, phenanthrene, fluoranthene, pyrene and chrysene, were quantified. At t=9 months, substantial loss of the five lowest molecular weight (LMW) PAHs had occurred and was attributed to natural attenuation. Phytoremediation did not enhance removal of the 5 LMW PAHs. During the first 9 months, below average precipitation delayed grass root development. Between t=9 and 12 months, above average precipitation was experienced and accelerated chrysene removal rates occurred in planted cells. Over the last 8 months of the study, minimum concentrations of acenaphthene, fluorene and phenanthrene were attained. Increased PAH aqueous solubility correlated to decreased constituent soil concentrations.

INTRODUCTION

Creosote, a compound containing 85% polycyclic aromatic hydrocarbons (PAHs) by weight, is commonly used as a wood preservative in the railroad industry and in other commercial applications (Mueller et al., 1989). Certain PAHs have been associated with toxic, mutagenic and/or carcinogenic effects in humans and aquatic organisms. The U.S. Environmental Protection Agency (U.S. EPA) has designated 16 PAHs as priority pollutants. In order to protect public health, remediation of PAH-contaminated hazardous waste sites is of fundamental importance.

Phytoremediation, an innovative form of bioremediation technology, involves the use of green plants to degrade contaminants. Phytoremediation has been reported to augment contaminant removal in two ways (Schnoor et al., 1995). Direct phytoremediation involves contaminant uptake into the root structure, subsequent plant metabolism and either volatilization and/or incorporation of the degradation by-products into the plant. Indirect phytoremediation, or rhizosphere remediation, involves augmentation of microbial populations capable of contaminant degradation in the plant root zone. Root exudates and decomposing root materials are comprised of readily degradable carbon-containing byproducts that support growth and activity of larger contaminant-degrading microbial communities. Aqueous phase contaminants are the most susceptible to plant or microbial utilization. Data from phytoremediation field studies at PAH or creosote-contaminated sites is limited.

Objective. The objective of this study was to assess the affects of phytoremediation on creosote-contaminated surface soils over a 17-month period.

Site Description. From the early 1950s until 1973, the site was used as a railroad cross-tie treatment facility. The facility consisted of an above ground creosote storage tank, a treatment unit and a holding pond. The treatment facility has been inactive since the 1970s. In October 1990, creosote seepage into a nearby creek bank was observed and reported to the state department of health and environment.

Over the next seven years, site assessment activities, including installation of monitoring wells and collection of soil, ground water and sediment samples, were performed. In 1991, eleven test pits were excavated to assist in site characterization and enable construction of a ground water collection trench. The contaminated soil was placed in three on-site piles. In 1997, the state granted permission to spread these piles over a test area to conduct a phytoremediation surface soil study.

Test Matrix. The 92 feet (28 m) by 32 feet (9.75 m) grass study area was subdivided into thirty-six, 4 foot (1.22 m) square cells. Cell matrix layout was based on a statistically random design with emphasis on control cells.

MATERIALS AND METHODS

The grass study was initiated in July 1997. One hundred and forty-two soil samples (t=0) were taken to allow adequate characterization of the study area. Fertilizer (10-10-10) and fescue and rye seed was dispersed to all cells and designated cells, respectively. Due to extremely dry soil conditions during the first 4 weeks of July, grass growth in planted cells was negligible. In late July, agitation of the top 3 inches (7.6 cm) of soil, addition of organic mix, reseeding of planted cells, watering and a final cover with bale straw occurred. Extremely saturated and dry subsurface conditions made sample collection impossible in February and September of 1998, respectively. Fifty-nine, sixty-three and ninety-nine samples were collected in March (t=9 mos.), June (t=12 mos.) and November (t=17 mos.) of 1998, respectively.

Soil Sampling and Analysis. Soil samples were obtained with a tube-type soil sampler from a depth of 6 to 8.5 inches (15 to 21 cm) below land surface (bls). During each sampling period, multiple samples were collected from each sampled cell in similar and random locations. A lab-validated procedure was used to extract the soil. Five grams of soil and 15 mL of methylene chloride were combined in a 40 mL amber vial. Extraction vials were agitated and vial supernatant was transferred to a 1.5 mL amber gas chromatography (GC) vial for analysis. Custom-made external PAH standards were used to quantify PAH concentrations.

Selection of Monitored PAHs. Prior to initiation of the grass study, PAH analytical results were reviewed from on-site soil borings collected in 1997. Soil concentrations of the 16 PAHs were quantified. Acenaphthene, fluorene, phenanthrene, fluoranthene, pyrene and chrysene represented the majority of total soil PAH contamination. In June 1997, soil samples were collected and extracted at Virginia Tech to confirm the prevalence of the 6 PAHs.

RESULTS AND DISCUSSION

PAH Concentrations. Box and probability plots were used to assess PAH concentration distributions for planted and unplanted cells at the four sampling periods. Fiftieth percentile concentrations were used to assess differences in PAH concentrations between sampling periods and treatments (control, fescue and rye). Acenaphthene and chrysene results will be discussed in detail.

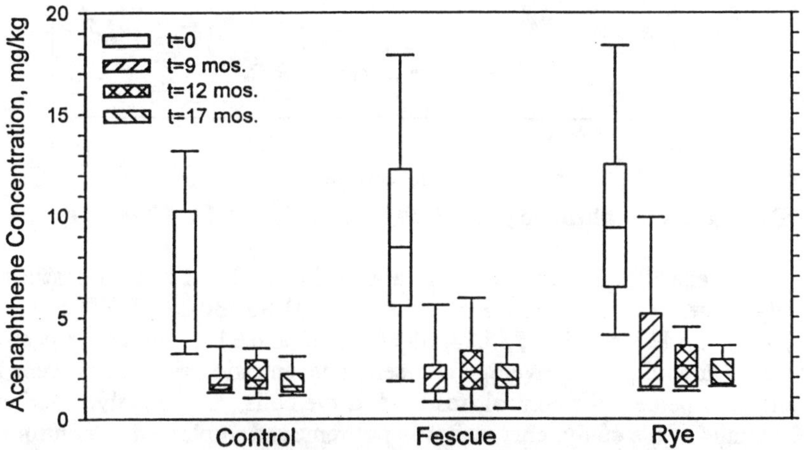

FIGURE 1. 5th/95th percentile box plot: acenaphthene concentrations at t=0, t=9, 12 and 17 months.

The acenaphthene box plot (Figure 1) indicates that concentration reductions occurred in all treatments over the study period. The largest concentration reduction occurred during the first 9 months. Fluorene, phenanthrene, fluoranthene and pyrene results were similar. The 50th percentile acenaphthene concentrations for control, fescue and rye plots at t=0, t=9, 12 and 17 months were 7.28, 1.69, 1.90 and 1.62; 8.35, 2.18, 2.27 and 1.91; 9.40, 2.58, 2.58 and 2.24 mg/kg, respectively. Overall acenaphthene concentration reductions were 77.8, 77.1 and 76.2 percent for control, fescue and rye plots, respectively. The data strongly suggests that phytoremediation did not augment acenaphthene removal. Natural attenuation processes should be attributed to acenaphthene removal. Fluoranthene and pyrene concentrations in control cells remained relatively constant for the remainder of the study, while planted cell concentrations slowly declined to levels present in control cells over the remaining 8 months.

Acenaphthene concentration trends are also clearly represented by the probability plot (Figure 2). Initial soil concentrations are clearly higher than concentrations at the other sampling periods. Over the last 3 sampling periods, a minimum concentration of acenaphthene was attained. This suggests that further acenaphthene removal may be limited. Similar results were experienced with fluorene and phenanthrene in planted and control cells.

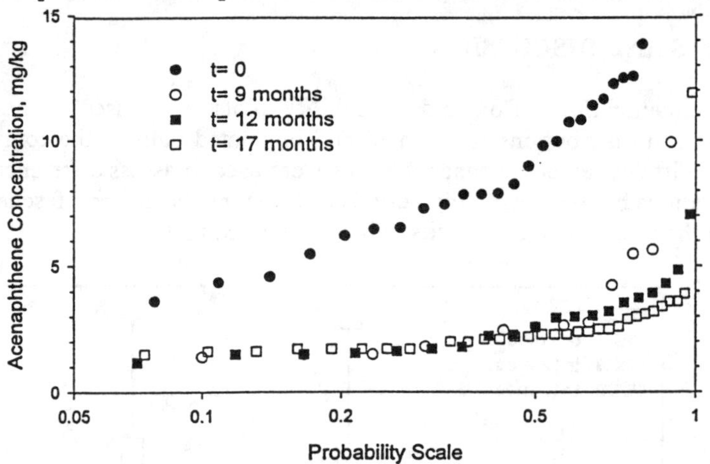

FIGURE 2. Acenaphthene probability plot:t=0, t=9, 12, 17 mos. rye data.

Chrysene 50th percentile concentrations (Figure 3) for control, fescue and rye cells at t=0, t=9, 12 and 17 months were 31.90, 30.28, 26.67 and 27.15; 31.57, 32.48, 24.86 and 25.71; 35.78, 35.66, 29.55 and 31.49 mg/kg, respectively. Over the 17-month period, average chrysene concentration reductions were 14.9, 18.6 and 12.0 percent for control, fescue and rye cells, respectively. During the first 9 months of the study, chrysene was not removed in planted or control cells. Over the next 3 months, chrysene concentration reductions were experienced in both planted and control cells. Planted cells experienced a slightly higher level of removal during this period. Higher removal rates coincided with increased precipitation (Figure 4). Increased levels of precipitation may have resulted in increased aqueous phase concentrations of chrysene. Stagnant soil pore water could have predicated mass transfer of chrysene from soil to the liquid phase potentially making the constituent available for plant or microbial uptake. At t=17 months, chrysene concentrations were stable, yet slightly elevated in planted and control cells.

PAH aqueous solubility is directly related to PAH soil concentration at t=17 months (Figure 5). The aqueous solubility of acenaphthene, fluorene, phenanthrene, fluoranthene, pyrene and chrysene is approximately 3.9, 1.8, 1.3, 0.2, 0.14 and 0.0015 mg/L, respectively. The relatively high aqueous solubility of acenaphthene and fluorene resulted in the lowest PAH soil concentrations after 17

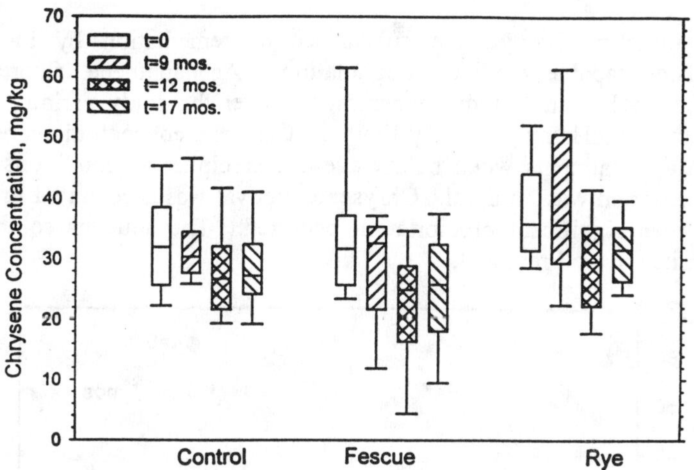

FIGURE 3. 5th/95th percentile box plot: chrysene concentrations at t=0, t=9, 12 and 17 months.

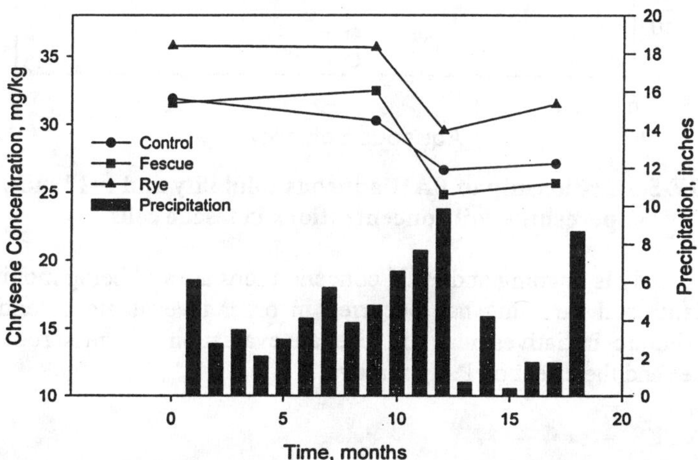

FIGURE 4. The influence of precipitation on chrysene concentration.

months. Conversely, chrysene removal was strongly impeded by a very low aqueous solubility. The t=17 mos. data clearly indicates that increasing PAH aqueous solubility correlates well with decreasing constituent soil concentrations.

CONCLUSIONS

Significant remediation of five of the six monitored PAHs, specifically acenaphthene, fluorene, phenanthrene, fluoranthene and pyrene, occurred in both planted and control cells. Phytoremediation did not augment LMW PAH removal. Natural attenuation processes were responsible for decreased LMW PAH concentrations. Removal of LMW PAHs can likely be attributed to constituent mobility through the subsurface or remediation by indigenous microorganisms capable of PAH degradation. Acenaphthene, fluorene and phenanthrene reached minimum concentrations over the study period. Further removal of these PAHs may be very limited. Chrysene concentrations remained the same in all treatments when below average precipitation conditions existed and root penetration was minimal. Chrysene removal was accelerated in planted cells only when significant precipitation occurred. The aqueous solubility of chrysene limited overall removal.

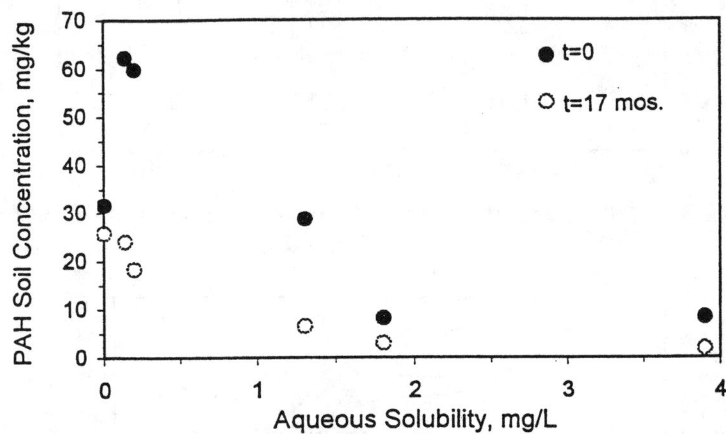

FIGURE 5. Relationship of PAH aqueous solubility and t=17 mos. 50th percentile soil concentrations in fescue cells

This study is ongoing and PAH concentrations are still being monitored. Multiple depth soil sampling has occurred in order to evaluate potential PAH transport. Future initiatives may include an evaluation of grass root organic matter cycles and the effect on PAH removal.

REFERENCES

Fetterolf, G. J. 1998. *Characterization of a creosote-contaminated tie yard site and the effects of phytoremediation.* M. S. Thesis. Virginia Polytechnic Institute & State University, Blacksburg, VA.

Mueller, J. G., P. J. Chapman, and P. H. Pritchard. 1989. "Creosote-contaminated Sites." *Environ. Sci. Technol. 25*(6): 1045-1054.

Schnoor, J. L., L. A. Licht, S. C. McCutcheon, N. L. Wolfe, and L. H. Carriera. 1995. "Phytoremediation of Organic and Nutrient Contaminants." *Environ. Sci. Technol. 29*(7): 318A-323A.

PHYTOTOXICITY AND PHYTOREMEDIATION STUDIES IN SOILS POLLUTED BY WEATHERED OIL

Augusto Porta, ***Nadia Filliat*** and Nadia Plata (Battelle, Geneva, Switzerland)

ABSTRACT: Following an oil well blowout that occurred near Trecate, Italy, phytotoxicity and phytoremediation studies were conducted. The purpose of these studies was to 1) understand the impact of hydrocarbon residues on local crops and vegetation and measure the uptake of polycyclic aromatic hydrocarbons (PAHs), and 2) assess whether specific plants contribute to degradation of hydrocarbons in the soil through direct uptake or by enhancement of microorganismal degradation. Greenhouse and on-site experiments were conducted on rice and corn crops for the phytotoxicity study. Greenhouse plants were sown on oil-contaminated soils containing different hydrocarbon levels. Total petroleum hydrocarbon (TPHC) content of the soil did not influence germination and growth of rice plants below 10,000 mg/kg. Corn plants showed signs of distress when cultivated in soil containing above 2,000 mg/kg TPHC. Total PAH concentrations were low in all parts of rice and corn plants analyzed. Because these total PAH levels were comparable to those in analyzed rice and corn plants taken from background (control) areas, we concluded that PAH uptake from the oil-contaminated soils had not occurred. The on-site phytoremediation study was conducted on soils that had been previously treated in a biopile to enhance hydrocarbon degradation. Four plant species, clover, lucern, rye-grass, and corn were sown and their biodegradation ability was compared to the one occurring in landfarmed soil. Corn and white clover exhibited the best biodegradation power, only slightly inferior to landfarming degradation.

INTRODUCTION

Following an oil well blowout that occurred in the area of Trecate, Italy, phytotoxicity and phytoremediation studies were conducted as part of a larger project to assess the impact of hydrocarbons in the area and elucidate possible methods of soil decontamination.

Our phytotoxicity studies were conducted on rice and corn plants grown both on the blowout site and in a greenhouse. Phytoremediation is the use of plants to enhance degradation of organic compounds in the soil. The phytoremediation study was conducted on-site. It entailed growing four plant species, clover, lucern, rye-grass, and corn, on soils that had been previously treated in a biopile to enhance hydrocarbon degradation, and comparing the biodegradation to the one occurring in the landfarmed soil.

The objectives of the phytotoxicity and phytoremediation studies were to evaluate the impact of hydrocarbons on rice and corn growth and to identify the potential interest to use phytoremediation to complement simple bioremediation.

MATERIALS AND METHODS

Phytotoxicity study. For the on-site study, rice was cultivated 2 and 3 years after the blowout on soils containing 20-500 mg/kg TPHC and 200-1000 ng/g PAH. 20 grain samples and 4 background ones were analyzed for PAH content.

Soils used for the greenhouse rice cultivation were taken from the oil-contaminated area. Soils used for the corn cultivation were taken from the biopile. Four levels of total petroleum hydrocarbons (TPHC) in soils [mg/kg] were used for each study: for rice, A: 400; B: 9,000; C: 15,000; D: 24,000; for corn, A: 200; B: 1,800; C: 2,700; D: 4,400. For both rice and corn studies, soil A was used as reference. Two cultivars of each crop were used in each experiment: Ballila and Thaïbonnet for rice, Starix and Tempra for corn. Plants were cultivated in pots containing 20 liters of soil. 5 pots were assigned to each cultivar and each TPHC level. One pot without plants was assigned to each level of TPHC content. Real cultivation conditions were approximated as nearly as possible. Soil samples were collected and analyzed prior to planting (t=0), during the experiment (t=1, 1.5 months after sowing [rice] and 1 month [corn]), and at the end of the experiment (t=2, 6 months [rice] and 4 months [corn]). Plant samples were collected and analyzed during and at the end of the experiment (t=1 and t=2, respectively). The germination, growth and maturation of the plants were tracked via periodic observation throughout the experiment. The details of the soil and plant analyzes are provided below.

Phytoremediation study. As previously mentioned, the experiment took place on soils that had been previously treated in a biopile. Seeds of different crops, rye-grass, corn, lucern and white clover, were sown and compared to landfarming according to the statistical agronomic model: 4 parcels with crops + 1 landfarmed parcel, with 5 repetitions, giving a total of 25 parcels. Crop development was followed through visual observation and periodic collection and analyses of plant and soil samples. The details of the soil and plant analyzes are provided below.

Chemical Analyses. For both the phytotoxicity and the phytoremediation studies, soils were analyzed for TPHCs and PAHs using the following methods. The extraction of TPHCs and PAHs were carried out with supercritical fluid (SFE, Isco SFX^{TM} 3560, CO_2, 8% methanol, 350 bar, 80°C, 2 ml/min, 50 min) or accelerated solvent extraction (Dionex ASE^{TM} 200, CH_2Cl_2/acetone (7/3: v/v) 100°C, 140 bar, 3x5 min). For both studies, the different plant parts were analyzed separately for PAHs. For the on-site rice, only the grains were analyzed. For the greenhouse rice, stalks and leaves, and grains were analyzed. For corn, stalks and leaves, grains and spathes (leaves protecting the spike) were analyzed. Extractions were conducted by SFE (Isco SFX^{TM} 3560, CO_2, 300 bar, 70°C, 60 min, 1.5 ml/min), ASE^{TM} (Dionex ASE^{TM} 200, CH_2Cl_2, 100°C, 140 bar, 3x10 min) or Soxhlet (CH_2Cl_2, 24 hours). TPHC analyses were carried out using GC/FID: (Perkin Elmer Autosytem, column BPX5 (SGE) 25 m x 0.32 mm; film 0.25 μm; carrier gas helium; inj. 300°C; oven 50°C (5 min) – 5°C/min → 320°C (10 min) – 10°C/min → 350°C; det. 300°C). PAH analyses were carried out using GC/MS: (HP 5890 series II, MSD 6972, column J&W DB5 MS 30 m x 0.32 mm; film

0.25 μm; carrier gas helium; inj.300°C; oven. 40°C (1 min) – 6°C/min → 290°C (28 min); and det. 280°C).

RESULTS AND DISCUSSION

Phytotoxicity

On site study. PAH concentrations in grains of rice grown in soils, whose TPHC and PAH concentrations were respectively 20-500 mg/kg and 200-1000 ng/g, ranged from 30 to 160 ng/g. This range of values was equal to or less than that of rice grains from the background area. Profile comparison did not show any correlation between soil and rice grain PAH concentrations, indicating that it was unlikely that PAHs had been taken up by these plants from the soil (Figure 1). PAHs found in the grains is likely due to adsorption of atmospheric PAHs (Nakajima et al., 1995; Simonich et al., 1994).

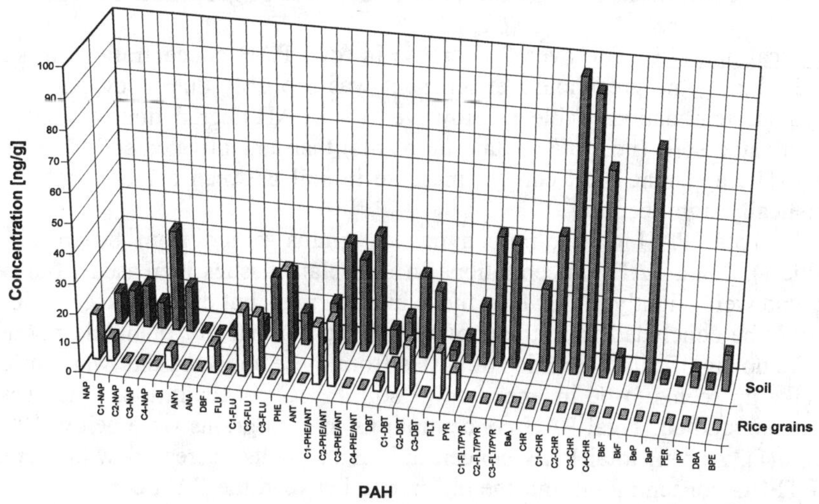

FIGURE 1. PAH profiles of a rice grain sample and the corresponding soil.

Greenhouse experiment

Growth observations. At TPHC contamination levels of 8,000-9,000 mg/kg (C) and 20,000 mg/kg (D), rice roots exhibited difficulties penetrating the soils. Algae and weed development was observed in reference soil (A). However, rice plants did not exhibit signs of stress during growth and maturation. No particular observations were made for soil B. Soil TPHC content influenced corn growth and had a marked effect on plant aerial biomass. The difference between the plant biomass for the different soils was statistically significant. The plants grown on soils uncontaminated by the oil well blowout (A) had a biomass twice that of the plants grown on the most polluted soils (D) (Figure 2). Hydrocarbons in the soils also led to a delay in flowering and an increase in protandry (condition where male elements mature and are shed before female elements mature).

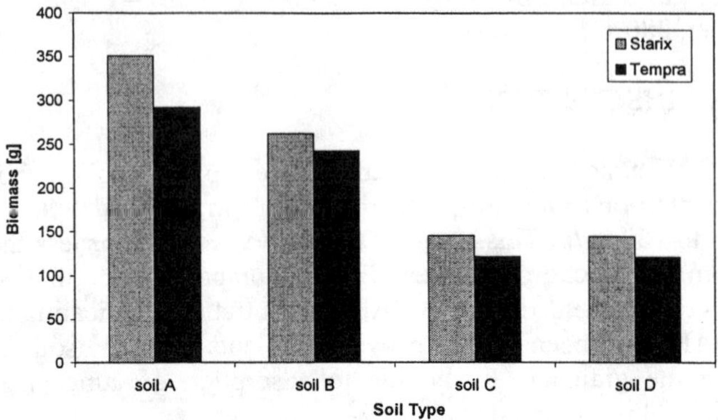

FIGURE 2. Average total aerial biomass of greenhouse corn.

Chemical analyses. The overall decrease in total PAH concentrations in soils C and D during the greenhouse experiment was probably due to evaporation and microbial degradation (Table 1). However, total PAH degradation was not higher in cultivated soils (i.e. pots containing plants) than uncultivated soils (i.e. control pots). The apparent PAH content increase in soil between t=1 and t=2 was not statistically significant.

Total PAH concentrations in rice grains varied from 80 to 115 ng/g (Table 1). Total PAH concentrations in rice plant tissues increased from t=1 to t=2, and were slightly higher than those in rice grains but did not exceed 180 ng/g. It could be concluded from the above observations that rice plants do not appear to take up significant quantities of PAHs from these soils and that what little they do take up seems to be independent of soil TPHC concentration, at least within the range studied. Total PAH concentrations in corn grains were below 100 ng/g (Table 1). At t=1, total PAH concentrations in plants decreased with increasing soil TPHC concentration, but the difference between the PAH contents was only statistically significant for Starix. However, at t=2, these values were constant, regardless of soil TPHC level, and lower than the t=1 values. At t=2, total PAH concentrations in plants and spathes ranged from 140 to 320 ng/g and from 120 to 220 ng/g, respectively. Comparison of PAH concentration profiles of plants, grains and soil did not show any correlation between the PAH content of rice and corn, and that of soil, indicating no apparent uptake. The PAH content in rice and corn plants is likely partly due to adsorption of atmospheric PAHs on aerial plant surfaces. Since the greenhouse is a relatively closed structure, it could favor maintenance of evaporated PAHs from the soils in the atmosphere. This phenomenon could explain the identical values found in rice and corn plants cultivated in the same airspace but in soils with differing TPHC levels.

TABLE 1. Total PAH concentrations [ng/g] in greenhouse crops and soils.

Rice	t = 0	t = 1		t = 2		
1996	soil	soil	plants	soil	plants	grains
Soil A						
Average Balilla	**624**	**15**	**39**	**1017**	**108**	**104**
Average Thaibonnet	**475**	**16**	**64**	**362**	**148**	**110**
Control (pot without plant)	560	5		1250		
Soil B						
Average Balilla	**38049**	**20019**	**46**	**21957**	**117**	**83**
Average Thaibonnet	**38068**	**32124**	**71**	**32209**	**143**	**98**
Control (pot without plant)	37342	8842		17091		
Soil C						
Average Balilla	**82995**	**30526**	**62**	**28520**	**156**	**82**
Average Thaibonnet	**84105**	**31832**	**57**	**36822**	**174**	**72**
Control (pot without plant)	86996	24197		18754		
Soil D						
Average Balilla	**260883**	**58120**	**54**	**77233**	**160**	**96**
Average Thaibonnet	**218670**	**55209**	**88**	**97986**	**163**	**113**
Control (pot without plant)	238612	29638		46608		

Corn	t = 0	t = 1		t = 2			
1997	soil	soil	plants	soil	plants	spathes	grains
Soil A							
Average Starix	**63**	**25**	**1544**	**117**	**218**	**178**	**99**
Average Tempra	**44**	**16**	**1037**	**66**	**259**	**191**	**54**
Control (pot without plant)	102	0		89			
Soil B							
Average Starix	**3947**	**4541**	**947**	**3775**	**138**	**167**	**80**
Average Tempra	**3835**	**3180**	**817**	**3480**	**148**	**170**	**72**
Control (pot without plant)	2387	4621		2501			
Soil C							
Average Starix	**4434**	**4267**	**631**	**4449**	**275**	**219**	**96**
Average Tempra	**5206**	**4464**	**680**	**4416**	**246**	**123**	**91**
Control (pot without plant)	4370	3087		2946			
Soil D							
Average Starix	**9524**	**5117**	**502**	**5341**	**323**	**145**	**38**
Average Tempra	**8156**	**7296**	**741**	**4881**	**267**	**132**	**65**
Control (pot without plant)	11248	6537		8187			

Phytoremediation

Growth observations. The observations made from the beginning showed reduced vegetative activity in some areas but without evident correlation with soil TPHC levels.

Chemical analyses. A global decrease in average TPHC concentration in soils over time was observed for all crops (Figure 3). However, average TPHC concentrations remained stable for lucern and increased slightly for white clover during the last period (June 16 – Sept. 15). Corn and white clover exhibited the best phytodegradation ability, slightly inferior to landfarming degradation.

PAH concentrations at maturation (September) were higher in white clover and rye-grass than in the corresponding background plants, indicating some uptake of PAHs (Table 2). In contrast, PAH concentrations in lucern and corn were inferior to those in the corresponding background plants. The background plants having been sampled only at maturation, no conclusion could be drawn for the uptake of the different crops before that time. The decrease in PAH concentration in all plants over time is likely a result of increased plant biomass.

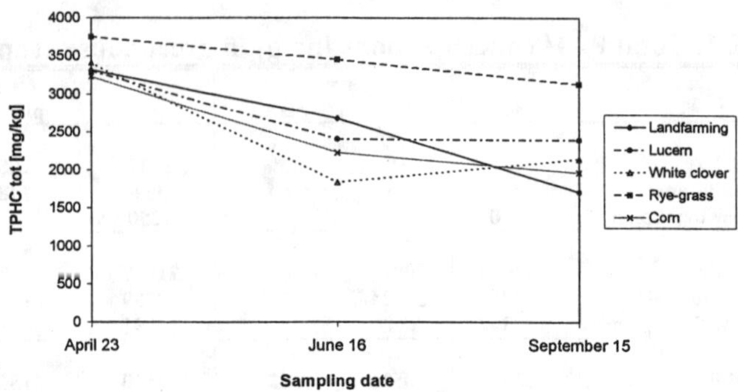

FIGURE 3. Evolution of TPHC concentrations in phytoremediation parcels.

TABLE 2. Evolution of PAH concentrations in plants.

Total PAH concentration [ng/g]	Crop	May 15	June 11	July 14	September 7
Background	White clover				493
Average	White clover		1148	1628	647
Background	Rye-grass				262
Average	Rye-grass	1962	758	681	603
Background	Lucern				495
Average	Lucern		1508	1126	388
Background	Corn				339
Average	Corn	1650	950	783	338

CONCLUSIONS

The phytotoxicity study conducted on corn and rice showed that the hydrocarbon content of the soil did not influence development of rice plants below 10,000 mg/kg TPHC, but corn plants showed signs of distress above 2,000 mg/kg TPHC. Total PAH concentrations were low in all parts of rice and corn plants and comparable to those in background rice and corn, indicating that probably no PAH uptake from the oil contaminated soils had occurred.

The phytoremediation study effectuated on corn, rye-grass, lucern and white clover showed that corn and white clover had a better degradation ability. The PAH content of white clover and rye-grass was higher than that in the background plants. These two plants, contrary to corn and rice, seem therefore to take up PAH from the contaminated soil.

Further research on phytoremediation is in progress on the site of the oil well blowout in order to understand the mechanisms occurring both in the rhizosphere area and inside the different plants.

REFERENCES

Nakajima, D., Yoshida, Y., Suzuki, J., Suzuki, S., 1995. "Seasonal Changes in the Concentration of Polycyclic Aromatic Hydrocarbons in Azalea Leaves and Relationship to Atmospheric Concentration." *Chemosphere,* 30:, 409-418.

Simonich, S.L. and Hiltes, R. 1994. "Vegetation-Atmospheric Partitioning of Polycyclic Aromatic Hydrocarbons." *Environ. Sci. Technol.,* 28: 939-943.

SHORELINE CLEANER EVALUATION ON A PETROLEUM IMPACTED WETLAND

Cydney Bizzell, Richard T. Townsend, James S. Bonner, Robin L. Autenrieth
(Texas A&M University, College Station, TX, USA)

ABSTRACT: The impact and performance of a shoreline cleaning agent, Corexit 9580 (EC9580), was evaluated by conducting a controlled release of weathered crude oil (Arabian-medium) in May 1998 at a well characterized and fully instrumented wetland research site. For this controlled oil release experiment, 21 plots were divided into three treatment regimes: six oiled, no-action control plots; six oiled plots, high application dose of Corexit; and six oiled plots, low application dose of Corexit. To determine the effectiveness of the Corexit applications, sediment samples from all plots were taken as a function of time from May, 1998, until October, 1998. These samples were analyzed for petroleum hydrocarbons (TEM and GC/MS), toxicity, microbial counts, nutrient concentrations and redox potential. Target hydrocarbon (aliphatic and polycyclic aromatic hydrocarbons (PAHs)) concentrations were normalized to 17α(H),21β(H)-hopane, a conservative biological marker, to account for physical petroleum losses and to reduce heterogeneity in the hydrocarbon data. Initial GC/MS results indicate that the Corexit 9580 did not significantly impact the removal of the petroleum from the test plots. Populations of hydrocarbon-degrading microorganisms (aliphatic- and PAH-degrading) increased following the oil application, but did not appear to be significantly affected by the application of the Corexit. Nutrient and toxicity data are currently being analyzed.

INTRODUCTION

Every year, millions of gallons of oil are introduced into marine environments from anthropogenic sources (NAS Report, 1985). Large spills resulting from the transport of petroleum products are perhaps the most devastating to the marine environment and are especially destructive to near-shore environments. Coastal wetlands are among these areas where oil damage can cause large economic losses and environmental damage. Crude oil removal will occur through physical, chemical and biological processes in wetland environments; however, rapid oil clean-up is often a difficult task. Physical clean-up methods involving large work crews and heavy equipment are difficult to implement in wetlands where they may cause further damage to sensitive plant and animal life. As a result, alternative remediation approaches are needed for wetland areas. One such alternative is the use of shoreline cleaners.

Shoreline cleaners are designed to release stranded oil from the shoreline substrates. Corexit 9580 is "a balanced formulation of selected biodegradable surfactants in a low-toxicity, highly-refined hydrocarbon solvent system" (Fiocco et al., 1991). The cleaner's high effectiveness and low toxicity was demonstrated by these researchers in both laboratory and field tests. Commonly these cleaners are used for oil removal from rocky shorelines; however, these cleaners may have uses in oil-contaminated wetlands as well.

Objective. The objective of this research was to evaluate the ability of the shoreline cleaner Corexit 9580 to enhance petroleum removal from a wetland environment.

MATERIALS AND METHODS

Site Description. The site used for the controlled crude oil application has been described previously (Mills *et al.*, 1997a) and was used for controlled oil applications in 1996 and 1997. In brief, the site is located on the lower San Jacinto River approximately 2 km upstream of the Interstate 10 bridge (approximately 29° 48' N and 95° 04' W). The site is a tidally-influenced estuarine wetland. The area fluctuates between no water coverage and complete inundation with intertidal flats exposed during low tide.

Experimental Design. A randomized complete block design was used to evaluate the three treatments: six oiled, no-action control plots; six oiled plots, high application dose of Corexit; and six oiled plots, low application dose of Corexit. The randomized block design has been used in previous experiments, similar in scale and nature, because inferences to the entire experimental system versus the specific experimental plots are possible (Venosa et al., 1996). Six blocks consisting of three plots each were oiled in this study. The treatments were randomly assigned within each block, and three plots were also maintained as unoiled controls. According to Venosa et al.(1996) this amount of replication is adequate for this type of experiment. The Arabian medium crude oil was artificially weathered by circulation in an open tank. This weathered crude oil was sprayed onto the plots in equal volumes (21 L) during a low tide (Day –1). Details concerning the oil application procedures can be found elsewhere (Mills, 1997).

Corexit Application. The Corexit was applied 24 hours after the oil application using a battery-powered garden sprayer. A high dose of Corexit 9580 (1.0 L/2.45 m^2) was applied to six plots referred to as Oiled High 9580. A low dose of Corexit 9580 (0.5 L/2.45 m^2) was applied to six plots referred to as Oiled Low 9580. The Corexit was flushed off the plots with approximately 25 gallons of water from the San Jacinto River thirty minutes after the application.

Sampling. Eleven sample sets were collected from May to October 1998. Three 5 cm diameter sediment cores were taken on a systematic, randomized sampling grid at each plot for each sampling event. The soil cores were extruded and the top 5 cm from each core was collected. These 5 cm sections were composited for each plot.

Microbial Analysis. To determine the effects of the Corexit treatments on microbial populations for both phases of this research, hydrocarbon-degrading microbial populations were monitored using a selective most-probable-number (MPN) technique. Based on the Screen Sheen technique (Brown and Braddock, 1990), this technique (Wrenn and Venosa, 1996; Haines et al., 1996) can be used to separately enumerate microorganisms capable of degrading aliphatic hydrocarbons and PAHs.

Hydrocarbon Analysis. The petroleum chemistry methods were based on those of Mills *et al.* (1997b). The petroleum chemistry methods consisted of specific target compound analysis by GC-MS. Target compound analysis allows for normalizing the hydrocarbon concentrations to an internal conservative marker within the petroleum. Interpretation of the data with less influence from the contaminants' spatial heterogeneity is also possible with normalization (Prince et al., 1994; Bragg et al., 1994). The target compound analysis performed by GC-MS quantified 27 saturate hydrocarbons, 37 polynuclear aromatic hydrocarbons, and the conservative biomarker, 17α(H), 21β(H)-hopane.

RESULTS AND DISCUSSION

Microbial Dynamics. The population dynamics of aliphatic-degrading and PAH-degrading microorganisms are found in Figures 1 and 2. The addition of crude oil (Day –1) initially increased populations of both groups of microorganisms over populations on unoiled plots. Aliphatic-degraders increased only slightly, while populations of PAH-degraders increased approximately one order of magnitude. These populations were consistently higher than populations on unoiled plots over the course of the experiment. The addition of Corexit on Day 0 did not appear to significantly affect the populations of aliphatic-degraders or PAH-degraders on any sample day.

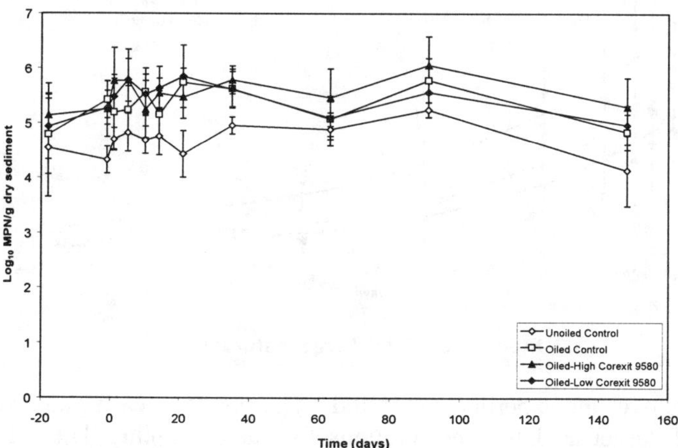

FIGURE 1. Population dynamics of aliphatic-degrading microorganisms in 0-5 cm plot segments.

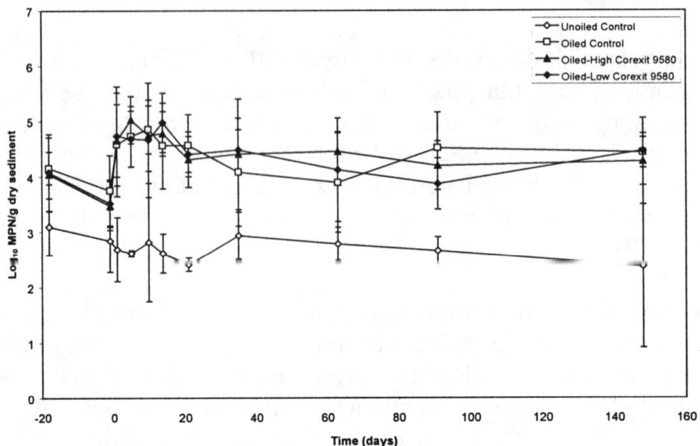

FIGURE 2. Population dynamics of PAH-degrading microorganisms in 0-5 cm plot segments.

Petroleum Chemistry. A first order differential equation was used to model the loss of the crude oil from the plots. This model has been used for previous bioremediation experiments at the site and complete model development can be found in Venosa et al., (1996). Figures 3 and 4 illustrate the removal history of the total target saturate hydrocarbons (sum of the hopane-normalized concentrations of n-C_{10}-C_{35} plus pristane and phytane) and the total target aromatic hydrocarbons (sum of the hopane-normalized aromatic concentrations)

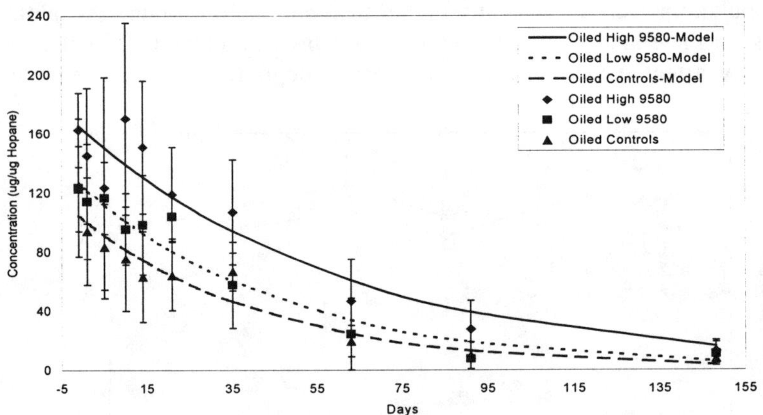

FIGURE 3. Total Target Saturates

from the plots over the experimental period. The points in each graph represent plot-averaged, target analyte concentrations for each sampling date. The solid and dashed lines are the first order models for each treatment. The addition of the

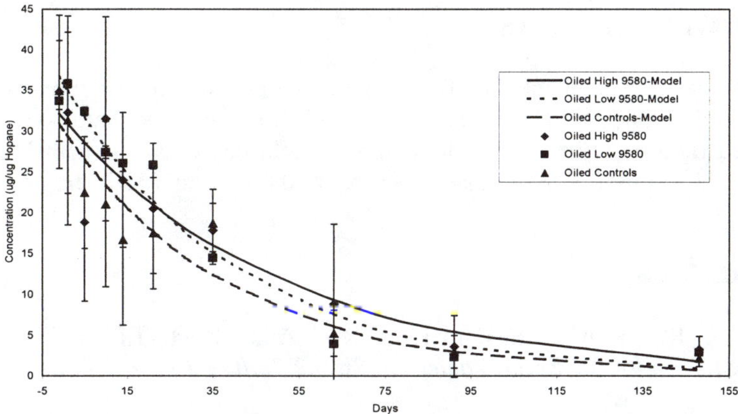

FIGURE 4. Total Target Aromatics

Corexit did not appear to remove a substantial amount of the analyzed hydrocarbons following its addition on Day 0. Higher concentrations of saturated aliphatic hydrocarbons were observed on plots receiving a Corexit applications, most likely the result of the numerous saturate compounds found in Corexit 9580 (Fiocco et al., 1991). Nonetheless, there were no significant differences between the k values (first order rate constants) for any of the treatments for either group of hydrocarbons. The site is located in a chronically exposed area, this exposure and the history of oil inputs and crude oil applications likely resulted in a rapid response to the oil's degradation. Biodegradation appeared to play a major role in removal of the petroleum hydrocarbons from the plots.

CONCLUSIONS

The removal of stranded crude oil from contaminated shorelines and coastal areas can be more deleterious to the environment than leaving the contaminant in place owing to the sensitivity of the wetlands to disturbances. A number of techniques have been examined for their potential to remove oil from such areas. Bioremediation has shown to be successful in some cases, most notably, following the *Exxon Valdez* spill in 1989 (Bragg et al., 1994). Despite its advantages, bioremediation can be relatively slow. Conversely, rapid oil clean-up of an oil spill is typically preferred. Shoreline cleaners such as Corexit 9580 offer a more direct, physical oil removal method. These products have shown some success on oiled, rocky shorelines (Fiocco et al., 1991); however, their potential has not been adequately assessed in wetland environments.

This research showed that Corexit 9580 did not significantly contribute to the removal of crude oil from wetland sediments. Natural attenuation appeared to play a major role in the loss of petroleum from the plots based on the decreases in the hopane-normalized concentrations of both saturated hydrocarbons and PAHs. Analysis of the toxicity data should further elucidate the effects of Corexit 9580 in an oil-impacted coastal wetland.

ACKNOWLEDGEMENTS

The authors wish to thank the Texas General Land Office for sponsoring and funding this project. The authors also wish to thank the hard-working and dedicated field crew including Mark Simon (field coordinator), Tom McDonald (analytical chemist), Cheryl Page, Merin Brodsky, Dan LaRiviere, and Jason Sparks.

REFERENCES

Bragg, J.R., R.C. Prince, J.E. Harner, and R.M. Atlas. 1994. "Effectiveness of Bioremediation for the *Exxon Valdez* Oil Spill." *Nature 368*: 413-418.

Fiocco, Robert, Canevari, G., Wilkinson, J., Jahns, H., Bock, J., Robbins, M., and Markarian, R. (1991) "Development of Corexit 9580 – A Chemical Beach Cleaner". From *Proceedings of the 1991 Oil Spill Conference*, p395-400.

Haines, J.R., B.A. Wrenn, E.L. Holder, K.L. Strohmeier, R.T. Herrington, and A.D. Venosa. 1996. "Measurement of Hydrocarbon-Degrading Microbial Populations by a 96-well Plate Most-Probable-Number Procedure." *Journal of Industrial Microbiology 16*: 36-41.

Mills, M. A. 1997. "Bioremediation of Petroleum Hydrocarbons in Aqueous and Sediment Environments." Ph.D. Dissertation, Texas A&M University, Civil Engineering Department, August 1997.

National Academy of Sciences (NAS). 1985. Oil in the Sea-Inputs, Fates, and Effects. National Academy Press,Washington, D.C.

Prince, R.C., D.L. Elmendorf, J.R. Lute, C.S. Hsu, C.E. Haith, J.D. Senlus, G.J. Dechert, G.S. Douglas, and E.L. Butler. 1994. "17α(H),21β(H)-Hopane as a Conserved Internal Marker for Estimating the Biodegradation of Crude Oil." *Environ. Sci. & Technol. 28*: 142-145.

Venosa, A.D., M.T. Suidan, B.A. Wrenn, K.L. Strohmeier, J.R. Haines, B.L. Eberhart, D. King, and E. Holder. 1996. "Bioremediation of an Experimental Oil Spill on the Shoreline of Delaware Bay." *Environ. Sci. & Technol. 30*: 1764-1775.

Wrenn, B.A. and A.D. Venosa. 1996. "Selective Enumeration of Aromatic and Aliphatic Degrading Bacteria by a Most-Probable-Number Technique." *Canadian. Journal of Microbiology 42*: 252-258.

MYCOREMEDIATION: A METHOD FOR TEST- TO PILOT-SCALE APPLICATION

S.A. Thomas, P. Becker, M.R. Pinza, and J.Q. Word (Battelle, Sequim, Washington)
P. Stamets (Fungi Perfecti, Olympia, Washington)

ABSTRACT: The Battelle Marine Sciences Laboratory (MSL) developed an approach called mycoremediation that begins with field collection of higher fungi from a contaminated vicinity of interest or a comparable site and that includes the steps of selection, culture, toxicity testing, screening, preconditioning, mesocosm-scale testing, and pilot-scale application. These steps result in development of proprietary fungal strains that are predisposed to remediate specific contaminants at increased efficiency under particular environmental regimes.

As an example, the MSL applied its method to remediate aged oil-contaminated soils that had been removed from sites administered by the Washington State Department of Transportation for treatment. In this pilot-scale study, the effectiveness of mycoremediation was compared with that of a bioremediation and an enhanced bacterial treatment to reduce the level of total petroleum hydrocarbons to the prescribed Washington State Department of Ecology Method A cleanup level.

The MSL also successfully applied its method to address organophosphates, such as dimethyl methylphosphonate and isopropyl methylphosphonic acid, alkaloids, estradiol, and other chemical contaminants at laboratory bench- and mesoscale. Proprietary strains were developed and shown to detect, to attack and destroy, or to inhibit the growth of bacterial contaminants, such as *Escherichia coli* and *Bacillus subtilis*, on agar and other substrata.

INTRODUCTION

There are several treatments available for the remediation of contaminated soils, such as chemical or mechanical methods, incineration, and application of biological materials. Each has advantages and disadvantages, and each may be appropriate under particular conditions or to suit particular needs. The application of fungi to break down certain soil contaminants has been successful in the speed and completeness, as well as in the economics of the treatment (Loske et al. 1990; Zazadril 1993; White and Lamar 1999). It is appropriate for remediating contaminated soil, water, and a variety of other substrata, and the end result is total, irreversible degradation of the chemical contaminants to carbon dioxide, water, and basic elements (White and Lamar 1999). In field applications, the deployment of proprietary, preconditioned fungal mycelia orchestrates the sequential participation of the native microbial community for efficient contaminant breakdown.

Based on their natural predisposition to degrade lignin, breaking it down in a stepwise depolymerization ultimately to carbon dioxide and water, the wood-degrading higher fungi can address complex compounds that are resistant to destruction by most microorganisms. They are particularly effective in degrading aromatic pollutants, as well as chlorinated and other aliphatics, and other classes of compounds. Wood-degrading fungi also are natural predators and competitors of microorganisms such as bacteria, nematodes, and rotifers, and can accordingly be adapted to aggressively remediate biological contaminants.

MYCOREMEDIATION METHOD

Working from the proven concept, Battelle Marine Sciences Laboratory (MSL) developed an approach called mycoremediation that begins with the field collection of higher fungi from a contaminated area of interest or a comparable site and that includes the steps of selection, culture, toxicity testing, screening, preconditioning, mesocosm-scale testing, and pilot-scale application. Selection begins by collecting from a field site that has contamination to be removed or other relevant characteristics, or alternatively, by choosing candidate strains from a library of previously collected and tested fungi. In the more than 20-year history of interest in using white-rot fungi in bioremediation, there has been an extraordinarily strong emphasis on a single fungal species, *Phanerochaete chrysosporium*, along with a few others. At the MSL, with our collaborator, Fungi Perfecti, we work with dozens of candidate higher fungal species, selected and cultured as proprietary strains suited to a wide range of environmental conditions and possible applications.

Toxicity testing for a contaminant of interest is accomplished through application of a relatively high concentration, generally 60% or higher. Strains that pass the toxicity test are then subjected to screening for other characteristics of interest to meet the requirements of a particular contamination event, site, or specific application. The selected strains are then preconditioned to increase their efficiency and effectiveness. In contrast to conclusions of some earlier work (Barr and Aust 1994), we have found that preconditioning does enhance the performance of selected strains. Further, neither nutrient deprivation nor nutrient supplementation improved the rate or extent of breakdown of contaminants such as organophosphates (Word et al. 1998; Thomas et al. 1999).

Mesoscale Testing. Mesoscale testing can be conducted under laboratory or field conditions, treating volumes from approximately one to many kilograms of substrate. For example, the MSL conducted both laboratory bench-scale and outdoor mesocosm experiments addressing remediation of petroleum hydrocarbons. In these experiments, mycoremediation was demonstrated to operate on a time-scale of weeks to months, and to be particularly effective in addressing the recalcitrant, mutagenic, and carcenogenic, higher-molecular-weight aromatic components of fresh petroleum products—that is, the four-ring PAHs, which tend to bioaccumulate and which are not readily degraded by bacterial systems. In the bench-scale test, a selected, preconditioned mycelial system in combination with native microbiota achieved 97% removal of PAHs from 2% Bunker C/diesel oil-contaminated soil over an 8-week period (Pinza et al. 1998).

In the followup mesocosm study, an applied mycelial system accomplished 93.5% removal of PAHs from 2% Bunker C/diesel oil-contaminated soil over a 12-week period (Pinza et al. 1998).

Pilot-Scale Experiment. One example of deployment to a pilot-scale technology is a study that was sponsored by Battelle and the Washington State Department of Transportation (WSDOT). The study was a 4-month field experiment to compare the efficacy of three approaches—mycoremediation, bioremediation, and enhanced bacterial remediation—to reduce the level of total petroleum hydrocarbons to below the prescribed Washington State Department of Ecology Method cleanup levels.

Objective. The objective of this study was to bring our proven laboratory-bench- and mesoscale biotechnology to a larger-scale field application that would be practical, low-maintenance, economical, and rapid, and that could yield as a byproduct a composted material with beneficial uses.

Site. The field experiment was conducted at the WSDOT maintenance yard at Bellingham, Washington (48 46.130'N, 122 28.052'W). The fenced yard is partially paved in asphalt, with several onsite storage sheds and other outbuildings. Several hundred cubic yards of scraped, contaminated soils were mounded and covered with tarpaulins for outdoor storage/treatment at the site.

Method. Three soils were excavated or scraped from contaminated sites and stored at the WSDOT maintenance yard: a) the earthen floor of a vehicle maintenance building that had operated for 30-40 years; b) a diesel-contaminated soil; and c) a soil described as gasoline-contaminated. On the basis of the Time 0 chemical analysis, it was determined that (c) was high in content of diesel and heavy oil. Each soil type was divided to create four 10-cu-yd mounds, one for each treatment type. A control mound of each was left untreated.

Deployment of the oil-preconditioned fungal system consisted of 20% by volume inoculation of the contaminated soils with alder sawdust fully grown out with the preconditioned fungal mycelium; after setup, there was no maintenance required for the duration of the test. WSDOT and its subcontractor prepared the bioremediation and enhanced bacterial treatments. WSDOT established the bioremediation regime, applying 12-lb nitrogen fertilizer (30-0-0) per 50 cu yd soil, assuming a contamination level of ~2000 mg/kg total petroleum hydrocarbons (TPH); maintenance consisted of monthly turning and addition of fertilizer. WSDOT contracted with PSCI Tank Services to conduct the enhanced bacterial treatment, which required initial and biweekly or monthly application of liquid fertilizer and bacterial inoculum, and biweekly or monthly turning of the soil. Cross-contamination was controlled by use of polyethylene underlayment; control and bacterial-treated mounds were also covered by polyethylene tarpaulins; and mycoremediated mounds were covered by 60% shade-cloth. The experiment took place in winter/early spring conditions for 16 weeks, March-July 1998.

Results. During the study, excellent growth of the introduced fungus was observed. After 4-5 weeks, and continuing monthly thereafter, fruiting was observed at the surface of the mycoremediated mounds of all three soils, and the penetration of the mycelium was throughout the mounds, to the bottom-most depths (4 ft). The smell of oil disappeared from these mounds, and the pockets of oil, tar, and other petroleum hydrocarbons were no longer apparent after the first few weeks. A vascular plant community began to develop on these mounds, and continued to become more diverse and abundant over time; by Week 12, there were secondary decomposer species of wild fungi fruiting.

The bioremediated and bacterially remediated soils, in contrast, did not show any visible evidence of change during the study. They retained their initial character of heavy clay composition, with an oil odor, and visible pockets of oil. These treatments usually require at least 1 year to be effective; therefore, the 4-month test termination may have been premature for optimal results.

Discussion. Based on results of the Northwest TPH-diesel extended (NWTPH-dx) method of chemical analysis prescribed by WSDOT, we could not claim with certainty that the cleanup criterion had been met in any of the treated or control soils during the test period. There were sampling and analytical limitations due largely to the nonhomogeneity of the soils; further, because the NWTPH-dx measures only the total of all oil components, we could not determine the specific ways in which the different treatments attacked the oils. For example, we would have expected from our prior studies (Pinza et al. 1998) that fungal mycelial activity would have broken down the higher-molecular-weight PAHs most effectively, and alkanes with nearly as high efficiency, but that perhaps as a consequence, it could have contributed a temporary increase in lower-molecular-weight compounds during the early stages of the process. It is possible that the test should have been extended to a longer duration because of the scale of the treatment and the low winter temperatures at the site, which could have slowed the biological activity.

***Toxicity tests*.** We conducted toxicity tests of the treated WSDOT Bellingham soils using earthworms, which showed no statistically significant difference among the treatments or controls for worm survival, although growth was slightly favored in the mycoremediated soils. Toxicity tests using Washington native plants used measures of plant growth and mortality to compare the three treatments. Initial results were inconclusive; more controlled testing would be necessary to determine the value of treated soils for beneficial uses.

OTHER APPLICATIONS

Organophosphates. We explored the effectiveness of selected, cultured, proprietary living fungal strains to degrade organophosphates such as dimethyl methylphonsphonate (DMMP) and isopropyl methylphosphonic acid (IMPA); further, we isolated specific fractions of the living fungal system that contain the enzymes we believe to catalyze the degradation. Some of the enzyme experiments were designed to study dephosphorylation, which we postulate to be the mechanism of the fungal breakdown of this group of compounds. Our focus was on the development of a fungal-technology for decontamination of equipment, personnel, and structures, and for wider-area remediation. We conducted preliminary laboratory bench-scale DMMP experiments to test our method and analytical techniques. One naive (not preconditioned) strain achieved about 40% removal of DMMP over a 4-week period. Toxicity and DMMP tolerance tests showed that there was robust growth in a range from 0.9 μg/g to 900 μg/g DMMP, and that there was no decline in growth with exposure to 100% DMMP. In bench-scale experiments, 4 of 17 naive strains degraded from 40% to 60% of DMMP in 12 weeks. At the mesoscale, preconditioned strains showed improvement, for example, by removing 80% of an original 10 μg/g concentration, and nearly 70% of an original 1000 μg/g concentration of DMMP in 8 weeks. In bench-scale tests with IMPA, 8 fungal strains removed >60% of the original 1000μg/g of the compound in 8 weeks; 16 strains removed from 32% to 59%; and 2 strains performed at a lower rate.

Bacterial Contaminants. Using the fecal coliform bacteria, *Escherichia coli* (ATCC 10798), and a well-studied bacillus, *B. subtilis* (ATCC 6633), to represent bacterial contaminants, we carried out experiments by which we documented the interactions of a number of proprietary fungal strains with the two bacterial species. Several types of interactions were observed with *E. coli*: four proprietary fungal strains completely *inhibited* the bacterial growth without actual contact between mycelium and bacteria, through the activity of fungal enzymes/exudates. All four are among species that change the pH of their substrate to a range of 6 to 6.5 (Becker et al., in preparation), and that are likely competitors of *E. coli,* a decomposer that grows in the same slightly acid range. Four others significantly *suppressed* the bacterial growth, without completely inhibiting it. Three strains *overwhelmed* the bacteria, growing rapidly to fill the agar plate with mycelium before the bacteria had started to develop. These strains change the substrate pH to a level less hospitable to *E. coli*, two to 8.5 and one to 4.5 (Becker et al., in preparation). Seven of the proprietary strains tested began to attack and *consume* the bacteria immediately upon contact. A pilot-scale experiment in a field setting involved the application of one wood-degrading fungal species that is predisposed to prey upon *E. coli*. In a straw-bed deployment, the fungus was able to reduce the fecal coliform burden of dairy runoff into a stream by 50% prior to the water's natural transport to commercial shellfish-growing beaches (Stamets 1993, with 1998 progress report).

In experiments with *B. subtilis,* which is a species known to produce fungicidal agents, bacterial spores were applied to culture plates with each of 11 fungal strains in three different experimental designs. Three strains showed highly significant ability to attack and digest germinated *B. subtilis* colonies. Other interactions are under investigation.

ECONOMICS

In our pilot-scale mycoremediation of petroleum hydrocarbon contamination, we determined the cost of commercial application to be under $50/cu yd of contaminated soil, including costs of bulk fungal spawn on sawdust for inoculation, shade cloth covering, and transportation, labor, and equipment for the application.

CONCLUSION

It is clear from these studies that the transition from bench scale to field scale poses some challenge. Nonetheless, we are confident in the efficacy of the method, and look forward to investigating the variables that could determine the success of larger-scale deployment. These could include among other factors the humidity and temperature within the substrate, bioavailability and distribution of contamination in soils, and duration of the treatment.

REFERENCES

Barr, D.P., and S.D. Aust. 1994. Mechanisms white rot fungi use to degrade pollutants. *Environ. Sci. Technol.*, 28(2): 78A-87A.

Becker, P., M.R. Pinza, and S.A. Thomas. 1999. Fungal influence on substrate pH. In preparation.

Loske, D., A. Hüttermann, A. Majcherczyk, F. Zadrazil, H. Loosen, and P. Waldinger. 1990. Use of white rot fungus for the clean-up of contaminated site, *pp. 311-321.* In M.P. Coughlan and M.T. Amaral-Collaco, eds., *Advances in Biological Treatment of Lignocellulosic Materials,* Elsevier Applied Science, Netherlands.

Pinza, M., P. Becker, S. Thomas, A. Drum, and J. Word. 1998. Bioremediation: mycofiltration study for the cleanup of oil-contaminated soil, pp. 16-19. PNNL-11860. *Laboratory Directed Research and Development Annual Report, Fiscal Year 1997.* Prepared for the U.S. Department of Energy by the Pacific Northwest National Laboratory, Richland, Washington.

Stamets, P. 1993. Mycofiltration of gray water run-off utilizing *Stropharia rugoso-annulata,* a white rot fungus. Research supported by a grant from the Mason County Water Conservation District, Shelton, Washington; and updated progress report March 1998.

Thomas, Becker, Pinza, Word. 1999. Adaptation of mycofiltration phenomena for wide-area and point-source decontamination of CW/BW agents. *Laboratory Directed Research and Development Annual Report, Fiscal Year 1998.* Prepared for the U.S. Department of Energy by the Pacific Northwest National Laboratory, Richland, Washington.

White, R.B., and R.T. Lamar. 1999. Degrading ability of white rot fungi: bioremediation of pentachlorophenol and lindane. *Soil & Groundwater Cleanup.* (December/January 1999): 11-16.

Word, J.Q., S.A. Thomas, A.S. Drum, P. Becker, M.R. Pinza, and T. Divine. 1998. Adaptation of mycofiltration phenomena for wide-area and point-source decontamination of CW/BW agents, pp. 16-19. PNNL-11860. *Laboratory Directed Research and Development Annual Report, Fiscal Year 1997.* Prepared for the U.S. Department of Energy by the Pacific Northwest National Laboratory, Richland, Washington.

Zadrazil, F. 1993. "Biological renewal of contaminated soils using basidiomycetes." *Mushroom Information* 4(1993-85): 13-23.

SELECTION OF LIGNINOLYTIC FUNGI FOR BIODEGRADATION OF ORGANOPOLLUTANTS

Václav Šašek (Institute of Microbiology, Prague, Czech Republic)
Cenek Novotný, Pavla Erbanová, Manish Bhatt, Tomáš Cajthaml, Alena Kubátová (Institute of Microbiology, Prague, Czech Republic)
Carlos Dosoretz (MIGAL-Galilee Technological Center, Kiryat Shmona, Israel)
Bhavin Rawal, Hans Peter Molitoris (University of Regensburg, Germany)

Abstract: Pleurotus ostreatus, Trametes versicolor, Irpex lacteus, and Bjerkandera adusta were screened from 95 strains of ligninolytic fungi using decolorization of synthetic dyes and further characterized for their growth and biodegradation of polycyclic aromatic hydrocarbons (PAHs) and synthetic dyes. All four species grew fast on agar media, however, only I. lacteus and P. ostreatus colonized soil. These two fungi significantly decolorized 9 out of 11 synthetic dyes tested on agar media. In liquid medium within 14 days, I. lacteus decolorized Methyl Red and Congo Red by 60%, and Remazol Brilliant Blue R, Copper phtalocyanine and Bromophenol Blue by more than 95%. In the soil contaminated with RBBR, I. lacteus removed 80% of the dye within 49 days. PAHs in liquid medium were significantly degraded only by T. versicolor and I. lacteus; after 28 days the removal rates were: anthracene 80 and 96%, phenanthrene 0 and 27%, pyrene 44 and 83%, and fluoranthene 53 and 67%, respectively. Using I. lacteus incubated for 3 months in soil contaminated with the same PAHs, the removal rates were: anthracene 96%, phenanthrene 60%, pyrene 83%, and fluoranthene 39%. I. lacteus thus showed to be a prospective fungus for further research of its application in soil bioremediation.

INTRODUCTION

White rot fungi have been demonstrated to attack a wide spectrum of organopollutants including polycyclic aromatic hydrocarbons (PAHs) (Bumpus, 1989; George and Neufield, 1989; Hammel et al., 1992) and different synthetic dyes (Glenn and Gold, 1983; Cripps et al., 1990; Kirby et al., 1995). However, most of the studies have been carried out using Phanerochaete chrysosporium and out of many hundreds of species possessing ligninolytic activity, only few more became research subjects (Field et al., 1992; Vyas and Molitoris, 1995; Heinfling et al., 1997; Sack and Fritsche, 1997; Martens and Zadrazil, 1998).

The purpose of our work was to select white rot fungi efficient in degradation of PAHs and synthetic dyes both in liquid and in soil. The first step of a screening of a large number of ligninolytic fungal strains was the decolorization of model synthetic dyes (Šašek et al., 1998) since this capability is supposed to be in correlation with degradation of other organopollutants (Field et al., 1993).

MATERIALS AND METHODS

Screening and Maintainance of Ligninolytic Fungi. A number of dyes belonging to several chemical groups (azo, diazo, metal complex, heterocyclic and triphenyl methane) were used to screen white rot fungi efficient in decolorizing synthetic dyes on nutritive agar media. The dye-agar plates were inoculated with fungal-grown agar plugs (9 mm diameter) and incubated at 26 °C. Decolorized zone around the fungal growth was measured. The selected fungi were maintained on malt extract glucose (MEG) agar (malt extract 10g/L, glucose 5g/L, agar 20g/L; pH 5.5).

Soil colonization. Soil collected from the garden of the Institute of Microbiology, Prague (native humidity 5.4%, pH 6.5, organic carbon 4.7% glucose equivalent) was sieved (<2mm). Both in native and tyndalized soil samples in test tubes different humidity (5, 10, 15 and 20%) was adjusted. On the top of the soil column covered with a layer of ground, moistened, sterilized straw, a plug of malt agar grown I. lacteus was added. The tubes were incubated at 26 °C and soil colonization was measured.

Dye Decolorization in Liquid Media and Soil. A volume of 20 ml of a dye containing (150 μg/ml) nitrogen limited mineral medium (NMM) (Tien and Kirk, 1988) in a 250-ml Erlenmeyer flasks were inoculated with I. lacteus-grown MEG agar plugs and kept at 22 °C. Every second day triplicate flasks were harvested and following filtration the culture liquid was collected and the absorbance of a corresponding dye at an optimum wavelength, after suitable dilution, was measured using a Perkin-Elmer λ 11 UV-VIS spectrophotometer.

In 100-ml Erlenmeyer flasks containing 8 g of the above soil, dried and tyndalized, the solution of Remazol Brilliant Blue R (RBBR) prepared in distilled water was spiked in such a way that the final concentration of RBBR and humidity in soil was 150 μg/ml and 20% respectively. Freshly prepared straw-grown I. lacteus (8 g) was mixed with the RBBR-spiked soil. Heat-killed control was prepared by adding the autoclaved inoculum. The flasks were incubated at 27 °C. The contaminated soil was extracted and analyzed in triplicates for RBBR recovery at different time intervals. RBBR was extracted from the spiked soil using a multisolvent system made up of chloroform, methanol and distilled water (1:1:1, V/V). The soil sample was then sonicated for 15 min and filtered through a fine nylon cloth to remove coarse particles. The filtrate was centrifuged and the supernatant was dried at 100 °C. The residuum was redissolved in 7 ml of distilled water and centrifuged. The supernatant was diluted 5-fold and the absorbance read at 578 nm using a Perkin-Elmer λ 11 UV-VIS spectrophotometer.

Removal of PAHs in Liquid Media and Soil. Erlenmeyer flasks (100 ml) with 20 ml MEG broth were inoculated with I. lacteus. The flasks were spiked with 200 μl acetone containing anthracene, phenanthrene, pyrene and fluoranthene making the final concentration of each PAH at 25 μg/ml. Heat killed controls were prepared by autoclaving the flasks before spiking with the PAHs. The cultures were incubated at 28 °C for 4 weeks. To extract PAHs, a total volume (20 ml) was sacrificed, first acidified to pH 4 and then extracted on a shaker with 10 ml of cyclohexane (5 times; 15 min each step, 300 strokes/min). HPLC detection was done using a 250 x 4 mm Merck LiChroCART column under isocratic conditions (methanol/water, 9:1, V/V) at a flow rate of 1 ml/min, at a pressure of 17.7 MPa and a temperature of 40 °C using a Perkin-Elmer 200 LC pump and a Waters 990 UV photodiode array detector.

For the study of degradation of PAHs in soil, 6g of tyndalized dry soil described above was contaminated with 2 g of sterilized sand spiked with the above PAHs (final concentration of 50 µg/g each). The soil was inoculated with straw-grown I. lacteus (15 g wet inoculum) and incubated for 12 weeks at 26 °C. PAHs were extracted using supercritical fluid extraction (SUPREX PrepMaster/AccuTrap/a variflow restrictor) performed in two fractions. A total sample of 12 g of soil, sand and straw was extracted, first at 400 atm, 150 °C, for 10 min, at a rate of 1 ml CO2/min, then at 450 atm, 150 °C, for 10 min, at a rate of 1 ml CO2/min using 10% (V/V) methanol as modifier. Cryogenic collection was used: glass beads, -20 °C. The HPLC detection was the same as above.

Chemicals. All PAHs (purity 〉 97%) were purchased from Aldrich-Chemie, Steinheim, Germany. Except methanol and choloform (technical grade) all solvents were of HPLC grade (purity 〉 98%). All dyes used were more than 90% pure.

RESULTS AND DISCUSSION

Decolorization of Dyes in Agar Media by I. lacteus and P. ostreatus. Using NMM and MEG agar media containing 200 µg/ml of the respective dye, four out of 95 strains of ligninolytic fungi were preselected and two of them were further studied for dye decolorization. Both I. lacteus and P. ostreatus completely deolorized all the dyes tested within 10 to 30 days. The individual dyes decolorized were: azo dyes - Methyl Orange, Methyl Red, Reactive Orange 16; diazo dyes - Reactive Black 5, Congo Red; anthraquinone based dyes - Remazol Brilliant Blue R, Disperse Blue 3; heterocyclic dyes - fluorescein, Methylene Blue; triphenyl methane dye - Bromophenol Blue.

Decolorization of Dyes in Liquid Media by I. lacteus. Further study of I. lacteus in NMM containing five different dyes (150 µg/ml), namely Methyl Red (MR), Congo Red (CR), Remazol Brilliant Blue R (RBBR), Copper Pthalocynine (CuPth, a metal complex dye) and Bromophenol Blue (BPB) showed 56, 58, 93, 98 and 99% decolorization rates within two weeks, respectively (Figure 1). Except for Congo Red, where the decolorization was continuous during the experiment, all decolorization were accomplished until 4-6 days.

Removal of RBBR from Soil by I. lacteus. Following the observation that I. lacteus colonized both tyndalized and untreated soil it was further studied for the removal of RBBR from soil. The fungus removed the dye gradually upto 80% of the original amount added (Figure 2).

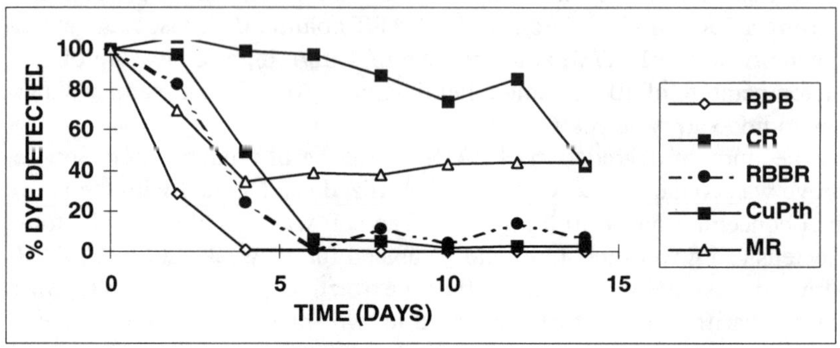

FIGURE 1. Decolorization of various dyes by I. lacteus in NMM.

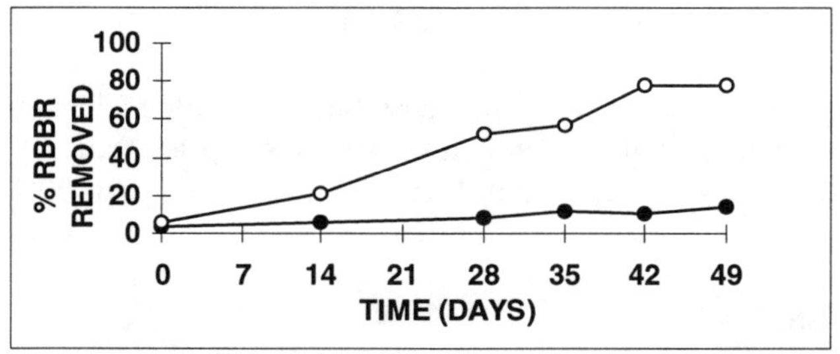

FIGURE 2. Removal of RBBR from artificially contaminated soil by I. lacteus (closed circles heat killed control, open circles live fungus).

Degradation of PAHs in Liquid Media by I. lacteus and T. versicolor. The efficiency of several white rot fungi to degrade polycyclic aromatic hydrocarbons (PAHs) was tested in liquid MEG medium containing phenanthrene, anthracene, fluoranthene and pyrene at 25 μg/mL The extraction and analysis of PAHs at the end of day 28 indicated that compared to their heat killed controls only I. lacteus and T. versicolor significantly degraded the PAHs showing removal of anthracene 96 and 80%, fluoranthene 67 and 53%, pyrene 83 and 44% and phenanthrene 27 and 0%, respectively. The live I. lacteus showed more than 80% presence of anthraquinone compared to the control while T. versicolor did not show such pattern.The results also showed that I. lacteus was more efficient in biodegradation of all PAHs tested in MEG medium, compared to other white rot fungi - Pleurotus tuber-regium, T. versicolor, P. ostreatus and Phanerochaete chrysosporium (Figure 3).

Degradation of PAHs in Soil by I. lacteus. The results of degradation of PAHs (50 μg/g) in soil by wheat straw grown I. lacteus were comparable to those in liquid cultures. After 12 week-incubation the fungus removed 74% phenanthrene, 96% anthracene, 61% fluoranthene and 88% pyrene compared to its heat-killed control. The experiment demonstrated that I. lacteus can be used for efficient removal of 3- and 4-ringed PAHs from both liquid and soil environment.

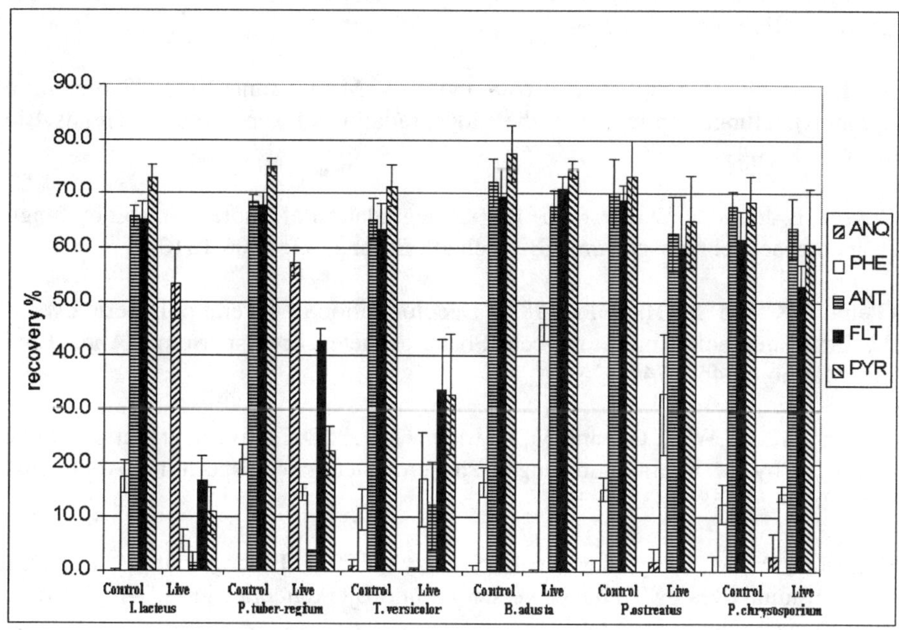

FIGURE 3. Degradation of PAHs in Liquid MEG Medium by White Rot Fungi after 28 days.

CONCLUSION

The screening of a set of cultures of white rot fungi resulted in selection of the strain I. lacteus that has a potential to be applied in remediation of soil contaminated with toxic organopollutants.

ACKNOWLEDGEMENTS

The research was supported by Grant No. TA-MOU-95-C15-190, U.S.-Israel Cooperative Development Research Program, Office of the Science Advisor, U.S. Agency for International Development and by the Program of Czech-German Bilateral Scientific Cooperation No. TSR-040-97.

REFERENCES

Bumpus, J. A. 1989. "Biodegradation of polycyclic aromatic hydrocarbons by Phanerochaete chrysosporium." Appl. Environ. Microbiol. 55: 154-158.

Cripps, C., J. A. Bumpus and S. D. Aust. 1990. "Biodegradation of azo and heterocyclic dyes by Phanerochaete chrysosporium." Appl. Environ. Microbiol. 56: 1114-1118.
Field, J. A., E. de Jong, G. F. Costa and J. A. M. de Bont. 1992. "Biodegradation of polycyclic aromatic hydrocarbons by new isolates of white rot fungi." Appl. Environ. Microbiol. 58: 2219-2226.

Field, J. A., E. de Jong, G. F. Costa and J. A. M. de Bont. 1993. "Screening of ligninolytic fungi applicable to the biodegradation of xenobiotics." Trends Biotechnol. 11: 44-49.

George, E. J. and R. D. Neufield. 1989. "Degradation of fluorene in soil by fungus Phanerochaete chrysosporium." Biotechnol. Bioeng. 33: 1306-1310.

Glenn, J. K. and M. H. Gold. 1983. "Decolorization of several polymeric dyes by the lignin degrading basidiomycete Phanerochaete chrysosporium." Appl. Environ. Microbiol. 45: 1741-1747.

Hammel, K. E., W. Z. Gai and M. A. Moen. 1992. "Oxidative degradation of phenanthrene by the ligninolytic fungus Phanerochaete chrysosporium." Appl. Environ. Microbiol. 58: 1832-1838.

Heinfling, A., M. Bergbauer and U. Szewzyk. 1997. "Biodegradation of azo and pthalocyanine dyes by Trametes versicolor and Bjerkandera adusta." Appl. Microbiol. Biotechnol. 48: 261-266.

Kirby, N., G. McMullan and R. Marchant. 1995. "Decolourisation of an artificial textile effluent by Phanerochaete chrysosporium." Biotechnol. Lett. 17: 761-764.

Martens, R. and F. Zadrazil. 1998. "Screening of white rot fungi for their ability to mineralize polycyclic aromatic hydrocarbons in soil." Folia Microbiol. 43: 97-103.

Sack, U. and W. Fritsche. 1997. "Enhancement of pyrene mineralization in soil by wood-decaying fungi." FEMS Microbiol. Ecol. 22: 77-83.

Šašek, V., È. Novotný, P. Vampola. 1998. "Screening for efficient organopollutant fungal degraders by decolorization." Czech Mycol. 50: 303-311.

Tien, M. and T. K. Kirk. 1988. "Lignin peroxidase of Phanerochaete chrysosporium." Methods Enzymol. 161: 238-249.

Vyas, B. R. M. and H. P. Molitoris. 1995. "Involvement of an extracellular H2O2-dependent ligninolytic activity of the white rot fungus Pleurotus ostreatus in the decolorization of Remazol Brilliant Blue R." Appl. Environ. Microbiol. 61: 3919-3927.

BIOTREATMENT OF hPAH-CONTAMINATED SOILS IN INTERMITTENTLY MIXED BATCH REACTORS (IMBRs)

Walter J. Weber, Jr., Han S. Kim, Andrei L. Barkovskii
Environmental and Water Resources Engineering Program
The University of Michigan, Ann Arbor MI 48109-2125 USA

ABSTRACT: The applicability of an in-situ intermittent slurry treatment system for remediation of heavy polycyclic aromatic hydrocarbon (hPAH) contaminated sites was examined. The biodegradation by indigenous microorganisms of selected PAH components of historically contaminated soils (PAHs ~100 mg/kg and chlorophenols ~50 mg/kg) that were collected from a Southern Maryland Wood Treatment (SMWT) site was evaluated in prototype lab-scale IMBRs. The treatment systems were augmented in a second series of experiments by addition of *Sphingomonas paucimoblis* EPA 505, a known hPAH degrader. Concentrations of phenanthrene, fluoranthene, and pyrene were monitored during the experiments, revealing nearly complete degradation of phenanthrene, 55~85% removal of fluoranthene and 65~67% removal of pyrene within 100 days. The results, coupled with preliminary cost analyses suggest that the IMBR technology may provide a means for efficient and cost effective remediation of hPAH contaminated sites.

INTRODUCTION

Soils, buried and lagooned deposits of dredge spoils, industrial sorbents and filtration media, and waste sludges contaminated with hydrophobic organic compounds (HOCs) such as polychlorinated biphenyls (PCBs) and hPAHs can often be found within a few feet of ground surface at old industrial and defense-related sites. These materials can pose serious risks to human health and the environment, especially if they are subject to leaching or some other forms of release that might lead to contamination of potable or recreational water resources. Natural attenuation of the contaminants in such deposits is usually negligible, and active treatment is thus necessary for effective and timely clean up. The limitations associated with conventional in-situ biotechnologies targeted at the removal of hPAHs center about the availability of the contaminants in a form conducive to microbial utilization, and the effective delivery of appropriate electron acceptors and other nutrients to contaminated zones. It is our hypothesis that in-situ intermittent slurry treatment systems can address these limitations, primarily by compressing the temporal and spatial scales of reactor mass transport processes to more closely match those of biotransformation processes. In our conceptualization of this type of system, it is envisioned that large diameter augers can be used to periodically mix the contaminated slurries while appropriate electron acceptors, nutrients, surfactants, bacteria, and co-substrates are simultaneously injected into the affected zone(s). The novel features of such a system would be: (i) elimination of the excavation and transportation of contaminated materials; (ii) reduction of operating costs through intermittent rather than continuous mixing; (iii) applicability to relatively deep contaminated zones (~30 ft depth); and, (iv) acceleration of in-situ biodegradation processes by increasing mass transport rates and through more thorough distributions of contaminants, nutrients, augmenting microorganisms, and appropriate reagents such as surfactants.

Objective. The objective of the study was to evaluate the efficacy of the concept by investigating hPAH degradation in a prototype IMBR. Experiments were conducted to: (i) examine differences in biodegradation patterns of target hPAHs during differently sequenced mixing of the contaminated soils; and, (ii) evaluate the need for and effects of microbial amendment with an identified hPAH-degrading strain.

MATERIALS AND METHODS

IMBR Design. Lab-scale bioreactors were constructed as prototypes for the conceptual in-situ IMBR using a 500-mL glass kettle covered with a glass head plate having sampling ports. A sketch of the laboratory reactor is provided in Figure 1. A 70-mm diameter stainless steel auger rotating at a constant speed (60 rpm) provided efficient mixing, with the SMWT soil at 66% solid content (mass basis) being completely mixed within a 2-minute period. Intermittent operation of the auger was achieved by controlling the drive motor with a programmable timer. No dead zones were observed in the reactor. A glass tube at the bottom of the reactor provided oxygen (mixed air) supply to the reactor.

Reactor Setup. Five IMBRs were operated at different mixing frequencies to provide a series of experiments involving reactors in which 2, 4, 6, 12, and 18 hours of mixing per day were provided. A continuously and completely mixed batch reactor (CMBR) and a reactor containing sterilized SMWT soils were used respectively as mixing and biological control systems. One series of experiments was conducted using only the indigenous microbial consortium. Another series was conducted with amendment by *Sphingomonas paucimoblis* EPA 505, an hPAH degrading strain, isolated and provided by Hap Pritchard, P. (Environmental Quality Service, U.S. Naval Research Lab, Code 6115). The number of *Sphingomonas paucimoblis* EPA 505 applied at the beginning of the latter experiments was 2×10^6 per mL of soil water by plate colony counting method. All systems were maintained at 25 ± 0.5°C.

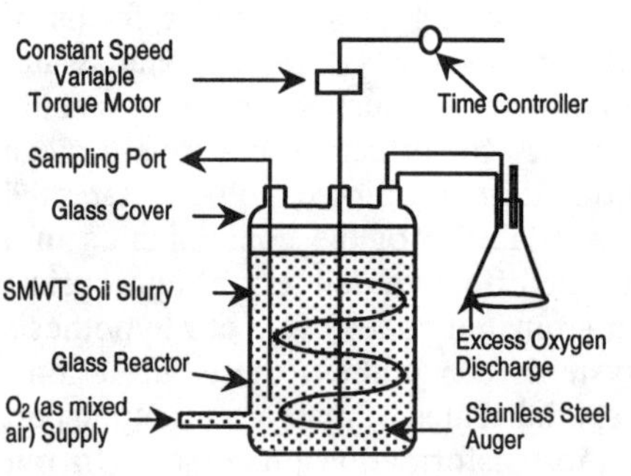

Figure 1. Schematic diagram of IMBR.

Extraction of hPAH from SMWT Soils. Samples of contaminated soil slurry were dewatered using 0.2 μm-microcentrifuge nylon filters. The water content of dewatered samples was 6.85 ± 1.24% (mass basis). Two extraction procedures (24-hour soxhlet and 4-hour sonication) and three different solvents (dichloromethane, n-butanol, and a mixture of dichloromethane and n-butanol (1:1 v/v)) were evaluated to determine an optimal tool for extracting hPAHs from the SMWT soil. Based on extraction efficiency and operation time, four-hour sonication with dichloromethane was selected.

Analytical Procedures. Triplicate samples of dichloromethane-extracted hPAHs were analyzed by gas chromatography (Hewlett-Packard 5890 Series II) employing a flame ionization detector (FID).

Cost Analysis. A preliminary cost analysis was performed using estimates for soil handling, treatment, equipment, and labor based on currently available information. To develop an appropriate basis for comparison, an imaginary contaminated site was developed and used for estimation of costs for the in-situ IMBR systems and alternative technologies. The in-situ IMBR system used for these cost estimates consisted of the following components: (i) a 10 foot-diameter and 20 foot-length auger; (ii) a crane for powering and moving the IMBR from one treatment column to another; (iii) injection equipments for such amendments as nutrients, microorganisms, surfactants, and water; (iv) an air supply system; and, (v) trapping equipment for dust and VOCs. The following clean-up conditions were stipulated: (i) approximately 90% degradation of petroleum hydrocarbons and total PAHs; (ii) an effective treatment depth of approximately 30 feet; (iii) a 50-day remediation time frame required for each auguring column; and, (iv) areas between auguring columns were assumed to be mixed and treated.

RESULTS AND DISCUSSION

Biodegradation of Target hPAHs by Indigenous Microbial Consortium. The indigenous microorganisms of the SMWT soil were found to successfully mineralize phenanthrene in 25 days in all reactors tested (Figure 2(a)). However, hPAHs such as fluoranthene and pyrene were not effectively degraded by the indigenous microflora (data not shown). This was not unexpected, given the extremely low aqueous solubilities of these compounds and the fact that only a limited number of microorganisms can mineralize them. In the CMBR experiment shown in Figure 2(a), degradation of phenanthrene was retarded during the first few days because oxygen was depleted and microbial activity was inhibited. After resuming the supply of oxygen, approximately 90% of phenanthrene was degraded within only five days.

Biodegradation of Target hPAHs by the Indigenous Microbial Consortium Amended with *Sphingomonas paucimoblis* EPA 505. The degradation pattern of phenanthrene in the presence of *Sphingomonas paucimoblis* EPA 505 (Figure 2(b)) was very similar to that produced by the indigenous consortium alone. Fluoranthene and pyrene, which showed no apparent degradation in the indigenous microbial systems, were degraded in the systems augmented with *Sphingomonas paucimoblis* EPA 505 . After 100 days of incubation, 55~85% of fluoranthene and 65~67% of pyrene disappeared (Figures 3(a), (b)). In the case of fluoranthene at least, residual contaminant concentrations corresponded to the cumulative mixing time.

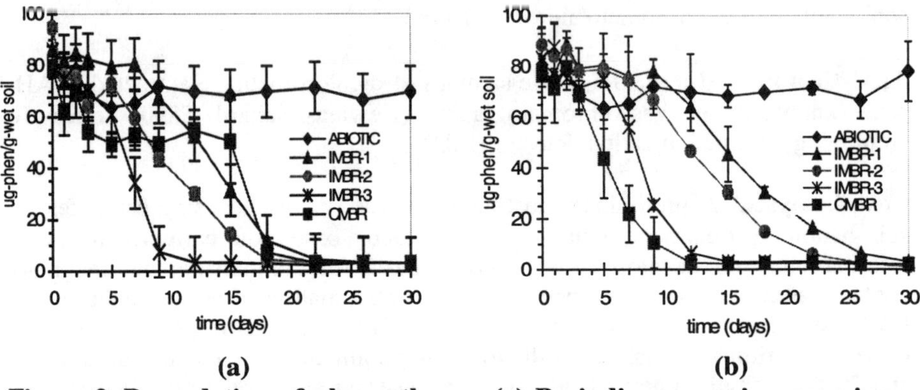

Figure 2. Degradation of phenanthrene. (a) By indigenous microorganisms. (b) Amendment with *Sphingomonas paucimoblis* EPA 505.

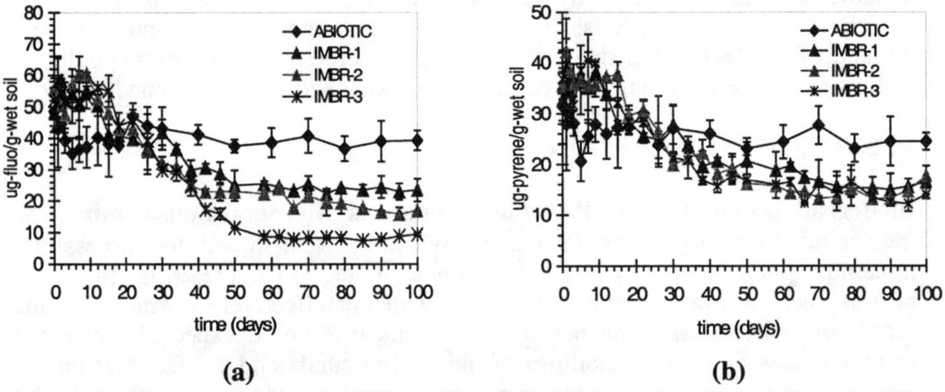

Figure 3. Degradation of (a) fluoranthene and (b) pyrene (Amendment with *Sphingomonas paucimoblis* EPA 505).

Legend: CMBR: Completely and continuously mixed batch reactor
ABIOTIC: Abiotic CMBR
IMBR-1: IMBR with 2hrs (0.5hr x 4times) per day
IMBR-2: IMBR with 4hrs (0.5hr x 8times) per day
IMBR-3: IMBR with 6hrs (0.5hr x 12times) per day

The Effects of Mixing Duration Time on the Mass Transfer and Biodegradability of hPAHs. It has been demonstrated in our laboratory

(Mukerji and Weber, 1998) that the rate of phenanthrene biodegradation associated with NAPL and soil phases is limited by mass transfer to the aqueous phase. It is reasonable to expect this to be at least an equally important limit on the rates of biodegradation of the less soluble hPAHs. The rate limiting process of mass transfer involves a linear concentration driving force, thus mimicking a first-order reaction having an associated rate coefficient. A non-linear relationship was observed between the first order rate coefficient for degradation of PAHs and cumulative reactor mixing time, as shown in Figure 4 with the coefficient approaching a limiting value as mixing time approaches the CMBR condition. This suggests that continuous mixing may not be necessary for effective biotreatment of hPAHs.

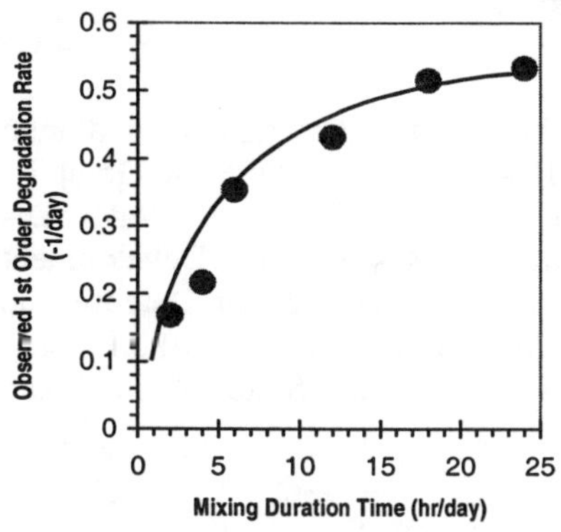

Figure 4. Correlation between observed 1st-order degradation rate coefficients and daily cumulative mixing time.

Cost Estimates. Figure 5 compares volume-normalized costs for various applicable remediation technologies. The costs for alternative conventional technologies used for comparison purposes were based on information from technical reports prepared by various federal agencies (See first three and last three publications listed in Reference section). These costs can significantly vary depending on the physical/chemical properties of the contaminated system involved. Soil flushing (SF) costs, for example, might increase by as much as an

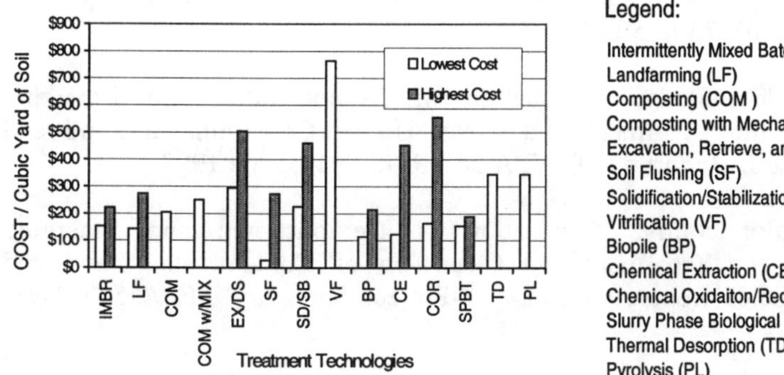

Figure 5. Cost comparison of various soil treatment technologies.

order of magnitude. As a rule of thumb, the greater the hydrophobicities of the contaminants (e.g. hPAHs), the slower are treatment rates and the higher are treatment costs for a given process. It is reasonable, therefore, for treatment of hPAHs and PCBs to compare the higher cost for each of the alternative technologies shown in Figure 5 with that of the in-situ IMBR system. It is evident that, on the basis of the evaluations described herein, in-situ IMBR systems compare favorably with alternative in-situ technologies with respect to operating cost, as well as with respect to expected removal efficiencies for heavier molecular weight hydrocarbons.

ACKNOWLEDGMENTS
Funding for this research was provided by the Great Lakes and Mid-Atlantic Center (GLMAC) for Hazardous Substance Research under USEPA Grant R-825539 through the Department of the Army Corps of Engineers Waterways Experiments Station and the Strategic Environmental Research and Development Program (SERDP), a consortium comprised by the Department of Defense, the Department of Energy, and the Environmental Projection Agency. Partial funding of the activities of GLMAC was also provided by the State of Michigan Department of Environmental Quality.

REFERENCES

AATDF Report TR-97-2, "AATDF Technology Practice Manual for Surfactants and Co-solvents", February 1997

"ATTIC Report", EPA/Office of Research and Development, Nov. 1998

The Federal Remediation Technology Roundtable (FRTR) Report, "Remediation Technologies Screening Matrix and Reference Guide", 3rd Ed., FRTR, Oct. 1997

RSMEANS, "Building Construction Cost Data", 55th Annual Ed., 1997

Mukerji, S. and Weber, Jr. W.J., 1998, "Mass Transfer Effects on Microbial Uptake of Naphthalene from Complex NAPLs", *Biotechnology and Bioengineering*, 60(6): 750-759.

NATO/CCMS Pilot Study, "Evaluation of Demonstrated and Emerging Technologies for the Treatment and Clean Up of Contaminated Land and Groundwater Phase II Report #219", EPA 542-R-98-001a, June 1998

NATO/CCMS Pilot Study, "Evaluation of Demonstrated and Emerging Technologies for the Treatment and Clean Up of Contaminated Land and Groundwater Phase II Report appendix IV-Project Summary", EPA 542-R-98-001c, June 1998

NATO/CCMS Pilot Study, "Evaluation of Demonstrated and Emerging Technologies for the Treatment and Clean Up of Contaminated Land and Groundwater Phase III 1998 Annual Report #218", EPA/542/R-98/002, May 1998

COMBINED INTRINSIC AND STIMULATED *IN SITU* BIODEGRADATION OF HEXACHLOROCYCLOHEXANE (HCH)

A.A.M. Langenhoff[1], H.C. van Liere[1], C.G.J.M. Pijls[2], G. Schraa[3], H.H.M. Rijnaarts[1]

[1] TNO Institute of Environmental Sciences, Energy Research and Process Innovation, Apeldoorn, The Netherlands; [2] Tauw Milieu Consultancy, The Netherlands; [3] Wageningen Agricultural University, The Netherlands

ABSTRACT The pollution of soil and groundwater with hexachlorocyclohexane (HCH) has caused serious environmental problems. Mineralisation of some HCH isomers can be expected under aerobic conditions whereas all isomers are known to be bioconverted to intermediate products under anaerobic conditions. Hence, redox conditions can be expected to have strong impact on the intrinsic degradation behaviour of HCH's is soil and groundwater. This was investigated at two HCH polluted sites in the Netherlands.

A field characterisation at the two polluted sites showed anaerobic conditions and a significant intrinsic biodegradation of HCH. Three breakdown products (monochlorobenzene, benzene and chlorophenol) were found in the core of the HCH-plume at one site, whereas the concentration of HCH seemed to decrease with time. The second site, an industrial site, showed high concentrations of HCH, and relatively low levels of intermediate products, indicating intrinsic biodegradation as insufficiently protective, and the need for a stimulation of the biodegradation processes.

To proof the feasibility of our bioremediation concept, the research will focus on two processes: I) fully activated anaerobic biodegradation and II) partial intrinsic anaerobic biodegradation followed by stimulated aerobic biodegradation. Both processes depend on the capacity of endogenous microorganisms, which will be tested in laboratory batch experiments with subsurface materials from the sites.

INTRODUCTION

The pollution of soil and groundwater with hexachlorocyclohexane (HCH) has caused serious environmental problems. Lindane (γ-HCH) is the best known and effective insecticide component of HCH, and only 17% of the HCH consists of this γ-isomer. The remaining HCH consists of α, ß- and δ-isomers, which have a low insecticide activity. These isomers were separated from γ-HCH, and dumped at waste sites resulting in polluted soils and groundwater. In contrast to α-, γ-, and δ-HCH, ß-HCH was for long known to be recalcitrant towards biodegradation under anaerobic and aerobic conditions, and the biological clean-up of soils was therefore assumed to be impossible (Bachmann et al. 1988). However, recent research has shown that ß-HCH can be microbiologically degraded to monochlorobenzene (MCB) and benzene (B) under anaerobic conditions (Middeldorp et al. 1996; Van Eekert et al. 1998). These compounds usually accumulate in anaerobic environments. Since biodegradation of monochlorobenzene, benzene and chlorophenol (CP) under aerobic conditions occurs easily, a complete mineralisation of all HCH isomers can be obtained by anaerobic conditions, followed by aerobic conditions (Figure 1).

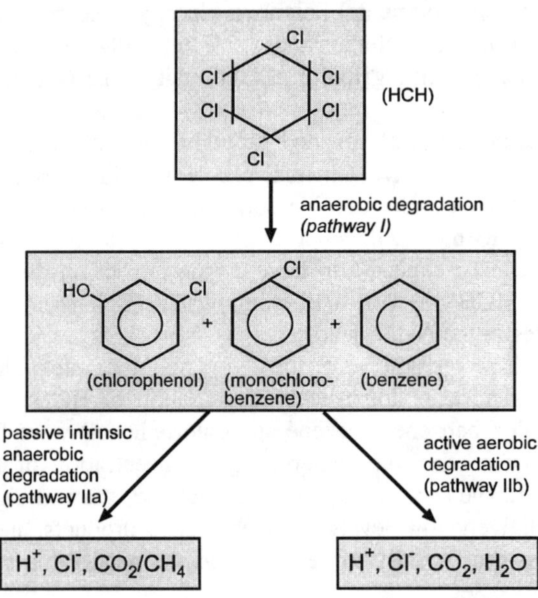

FIGURE 1. Two step HCH biodegradation pathways: I) anaerobic biodegradation of HCH to MCB, B, and CP and II) mineralisation of intermediates via aerobic (IIa) or anaerobic (IIb) biodegradation.

In this paper the feasibility of anaerobic and aerobic biodegradation of HCH and its intermediates will be discussed, and the concept of a combined intrinsic and stimulated *in situ* bioremediation will be evaluated for HCH contaminated sites. using data obtained at the industrial site.

MATERIALS AND METHODS

Site description. Lindane has been produced in the eastern part of the Netherlands from 1948 until 1954. Nowadays it is imported to manufacture lindane containing products (Slooff and Matthijsen 1990). HCH waste isomers were at that time amended with lime and stored at the industrial production site. A portion of this material spread over the region (province) because it was used as backfilling material in roads and farm yards. The top soil layer of most of these sites contained high concentrations of HCH and was removed a few years ago. However, the combination of rainfall and varying groundwater levels have caused heterogeneous groundwater contamination in the years before the removal of the top soil layer. One of the sites studied is such a location with a removed toplayer. The other site is an industrial facility with an HCH-contaminated area of 300m by 100m, where dissolved HCH compounds have formed a contaminated plume. This plume is moving away from the contaminated area into the direction of a nearby canal. All data presented and discussed here are obtained at that industrial site.

Field characterisation. The geohydrological situation of the industrial site has been characterised. Ten sampling wells have been installed at various depths in a selected area to analyse the soil and groundwater on the following parameters:

1) organic compounds; hexachlorocyclohexane isomers (HCH), chlorobenzene (CB), benzene (B) and chlorophenols (CP)
2) redox parameters; O_2, Mn(IV)/Mn(II), NO_3^-/NO_2^-, Fe(III)/Fe(II), SO_4^{2-}/S^{2-} and HCO_3^-/CH_4
3) electron donor capacity; DOC, TOC and N-Kjehldal

This selected area forms a small part of the total contaminated industrial site and will serve as the future pilot scale research area.

Batch experiments. Batch experiments are underway with sediment and groundwater from the site. The stimulation of the degradation of HCH is tested by the addition of lactate, landfill percolate, and compost percolate. The natural anaerobic degradation capacity of the site is tested for monochlorobenzene and benzene as well as the aerobic degradation, via the addition of oxygen.

RESULTS AND DISCUSSION

Field characterisation. A schematic overview of the site is shown in Figure 2. Three defined sandy aquifers can be defined down to a depth of 25 meter, and are separated by a peat-clay layer. The groundwater velocities are 7.5, 15 and 30-60 $m.yr^{-1}$ in the first, second and third aquifer, respectively. The groundwater direction (north-north-east) is towards a freshwater system, which forms the natural boundary of the site.

MCB, B, and CP were found as breakdown products of HCH in the core of the plume. In the first aquifer, benzene and chlorophenol appear to be removed during downstream transport in the direction of the water system, whereas the monochlorobenzene concentration remained fairly constant. In the second aquifer, the concentration of MCB, B, and CP showed no decline.

The redox potential characterisation of the source area of the first aquifer showed indications of sulphate reducing and methanogenic conditions. These low redox potential conditions are reported to be favourable for reductive dechlorination. In the downstream part of the plume, iron reducing conditions were found to be predominant. The redox conditions in the second aquifer were less clearly defined, but can be characterised as reducing conditions.

Both aquifers showed an average DOC concentration of 27 $mg.l^{-1}$, indicating the presence of sufficient electron donor capacity for reductive dechlorination.

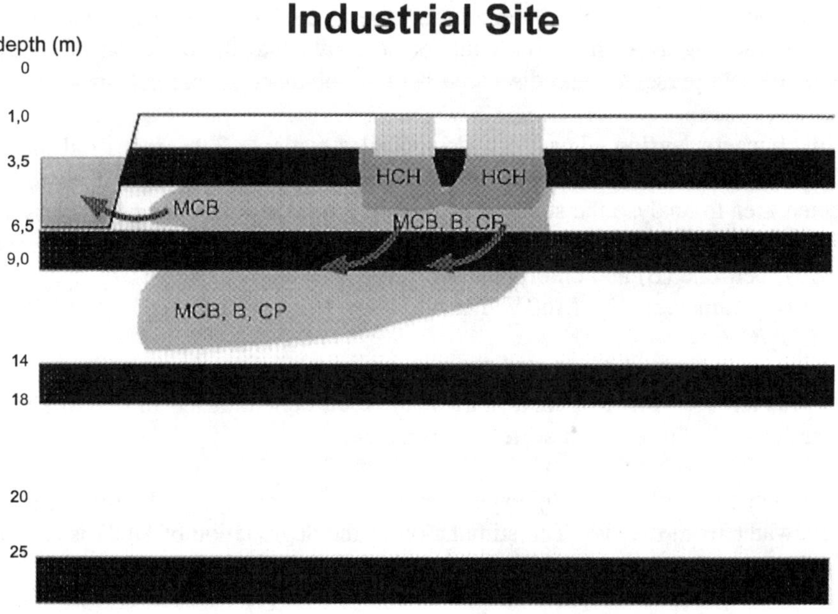

FIGURE 2. An overview of an industrial site in the eastern part of the Netherlands. Groundwater moves towards a canal.

HCH bioremediation concepts. Theoretically, four bioremediation processes are possible at this industrial site (Figure 3).

I. HCH is completely degraded to carbondioxide and methane via intrinsic processes (natural attenuation), and monitoring the degradation process is the only activity needed.

II. The intrinsic degradation capacity of the soil is insufficient, and the anaerobic degradation of HCH to its intermediates monochlorobenzene, benzene, and chlorophenol needs to be stimulated. The intermediates can then be mineralised via intrinsic degradation.

III. The intermediates monochlorobenzene, benzene, and chlorophenol are formed via intrinsic degradation, but no further anaerobic degradation occurs through intrinsic processes. Stimulation is needed to mineralise monochlorobenzene, benzene, and chlorophenol aerobically.

IV. A combination of anaerobic and aerobic stimulation (IV) can be successful if both the natural degradation of HCH to its intermediates, and the degradation of the intermediates are insufficient.

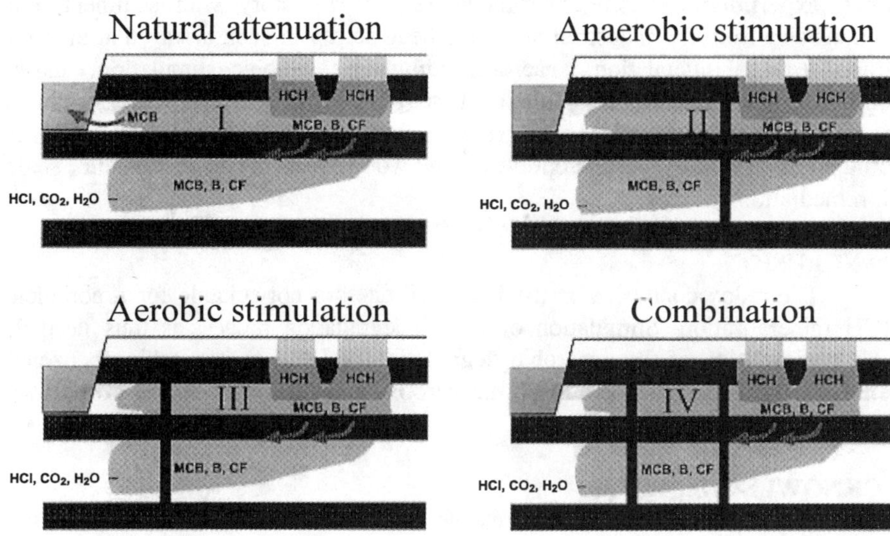

FIGURE 3. Four HCH contaminant bioremediation concepts: I) Natural attenuation, II) Anaerobic stimulation, III) Aerobic stimulation, IV) Combination of anaerobic and aerobic stimulation.

Analyses of the aquifer solids and groundwater has shown the presence of HCH, MCB, B, and CP. In the first aquifer, HCH is degraded into MCB, B and CP under sulphate reducing and methanogenic conditions near the core of the HCH plume. This indicates that HCH is degraded as previously demonstrated (Haider and Jagnow 1975; Jagnow et al. 1977; Middeldorp et al. 1996; Van Eekert et al. 1998), but not fast enough, as HCH is still found in high concentrations (400 mg/l) in the first aquifer.

MCB, B and CP are formed in the first aquifer under sulphate-reducing and methanogenic conditions and partly degraded. Further downstream, iron reducing conditions are present, and B and CP are also degraded, but MCB seems to persist under these conditions. Both B and CP can be degraded under anaerobic conditions (Lovley 1997; Van Agteren et al. 1998), but so far MCB seems to be recalcitrant to anaerobic biodegradation (Van Agteren et al. 1998). In the second aquifer MCB, B and CP do not seem to be degraded under the existing redox conditions. This shows that the aquifer and the present redox conditions greatly influence the degradation of the intermediates; some can be degraded, whereas others accumulate, especially MCB in the first aquifer. Stimulation of the biodegradation of all intermediates can be achieved with the addition of oxygen as all intermediates can be degraded aerobically.

Our data show that the fourth bioremediation option will be the most effective process: stimulated anaerobic transformation of HCH to monochlorobenzene, benzene, and chlorophenol via the addition of a suitable electron donor, followed by a stimulated aerobic mineralisation of these products.

Batch experiments. Batch experiments in the laboratory with sediment and groundwater from the industrial site have been set up in such a way that all four options (natural attenuation, anaerobic stimulation, aerobic stimulation, and a combination) are tested. The results of these experiments will give more insight in the biodegradation processes occurring at the industrial site and the need for stimulation. The most feasible option will be worked out into an *in situ* pilot scale bioremediation concept.

CONCLUSION

The redox conditions at the industrial site are not suitable for a complete HCH mineralisation. Stimulation of the biodegradation process is thus needed; both a stimulation of the anaerobic degradation of HCH to monochlorobenzene, benzene, and chlorophenol and of the aerobic mineralisation of the breakdown products.

ACKNOWLEDGEMENTS

Major parts of the work presented have been financed by NOBIS, the Dutch Research Programme Biotechnological In-Situ Remediation.

REFERENCES

Bachmann, A., Walet, P., Wijnen, P., de Bruin, W., Huntjens, J. L. M., Roelofsen, W. and Zehnder, A. J. B. 1988. "Biodegradation of a-, ß-hexachlorocyclohexane in a soil slurry under different redox conditions." *Appl. Environ. Microbiol.*, 54: 143-149.

Haider, K. and Jagnow, G. 1975. "Abbau von ^{14}C-, ^{3}H- und ^{36}Cl-markierten gamma-Hexachlorocyclohexan durch anaerobe Bodenmikroorganismen." *Arch. Microbiol.*, 104: 113-121.

Jagnow, G., K., H. and P.-C., E. 1977. "Anaerobic dechlorination and degradation of hexachlorocyclohexane isomers by anaerobic and facultative anaerobic bacteria." *Arch. Microbiol.*, 115: 285-292.

Lovley, D. R. 1997. "Potential for anaerobic bioremediation of BTEX in petroleum- contaminated aquifers." *J. Ind. Microbiol. Biotechn.*, 18(2-3): 75-81.

Middeldorp, P. J. M., Jaspers, M., Zehnder, A. J. B. and Schraa, G. 1996. "Biotransformation of a-, ß-, γ-, and δ--hexachlorocyclohexane under methanogenic conditions." *Env. Sci. Technol.*, 30(7): 2345-2349.

Slooff, W. and Matthijsen, A. J. M. C. 1990. "Basisdocument hexachloorcyclohexanen (en advies Gezondheidsraad)." *Rapport nr. 7*, RIVM, Bilthoven.

Van Agteren, M. H., Keuning, S. and Janssen, D. B. 1998. *Handbook on biodegradation and biological treatment of hazardous organic compounds*, Kluwer Academic Publisher, Dordrecht.

Van Eekert, M. H. A., Van Ras, N. J. P., Mentink, G. H., Rijnaarts, H. H. M., Stams, A. J. M., Field, J. A. and Schraa, G. 1998. "Anaerobic transformation of ß-HCH by methanogenic granular sludge and soil microflora *Env. Sci. Technol.*, 32: 3299-3304.

ANAEROBIC BIOREMEDIATION OF TOXAPHENE-CONTAMINATED SOIL

Harry L. Allen (U.S. EPA, Edison, NJ USA)
Robert M. Mandel (U.S. EPA, San Francisco, CA USA)
Michael Torres (U.S. EPA, Dallas, TX USA)
Daniel G. Crouse, T. Ferrell Miller (Roy F. Weston, Inc., Edison, NJ, USA)

ABSTRACT: For seven years, studies have been conducted to develop an anaerobic solid-phase bioremediation process for removal of toxaphene from soil. Medium development studies in bench-scale reactors indicated that blood meal promoted the rapid degradation of toxaphene under anaerobic conditions. Recipes were developed and evaluated in field studies at three sites in Arizona and New Mexico.

In field studies conducted at the Navajo Vats site, over 75% of the toxaphene residues (291 mg/kg) was degraded in as little as 33 days in nutrient-amended reactors, while there was no change in toxaphene concentration in unamended reactors after 300 days. Additional field studies were conducted at other sites to clean up toxaphene residues. In studies at sites where initial concentrations were 20 mg/kg or higher, toxaphene reduction ranged from 58% to 86%.

Similar results were found in field studies at the Sanders Aviation site. Toxaphene concentrations ranging from 930 to 1,530 mg/kg were reduced by 94% to 95% in 216 days. In recent field studies at the Ojo Caliente Dip Vat site, toxaphene levels (14 mg/kg) were reduced by 70% in as little as 14 days.

INTRODUCTION

Toxaphene is a major broad-spectrum insecticide which has been used in the United States since 1947. It is primarily used to control pests on cotton and other food crops and ectoparasites on livestock (Korte, Scheunert, and Parlor, 1979). Over a 25 year period, it was used at a rate of approximately 40 million pounds (18.2 million kg) per year (Guyer et al, 1971). The insecticide, produced by chlorinating camphene, is actually a mixture of over 177 chlorinated derivatives, none of which account for more than a few percent of the total. Most of the derivatives are isomeric hepta-, octa-, and nonachlorobornanes with an overall elemental composition of $C_{10}H_{10}Cl_8$ (Casida et al., 1974; Holmstead et al, 1974). Several of these components have shown mammalian and fish toxicity as well as mutagenic and carcinogenic properties. The structure of one of the more toxic components in toxaphene is shown in Figure 1.

Toxaphene can persist in soil for years. Because of its persistence and toxicity, it was taken off the market in the United States in 1982, leaving contaminated soil wherever it was disposed of or used. Fifty-eight sites were included on the National Priority List by the United States Environmental Protection Agency (U.S. EPA).

A number of studies have shown that toxaphene is biodegradable, especially by anaerobic processes (Mirsatari et al, 1987, Parr and Smith, 1976, and Smith and Willis, 1978). The U S. EPA Environmental Response Team Center (ERTC) and Response Engineering and Analytical Contract (REAC) personnel have been

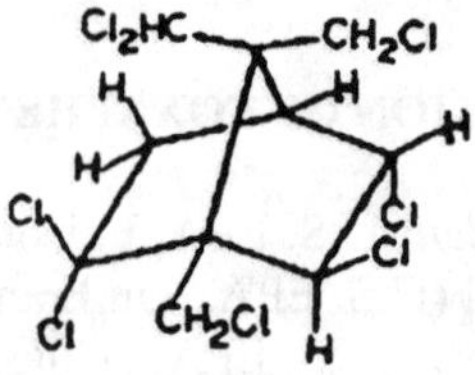

FIGURE 1. Structure of Toxaphene (Toxicant A-2, Turner et al., 1975)

conducting studies on removal of toxaphene using anaerobic bioremediation technology since 1991. A summary of results from three sites is presented.

Objective. The objective of these studies was to assess the feasibility of anaerobic bioremediation technology in removing toxaphene residues from contaminated soil. Bench-scale studies were initially conducted, and a recipe, containing blood meal, limestone, and sodium phosphate, was developed which consistently promoted the rapid degradation of toxaphene. Once recipes were developed, the anaerobic process was standardized in bench-scale reactors and then evaluated in field studies on test sites. The extent of toxaphene removal was monitored over the duration of each study. All of the sites were contaminated with other pesticides in addition to toxaphene. These pesticides were also monitored throughout the study.

Site Description. Studies were conducted at three sites: (1) Navajo Vats site, (2) Sanders Aviation site, and (3) Ojo Caliente Dip Vat site. The Navajo Vats site is actually a series of livestock dipping facilities located throughout the Navajo Nation. Approximately 170 to 250 concrete dip vats were used to dip livestock to control insect pests. Each dip vat was filled with about 2,000 gallons (7,560 liters) of pesticide formulation containing toxaphene and lindane. After the animals passed through the dip vats, the pesticide formulation was pumped out of the vats into unlined earthen pits. Of approximately 133 sites that have been screened to date for toxaphene and lindane, toxaphene concentrations ranged from non-detectable to 33,000 mg/kg, while the lindane concentration ranged from 2 to 309 mg/kg. There are roughly 70 sites remaining to be screened (Roy F. Weston, Inc., 1992).

The Sanders Aviation site is a former aerial pesticide applicator business, which operated from 1951 to 1984. Improper handling procedures resulted in the contamination of soil and groundwater with toxaphene and DDT (1,1,1-trichloro-2,2-bis(p-chlorophenyl)ethane). Approximately 175,000 square feet (16,275 square meters) of soil was contaminated with toxaphene, with contamination levels as high as 9,500 mg/kg.

The Ojo Caliente Dip Vat site is a former livestock dipping facility located on the Zuni Reservation south of Gallup, New Mexico. Until recently, the site was in operation as a working facility, with malathion replacing toxaphene as the active pesticide. Tribal plans to reestablish a community near the site led to the need for a priority cleanup.

MATERIALS AND METHODS

Navajo Vats site. A field study was conducted to evaluate recipes in promoting the removal of toxaphene in soil. The studies were conducted at the Nazlini and Whippoorwill sites. A set of pilot reactors, consisting of four 325-gallon (1,228-liter) plastic tanks, were buried in-ground at each site (Roy F. Weston, Inc., 1994). Each reactor was then charged with 3,500 pounds (1,591 kg) of contaminated soil. The soil toxaphene contamination levels at these two sites were approximately 300 mg/kg and 40 mg/kg, respectively. Two of the four tanks were used as controls and contained contaminated soil and water only, while the other two tanks were filled with contaminated soil, and amended with sheep manure, limestone, and blood meal at a rate of approximately 100 g/kg. Sheep manure was added as a bulking agent and to ensure sufficient organic matter to maintain anaerobic conditions in the reactor. Each tank was then filled with 200 gallons (756 liters) of water. Nine sampling ports were cut in the top of each tank. Core samples were collected and composited from each of three sets of three randomly selected sampling ports; a total of three composite samples were generated and analyzed for toxaphene. Samples were periodically collected over a 16-month period.

Following the success of the initial field studies, additional studies were conducted at 22 other sites. In these studies, pits were dug and lined with a high-density polyethylene (HDPE) liner. Contaminated soil, nutrient supplements (blood meal, limestone, and sodium phosphate), and water were slurried in a Maxcrete ® mixer and added to the test pits. Other nutrients were evaluated, but were discontinued due to inconsistent performance. The pits were covered with an HDPE liner with precautions taken to allow for venting of gases generated from biodegradative processes. Core samples were taken in triplicate at each sampling event and analyzed for toxaphene. Study periods ranged from 4 to 21 months.

Sanders Aviation site. In 1995, field studies were conducted at the Sanders Aviation site in Tempe, Arizona (Roy F. Weston, Inc., March 1995). Three test cells (pits), with dimensions of 5 feet (1.5 meters) square by 1.5 feet (0.46 meters) deep, were constructed and lined with plastic liners. Contaminated soil was sieved through a ¼-inch (6.35 mm) screen prior to loading into the test cells. Approximately 1,800 to 2,300 pounds of soil (818 to1,045 kg) was loaded into each cell. The control cell was filled with 100 gallons (378 liters) of water. In the remaining cells, contaminated soil was mixed dry with varying rates of blood meal and with limestone in a cement mixer and added to the cell. These cells were also filled with 100 gallons (378 liters) of water. Sodium phosphate was added later to the nutrient-amended cells to buffer the soil. The rate of nutrient dosage was: blood meal, 10 g/kg or 25 g/kg; limestone, 50 g/kg; and sodium phosphate, approximately 1 g/kg. Each cell was covered with a black plastic cover. Sediment cores were collected in triplicate from each cell and analyzed for toxaphene. The study lasted for approximately 17 months.

Ojo Caliente Dip Vat site. Approximately 160 cubic yards (122 cubic meters) of contaminated soil was excavated and stockpiled. A pit having dimensions of 60 ft (18 meters) long, 18 feet (5.5 meters), and 5 feet (1.5 meters) deep was constructed and lined with a plastic liner. Contaminated soil was mixed with a backhoe prior to adding to the pit. Blood meal and limestone were added at rates of 25 g/kg, and sodium phosphate at a rate of 10 g/kg. Sodium phosphate was added as a mixture of dibasic and monobasic salts at a ratio of 2.2:1 as a pH buffer as well as a nutrient source. Contaminated soil, blood meal, limestone, and sodium phosphate were added in layers, with each layer being hydrated with water. Once the pit was loaded, it was covered with a plastic liner. Sample cores were taken in triplicate and analyzed for toxaphene over a 40-day period.

Analytical Procedure. Toxaphene was analyzed by a modified gas chromatography/electron capture detector (GC/ECD) method developed at REAC laboratories (EPA/REAC 1994).

RESULTS AND DISCUSSION

Navajo Vats and Ojo Caliente Dip Vat sites. Table 1 summarizes results obtained from selected sites at the Navajo Vats and Ojo Caliente Dip Vat sites. Toxaphene reduction levels ranged from 58% to 86% over a 3- to 12-month period in soils having contamination levels ranging from approximately 20 to 300 mg/kg. High degradation levels were also found in Ojo Caliente Dip Vat soil, with over 70% removed in as little as 14 days.

TABLE 1. Toxaphene Degradation in Cells Using Blood Meal at the Navajo Vats and Ojo Caliente Dip Vat Sites

Site Name	Initial Conc. (mg/kg)	Final Conc. (mg/kg)	Time (Days)	Degradation (%)
Navajo Vats				
Nazlini	291	71	108	75.6
Whippoorwill	40	17	110	57.5
Blue Canyon Rd.	100	17	106	83.0
Jeddito Island	22	3	76	86.4
Poverty Tank	33	8	345	75.8
Ojo Caliente Dip Vat	14	4	14	71.4

Sanders Aviation site. Results of the pilot-scale study at the Sanders Aviation site are summarized in Figure 2. From initial concentrations ranging from 930 to 1,530 mg/kg, toxaphene levels decreased rapidly in reactors amended with blood meal at rates of 10 g/kg (cell AN2) or 25 g/kg (cell AN3). After seven months (day 216), toxaphene levels were reduced by 95% and 94%, respectively. Toxaphene levels in the control reactor (cell AN1), which was not amended with blood meal, were reduced by approximately 19% over the same time period.

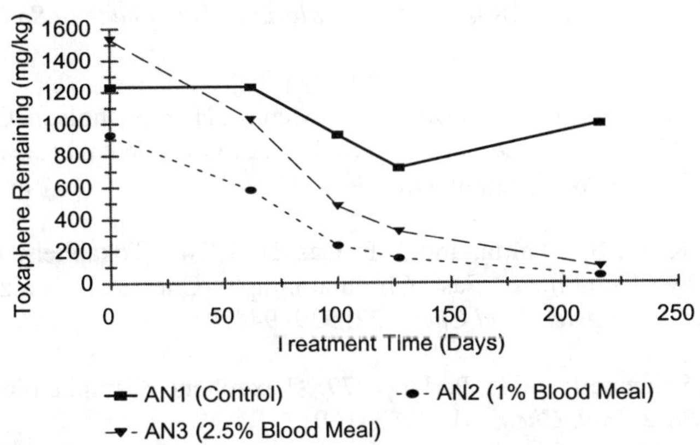

FIGURE 2. Toxaphene Degradation in Sanders Aviation Site Soil

Similarly designed studies (Camacho, et al., 1997) have shown that the use of a blood meal, limestone, and phosphate recipe promoted the rapid removal of toxaphene from soil by anaerobic processes. Studies at the Chem Air Spray site near Pahokee, Florida showed that 85 to 92% of toxaphene residues were removed from highly contaminated (1,200 mg/kg) soil in 6 months.

CONCLUSIONS

An anaerobic process has been developed which has demonstrated rapid removal of toxaphene from contaminated soil. The process has shown considerable versatility, with excellent performance demonstrated on three different sites having initial toxaphene levels ranging from 20 to 1500 mg/kg. Degradation levels ranged from 58 to 95% in 3 to 12 months.

Additional full-scale cleanups are planned for the Navajo Nation; the Zuni are considering additional site cleanups as well. Bench-scale feasibility studies are becoming standardized for screening toxaphene sites for bioremediation.

ACKNOWLEDGMENT

The authors wish to thank Nancy McGuire of Roy F. Weston, Inc. for her assistance in the preparation of this paper.

REFERENCES

Camacho, J. M., W. Joyner, R. Vemuri, and B. Holderness. 1997. "Anaerobic Degradation of Toxaphene in Soils: A Pilot Study." In B. C. Alleman and A. Leeson, Symposium Chairs, *Fourth International In Situ and On-Site Bioremediation Symposium*, 251–256, Vol. 4(2), Battelle Press, Columbus, Ohio.

Casida, J. E., R. L. Holmstead, S. Khalifa, J. R. Knox, T. Ohsawa, K. J. Palmer, and R. Y. Wong. 1974. "Toxaphene Insecticide: A Complex Biodegradable Mixture." *Science* 18: 520–521.

EPA/REAC. 1994. SOP #1802 *Routine Analysis of Toxaphene in Soil/Sediment by GC/ECD.*

Guyer, G. E., P. L. Adkisson, K. DuBios, C. Menzie, H. P. Nicholson, G. Zweig, and C. L. Dunn. 1971. *Toxaphene Status Report.* U. S. Environmental Protection Agency, EPA 540/9-71-005, Washington, D.C.

Holmstead, R. L., S. Khalifa, and J. E. Casida. 1974. "Toxaphene Composition Analyzed by Combined Gas Chromatography-Chemical Ionization Mass Spectrometry." *J. Agric. Food Chem.* 22: 939–944.

Korte, F., I. Scheunert, and H. Parlar. 1979. "Toxaphene (Camphethlor). A Special Report." *J. Pure Appl. Chem.* 51: 1583–1601.

Mirsatari, S. G., M. M. McChesney, A. C. Craigmill, W. L. Winterlin, and J. N. Seiber. 1987. "Anaerobic Microbial Dechlorination: An Approach to On-Site Treatment of Toxaphene-Contaminated Soil." *J. Environ. Sci. Health.* B22 (6): 663–690.

Parr, J. F., and S. Smith. 1976. "Degradation of Toxaphene in Selected Anaerobic Soil Environments." *Soil Sci.* 121(1): 52–57.

Roy F. Weston, Inc. 31 March 1992. "Final Report — Treatability Study, Navajo Vats, Window Rock, Arizona." Prepared for U.S. Environmental Protection Agency/Environmental Response Team under Contract Number 68-03-3482. Edison, NJ.

Roy F. Weston, Inc. 4 January 1994. "Field-Scale Treatability Study, Navajo Vats, Window Rock, Arizona — Status Report." Prepared for U.S. Environmental Protection Agency/Environmental Response Team under Contract Number 68-03-3482. Edison, NJ.

Roy F. Weston, Inc. 13 March 1995. "Sanders Aviation Company Site, Tempe, Arizona — Interim Report." Prepared for U.S. Environmental Protection Agency/ Environmental Response Team under Contract Number 68-03-3482. Edison, NJ.

Smith, S., and G. H. Willis. 1978. "Disappearance of Residual Toxaphene in a Mississippi Delta Soil." *Soil Sci.* 126(2): 87–93.

Turner, W.V., S. Khalifa, and J.E. Casida. 1975. "Toxaphene Toxicant A. Mixture of 2,2,5-endo,6-exo,8,8,9,10-octachlorobornane and 2,2,5-endo,6-exo,8,9,9,10-octachlorobornane." *J. Agric. Food Chem.* 23(5): 991–994.

INTERNATIONAL ACTIVITIES IN PHYTOREMEDIATION: INDUSTRY AND MARKET OVERVIEW

David J. Glass, D. Glass Associates, Inc., Needham, Mass., U.S.A.

Abstract: Phytoremediation, the use of plants to remove, destroy or sequester hazardous substances from the environment, is an innovative technology that has quickly gained a foothold in the United States market, while also attracting substantial interest outside the U.S. Several dozen companies and numerous non-profit research groups are conducting research on phytoremediation or are using it commercially, with perhaps 200 or more field remediations or demonstrations completed or underway around the world. The 1998 U.S. market of $16.5-29.5 million could grow to $214-370 million by 2005.

OVERVIEW OF COMMERCIAL USES OF PHYTOREMEDIATION

Several different types of phytoremediation are being used commercially, or are in advanced stages of research and development (Schnoor 1997). The most common applications rely on plants' ability to accumulate large quantities of contaminants from soils into their roots or to take up and transpire large amounts of contaminated groundwater. Once inside the plant, the contaminants might be degraded, metabolized to a volatile form and transpired, or may simply be sequestered inside plant biomass. Contaminants can also be removed from aqueous wastestreams by absorption onto plant roots. In other applications, the plants indirectly exert their impact on the soil, using secreted plant enzymes to degrade pollutants; by stimulation of microbial biodegradative activity in the rhizosphere; or by immobilizing contaminants in the soil using plant exudates.

Phytoremediation can be used in a number of environmental applications, specifically including the remediation of organics, metals and radionuclides from soils and water; municipal and industrial wastewater treatment; and-the use of plants as vegetative caps for control of landfill leachate or barriers in riparian zone buffers. Together, these applications offer U.S. markets totaling tens of billions of dollars, and intriguing opportunities overseas (Glass 1998).

U.S. PHYTOREMEDIATION COMPANIES

Dedicated Phytoremediation Companies. These are companies whose sole or primary remediation technology is phytoremediation. United States dedicated phytoremediation companies are shown in Table 1. The longer-established companies include several that use deep-rooted trees such as poplar, willow and cottonwood to remediate or control contaminated groundwater or aqueous runoff through the withdrawal of large amounts of water from the subsurface. Included in this group are Ecolotree, Applied Natural Sciences, Phytokinetics, Thomas Consultants and Verdant Technologies. Several companies, notably Phytokinetics and EarthCare, also specialize in the use of different plants, trees and grasses to stimulate microbial degradation of organic contaminants in soil.

Other dedicated companies are targeting inorganic contaminants, principally heavy metals or radionuclides. Phytotech is focused on removing lead and radionuclides from soil and aqueous media, using plants like *Brassica juncea* (Indian mustard) or sunflowers. PhytoWorks is developing transgenic plants for remediation of elemental mercury from soils by enzymatic reduction to less toxic forms and volatilization (PhytoWorks is also targeting organic contaminants, through the characterization of appropriate plant biodegradative pathways).

Diversifying Specialty Firms. Also shown in Table 1 are several companies whose specialties are in related fields or technologies, but which have begun to offer phytoremediation services. Many of these companies, including Ecoscience, Wolverton Environmental, and Sustainable Strategies, have expertise in constructed wetlands or artificial ecosystems to purify domestic, municipal or industrial wastewater. Other companies such as The Bioengineering Group and Bitterroot Restoration have expertise in revegetation for erosion control.

Diversified Consulting/Engineering Firms. A number of national or regional consulting/engineering firms have developed phytoremediation capabilities. As shown in Table 1, the firms having the most experience to date, mostly involving groundwater treatment or landfill leachate control, are ARCADIS Geraghty & Miller, CH2M Hill, Roy F. Weston, and a group at Geosyntec Consultants which relocated from Canada's Beak Consultants in 1998. A growing number of other c/e firms today have or claim to have phytoremediation expertise.

Industrial Companies. Numerous large industrial companies, principally in the oil, gas or chemicals industries, are also active in phytoremediation research (Table 1), mostly for possible use on their own contaminated sites. The most active companies to date include Chevron, Exxon, Amoco, DuPont and Occidental Chemical. These companies and others are participating in consortia that have

TABLE 1. U.S. Companies with Phytoremediation Experience.

Dedicated Phytoremediation Firms	*Diversifying Specialty Firms*	*Diversified Consulting/ Engineering Firms*	*Industrial Companies*
Applied Natural Sciences	Azurea	ARCADIS Geraghty & Miller	Alcoa
EarthCare	Bitterroot Restoration	Braun Intertec	Amoco
Ecolotree	Ecoscience	CH2M Hill	Chevron
Phytokinetics	Living Technologies	Geosyntec Consultants	DuPont
Phytotech	Soil Enrichment Systems	MSE Technology Applications	Exxon
PhytoWorks	Sustainable Strategies	Parsons Engineering	Occidental Chemical
Thomas Consultants	The Bioengineering Group	SoilQuest International	Shell
Verdant Technologies	Wolverton Environmental	Roy F. Weston	Union Carbide

supported phytoremediation research: at this writing two such consortia are in the process of merging: one established under the auspices of the Petroleum Environmental Research Forum and the other the Petroleum Hydrocarbon subgroup of the Phytoremediation of Organics Action Team of the Remediation Technologies Development Forum (a government-industry-academia consortium led by the U.S. Environmental Protection Agency).

PHYTOREMEDIATION FIRMS OUTSIDE THE UNITED STATES

Table 2 shows several companies in Europe and Canada known to be developing or commercializing phytoremediation. Slater (UK) Ltd., one of the few to be founded specifically to develop phytoremediation, is pursuing plant-based processes to remediate land contaminated with metals and recalcitrant organic pollutants, and is developing chelator-enhanced metal phytoremediation and a combined biotic/abiotic denitrification system. The company hopes to begin commercial use of its technologies by 2000.

TABLE 2. Non-U.S. Companies with Phytoremediation Expertise.

Aquaphyte Remediation (Canada)
BioPlanta (Germany)
Consulagri (Italy)
Ecobios (Italy)
Euramtecna (Portugal)
OEEL (U.K.)
Piccoplant (Germany)
Plantechno (Italy)
Slater (U.K.)

Most non-U.S. companies fall into the "diversifying specialty company" category, and maintain expertise in other commercial uses of plants such as constructed wetlands (reed beds) or agricultural technologies like micropropagation. Oceans Environmental Engineering Ltd. of the United Kingdom specializes in the construction of reed beds to treat industrial and municipal wastewaters, but is involved in soil phytoremediation in conjunction with its sister company ESU Services, Ltd. Aquaphyte Remediation (Canada) also focuses on the use of aquatic plants for wastewater treatment. Euramtecna (Portugal) and its sister company Cobelgal (Belgium/Portugal) commercialize constructed wetlands for wastewater treatment and sludge dewatering, and also coordinate a European cooperative project on new industrial uses for bamboo.

BioPlanta (Germany) is developing processes for soil rehabilitation, water cleaning, sewage water disposal, rehabilitation of rivers and other surface waters using specially optimized plants and constructed wetlands. Also in Germany is Piccoplant, founded in 1986, which specializes in *in vitro* propagation of woody and decorative plants, grasses and perennials, and conducts research on selection of plants for reconstruction, cleaning of contaminated sites, and erosion protection. Consulagri (Italy) is a service company which maintains expertise in plant systems to accumulate metals from soils and the use of plants for biomonitoring of pollution. Another Italian firm, Plantechno, is a plant biotechnology company founded in 1995 which is working to engineer an aquatic plant species for improved heavy metal uptake from wastewater.

PHYTOREMEDIATION IN THE FIELD: COMMERCIAL USES AND DEMONSTRATION PROJECTS

Phytoremediation has been the subject of a great deal of research not only in North America and Europe, but also in Australia, New Zealand, China and elsewhere in the world. There have also been numerous commercial and pilot-scale field projects conducted around the world, several of which are summarized below.

United States. Phytoremediation has been used in the field at dozens of sites in the U.S., including pilot-scale and commercial projects by commercial firms and a number of demonstration projects funded or supported by the U.S. government. Most of these projects have involved treatment of landfill leachate, petroleum hydrocarbon clean-up, removal of TCE or TNT from groundwater, and removal of metals from soils. These and other projects are described in more detail elsewhere (Chappell 1997, Schnoor 1997, Glass 1998). At this writing, the U.S. EPA is in the process of establishing a database on the World Wide Web that will list over 160 phytoremediation research projects (S. Rock, personal communication).

Canada. Canada has seen the involvement of several government agencies and a number of field demonstrations. In particular, a division of Environment Canada has initiated a program to facilitate and explore phytoremediation's potential, that includes research sponsorship, information gathering, and regulatory clarification (T. McIntyre, personal communication). There are ongoing demonstrations in British Columbia (a mine site evaluating the use of plants for Cd and Zn uptake); in Saskatchewan (a refinery evaluating poplars and willows for petroleum hydrocarbon remediation); in Ontario (two sites using scented geraniums for uptake of heavy metals and one project involving the use of rye grasses and fescues at a site contaminated with petroleum hydrocarbons); in Quebec (aquatic plant uptake of heavy metals); and the Atlantic Provinces (remediation of PCBs and petroleum hydrocarbons) (T. McIntyre, personal communication).

Europe. There has also been a good deal of phytoremediation activity in Europe. A group from Florida State University is managing a phytoremediation demonstration project at a contaminated soil refinery in Katowice, Poland, as part of a larger joint venture between U.S. and Polish entities, including Phytotech. Three plant species (corn, *Brassica* and *Brachinia*) are being tested for their ability to remove lead from the soil. The project began in 1997, and preliminary results from the 1997 harvest showed encouraging reductions in lead concentrations in soil (Kuperberg, et al. 1998).

Phytotech is also involved in two phytoremediation demonstrations at the site of the Chernobyl nuclear disaster in the Ukraine. The company is collaborating with U.S.-based Consolidated Growers and Processors and Ukraine's Institute of Bast Fibers to test the ability of industrial hemp to remove radionuclides from contaminated soil near the power plant. A pilot project took place in 1998, with full-scale trials planned for 1999. Phytotech has previously tested sunflowers on uranium-contaminated water at the Chernobyl plant.

There has been an application of phytostabilization in Belgium, on a site formerly polluted by huge amounts of heavy metals, carried out by Van Gronsveld, et al. from the University of Limburg (H.C. Dubourguier, personal communication). Projects to investigate the removal of heavy metals from dredged river sediments and the ability of phytoremediation to treat PAHs from soil were scheduled to begin in 1998 (H.C. Dubourguier, personal communication).

There are also several ongoing phytoremediation initiatives by European governments or research consortia. A group of academic scientists has proposed a phytoremediation study to be funded by the European Science Foundation. Another project is being carried out under the COST (European Cooperation in the field of Scientific and Technical Research) program. COST provides a framework for scientific and technical cooperation and coordination of research in different European countries. The COST phytoremediation project will study plant biotechnology for the removal of organic pollutants and toxic metals from wastewaters and contaminated sites (J.P. Schwitzguébel, personal communication).

WORLD PHYTOREMEDIATION MARKETS

Phytoremediation, which, as a new technology had almost no market share five years ago, has made considerable strides in establishing itself in the marketplace, particularly in the U.S. Any innovative remediation technology must compete on its merits, on a site-by-site basis, with many other available technologies. As a low-cost, permanent, *in situ* technology, phytoremediation tends to stack up well against alternative techniques, and should also fare well in the changing U.S. marketplace, where economic factors are becoming more important market drivers than are regulatory factors (Glass 1997, 1998).

United States Market. As shown in Table 3, we estimate that the largest 1998 U.S. markets for phytoremediation were for treatment of organic contaminants in groundwater, control of landfill leachate, and remediation of metals from soil. In general, the markets involving organic contaminants should see strong, steady growth in the coming years,. Markets for phytoremediation of metals or radionuclides are capable of dramatic growth as efficacy becomes better established, because there are few cost-effective alternatives, and because these sites tend to be expensive to treat. We estimate total U.S. phytoremediation revenues for 1998 were $16.5-29.5 million, and that the market will grow to $55-103 million by 2000 and $214-370 million by 2005. Uses of phytoremediation for organic pollutants should continue to dominate the market in the short term, with applications for metals becoming more prominent later.

Markets Outside the United States. The European market offers the next largest single remediation market. The market is currently dominated by remediation of hydrocarbon contamination and other easily-treatable pollutants, but metal-contaminated sites will ultimately begin to be remediated as well, once cost-effective technology is available. Certain regions of Europe, particularly the former Soviet bloc countries, have significant problems of improper landfilling and wastewater treatment, and these may present opportunities for phytoremediation.

TABLE 3. Estimated 1998 U.S. Phytoremediation Markets.

Organics in groundwater	$5-10 million
Organics in soil	$2-3 million
Metals in groundwater	$0.1-0.2 million
Metals in soil	$3-5 million
Radionuclides	$0.5-1.0 million
Landfill Leachate	$3-5 million
Organics in Wastewater	$2-3 million
Metals in Wastewater	$0.1-0.3 million
Other	$0.8-2.1 million
Total	$16.5-29.5 million

The European market has proven to be receptive to the use of innovative technologies. In view of the history of use of constructed wetlands in many European countries, we believe that Europe represents a major short-term (5-year) opportunity for phytoremediation.

Longer-term opportunities will be found in the developing nations, particularly in Asia and Latin America. Although site remediation is generally not a priority in these countries, phytoremediation is capable of addressing certain of these nation's high priority issues such as wastewater treatment and control of leachate from uncontrolled dumping sites. These markets might offer opportunities for market growth later in the first decade of the 21st Century.

The phytoremediation market outside the U.S. is currently very small, perhaps U.S. $1-2 million in 1998 revenues. We are cautiously optimistic that these markets will grow at rates such as those seen in the early days of the U.S. market, so that phytoremediation will ultimately become an important tool for several key environmental uses throughout the world in the 21st Century.

REFERENCES

Chappell, J. 1997. *Phytoremediation of TCE using Populus*. U.S. Environmental Protection Agency, Washington, DC. Available at http://clu-in.com/phytotce.htm.

Glass, D. J. 1997. "Evaluating Phytoremediation's Potential Share of the Hazardous Site Remediation Market." in *Phytoremediation, Proceedings of 2nd Annual IBC Phytoremediation Conference*. IBC Library Series, Southborough, MA.

Glass, D. J. 1998. *The 1998 United States Market for Phytoremediation*. D. Glass Associates, Inc., Needham, MA.

Kuperberg, J. M., R. Kucharski., P. Richter, and M. Kolta. 1998. "Phytoremediation in Poland: Central and Eastern Europe Technology Identification and Development Program." Poster presented at 3rd Annual IBC Phytoremediation Meeting, Houston, TX.

Schnoor, J. L. 1997. *Phytoremediation: Technology Evaluation Report*. Ground-Water Remediation Technologies Analysis Center, TE-98-01. Available at http://www.gwrtac.org/html/tech_eval.html#PHYTO.

PHYTOREMEDIATION OF TRICHLOROETHENE (TCE) USING COTTONWOOD TREES

S.A. Jones (U.S. Geological Survey, Austin, Texas)
R.W. Lee (U.S. Geological Survey, Dallas, Texas)
E.L. Kuniansky (U.S. Geological Survey, Atlanta, Georgia)

ABSTRACT: Phytoremediation uses the natural ability of plants to degrade contaminants in ground water. A field demonstration designed to remediate aerobic shallow ground water that contains trichloroethene began in April 1996 with the planting of cottonwood trees over an approximately 0.2-hectare area at the Naval Air Station, Fort Worth, Tex. Ground water was sampled in July 1997, November 1997, February 1998, and June 1998. Analyses from samples indicate that tree roots have the potential to create anaerobic conditions in the ground water that will facilitate degradation of trichloroethene by microbially mediated reductive dechlorination. Dissolved oxygen concentrations, which varied across the site, were smallest near a mature cottonwood tree (about 20-years old) 60 meters southwest of the cottonwood plantings. Reduction of dissolved oxygen is the primary microbially mediated reaction occurring in the ground water beneath the planted trees, whereas near the mature cottonwood tree, data indicate that methanogenesis is the most probable reaction occurring. Reductive dechlorination either is not occurring or is not a primary process away from the mature tree. On the basis of isotopic analyses of carbon-13 at locations away from the mature tree, trichloroethene concentration is controlled by volatilization.

INTRODUCTION

The terrace alluvial aquifer that underlies U.S. Air Force Plant 4 (AFP4) and the adjacent Naval Air Station (NAS), Fort Worth, Tex., is contaminated with trichloroethene (TCE) (Rust Geotech, 1995). The site has been designated by the North Atlantic Treaty Organization as a demonstration of innovative remedial technology (G.J. Harvey, U.S. Air Force, Aeronautical Systems Center/Environmental Management Directorate, oral commun., 1997). Phreatophytes (cottonwood trees) have been planted at a test site near Carswell Golf Course (Figure 1) to determine if the tree roots create anaerobic conditions in the ground water that will facilitate degradation of TCE by reductive dechlorination. The efficiency of TCE degradation varies depending on microbially mediated redox reactions (most efficient to least efficient—methanogenesis, sulfate reduction, iron (III) reduction, oxidation) (Chapelle, 1996). Microbial processes leading to TCE degradation require organic carbon as an electron donor. Naturally occurring organic carbon (CH_2O) is scarce in the aquifer, but decaying roots or root exudates containing organic carbon could be introduced by mature trees.

The study area is located on the northwest side of Fort Worth. The project area comprises 10 hectares northeast of the Carswell Golf Course (Figure 1). Two

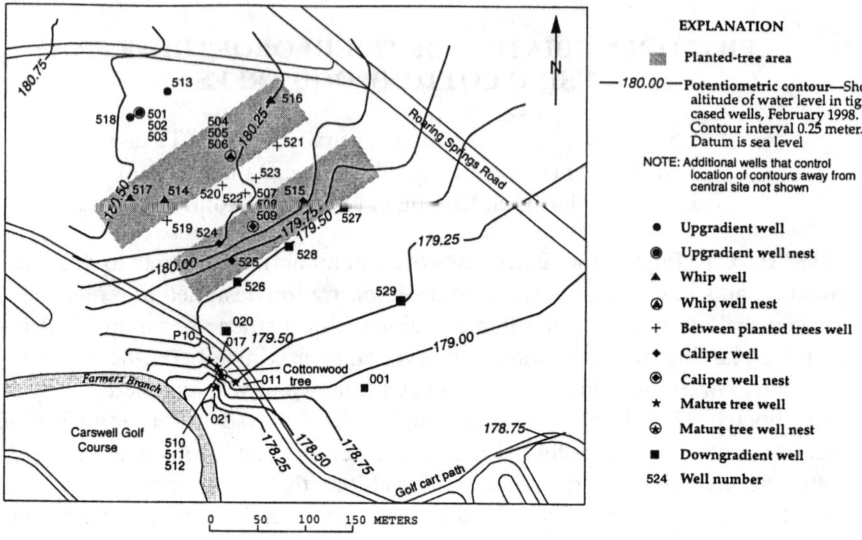

FIGURE 1. Locations of wells sampled at phreatophyte demonstration site, Fort Worth, Tex.

cottonwood tree plots were prepared and planted in April 1996 approximately orthogonal to ground-water flow. The plots are about 15 by 77 meters (m) each, consisting of 7 rows of 50 equally spaced plantings. The westernmost plot was planted with cottonwood whips (cuttings). The second plot was planted with 2.5- to 3.8-centimeter (cm) caliper (trunk diameter) cottonwood trees. A mature cottonwood tree (about 20-years old) is located 60 m southwest of the caliper plantings.

The terrace alluvium is composed of poorly sorted, cross-bedded unconsolidated gravel, sand, silt, and clay. The terrace alluvial deposits range from 1.8 to 4.6 m thick at the site. Saturated thickness in the unconfined terrace alluvial aquifer varies but normally ranges from 0.3 to 1.5 m above the confining unit beneath the aquifer. Slug tests were done at 11 wells at the site. Estimates of hydraulic conductivity based on the slug tests using the Bouwer and Rice method (Bouwer and Rice, 1976) ranged from 0.9 to 30 m/day. From the tests, the mean and median hydraulic conductivity of the terrace alluvium at the site is 9 m/day.

METHODS

Since April 1996, 52 wells have been installed near the planted trees. Selected wells were sampled in July 1997, November 1997, February 1998, and

June 1998, and the samples analyzed for selected chemical properties and constituents to define the geochemical environments in ground water at the site. Samples also were analyzed for volatile organic compounds (VOCs), dissolved organic carbon (DOC), and carbon-13 (• ^{13}C) isotopes in TCE. To determine the geochemical environments that could affect degradation of TCE, the constituents listed in Table 1 were analyzed on-site or in the laboratory, as indicated.

To determine if trees were influencing geochemical and microbial environments, the site was divided into areas on the basis of the presence or absence of trees. The areas are

1. Upgradient (of planted trees)
2. Whip plantings
3. Between planted trees
4. Caliper plantings
5. Mature tree
6. Downgradient (of planted trees)

Wells in the six areas were sampled synoptically in four 2-week periods from July 1997 through June 1998 to characterize changes in geochemical environments.

RESULTS

Reduction of dissolved oxygen (DO) is the primary microbial process in ground water beneath the planted trees. Concentrations of DO in ground water showed some spatial variation. Median concentrations of DO upgradient, in the whip plantings, between the planted trees, in the caliper plantings, and downgradient are all about 3 milligrams per liter (mg/L) (Figure 2). Concentrations from wells in the whip plantings vary more than those of other areas. At the southern end of the whip plantings, DO concentrations among four samples from well 514 ranged from 0.7 to 2.5 mg/L (Table 2), whereas concentrations among 16 samples from four other whip wells ranged from 3.0 to 4.8 mg/L. Table 2 lists selected chemical data from wells (one in each area). DO concentrations decreased in water from well 514 from July 1997 through February 1998, which indicates some oxygen consumption and the potential for establishment of reducing conditions near this well. However, DO concentrations were larger at well 514 in June 1998, reversing the trend. (April–June was a drought period in the area.) A similar trend in DO concentrations occurred in water from well 515—a decrease during July 1997–February 1998 followed by an increase in June 1998, indicating the potential to establish reducing conditions in the aquifer. DO concentrations were smallest near the mature cottonwood tree (median of 25 samples 1.0 mg/L). The relatively small DO concentrations indicate that oxygen likely is almost completely consumed by microbial oxidation of organic matter.

Dissolved sulfide and dissolved total iron concentrations typically were less than 0.001 and 0.1 mg/L, respectively, for most of the wells sampled. However, dissolved sulfide was detected in November 1997 and February 1998 in wells 511 and 514, indicating a trend toward reducing conditions consistent with small (well 511) or decreasing (well 514) DO concentrations in those months. Well 511, near the mature cottonwood tree, had consistently large but variable dissolved iron concentrations during the period of sampling, ranging from 3.9 to 7.7 mg/L, which also is consistent with anaerobic conditions.

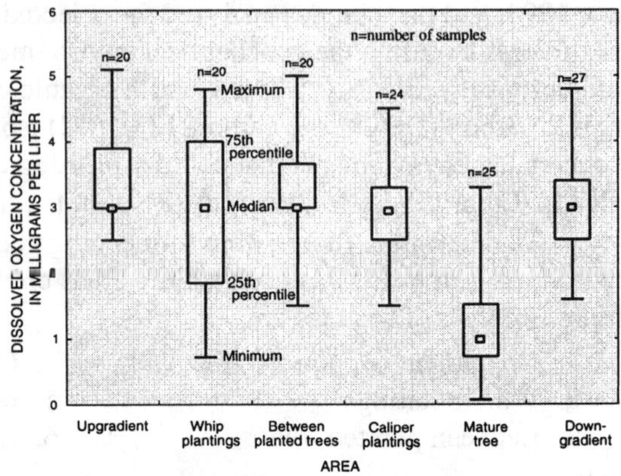

FIGURE 2. Boxplots showing dissolved oxygen concentrations for the six areas.

Molecular hydrogen and methane concentrations typically were less than 0.05 nanomolar (nM) and 0.1 micromolar (• M), respectively, at the site, although measurable concentrations of molecular hydrogen were observed at some wells. Measurable molecular hydrogen concentrations indicate that microbial activity could be developing in some locations capable of establishing reducing environments in the aquifer. Wells 511 and 514 had molecular hydrogen concentrations as large as 0.9 and 12.2 nM, respectively. Methane also was present in well 511 each time it was sampled, with concentrations ranging from 5.1 to 24 • M, which indicates some local methane production and possibly methanogenesis.

TABLE 1. Chemical analyses used to evaluate reductive dechlorination at the phreatophyte demonstration site, Fort Worth, Tex.

[mg/L; milligrams per liter; VOC, volatile organic compound; GCMS, gas chromatograph/mass spectrometer; DOC, dissolved organic carbon; µm, micrometer; °C, degrees Celsius; GC/C/IRMS, gas chromatograph/combustion/isotope ratio mass spectometer]

Constituent	Method	Procedure
Oxygen	Chemets[1]	On-site colorimetric (0–10 mg/L)
Fe^{+2}/total iron	Hach[1]	On-site colorimetric methods
Sulfide	Hach[1]/Chemets	On-site colorimetric analysis
Carbon dioxide, methane	Gas chromatography	On-site method, headspace, thermal conductivity
Hydrogen	Gas chromatography	On-site method, headspace analysis, (0.05–30 nM)
VOC	GCMS/gas chromatography	Laboratory analysis/on-site gas chromatography
DOC fractions	Lab separation	0.45-µm silver filter to glass bottle, chilled to 4 °C
• $^{13}C_{TCE}$	Compound specific isotopic analysis	Laboratory analysis by GC/C/IRMS per Slater et al. (1998)

[1]Any use of trade, product, or firm names is for descriptive purposes only and does not imply endorsement by the U.S. Government.

TABLE 2. Selected chemical data.

[mg/L, milligrams per liter; nM, nanomolar; µM, micromolar; well 501 is upgradient, well 511 is near the mature cottonwood tree, well 514 is in the whip plantings, well 515 is in the caliper plantings, well 523 is between the planted trees, and well 529 is downgradient]

Well	Dissolved oxygen (mg/L)				Total iron (mg/L)				Sulfide (mg/L)			
	Jul	Nov	Feb	Jun	Jul	Nov	Feb	Jun	Jul	Nov	Feb	Jun
501	3.5	3.0	3.0	4.7	0.1	<0.1	<0.1	<0.1	<0.001	<0.001	<0.001	<0.001
511	1.1	.7	.9	.8	4.9	7.7	3.9	5.5	<.001	.005	.007	<.001
514	2.5	1.2	.7	1.7	<.1	<.1	.1	.2	<.001	.120	.056	<.001
515	3.0	2.5	1.5	2.9	.1	<.1	.1	<.1	<.001	<.001	<.001	<.001
523	3.5	3.5	3.0	4.5	<.1	<.1	<.1	<.1	<.001	<.001	<.001	<.001
529	3.5	4.0	3.0	2.7	<.1	<.1	<.1	<.1	<.001	<.001	<.001	<.001

Well	Hydrogen (nM)				Methane (µM)			
	Jul	Nov	Feb	Jun	Jul	Nov	Feb	Jun
501	<0.05	<0.05	<0.05	0.3	<0.1	<0.1	<0.1	<0.1
511	<.05	<.05	.1	.9	5.1	7.5	24	15
514	<.05	12.2	.7	.5	<.1	<.1	<.1	<.1
515	<.05	.8	<.05	.1	<.1	<.1	<.1	<.1
523	.47	<.05	<.05	.2	<.1	<.1	<.1	<.1
529	<.05	<.05	<.05	.5	<.1	<.1	<.1	<.1

DOC was analyzed in samples collected in November 1997 from nine wells near the newly planted cottonwood trees and from well 511 adjacent to the mature cottonwood tree. Concentrations ranged from 0.8 to 1.8 mg/L. Wells 511 (mature tree) and 514 (whip plantings) had the largest concentrations of DOC, 1.7 and 1.8 mg/L, respectively. Analyses of acid, base, and neutral fractions of hydrophobic and hydrophilic organic compounds composing the dissolved organic carbon (Leenheer and Huffman, 1979) show that the principal compound classes at the site are hydrophobic neutral, hydrophobic acid, and hydrophilic acid.

Hydrophilic acids (containing less than 5 carbon atoms) typically are readily consumed by microbes. The largest measured concentrations of hydrophilic acids in ground water were 0.6 mg/L at well 511 (near the mature cottonwood) and 0.9 mg/L at well 514 (whip plantings). The data indicate that root systems under the mature cottonwood tree and under part of the planted cottonwoods produce hydrophilic acids that can be consumed by microbes, which results in increased DO consumption.

Concentrations of TCE were very small (less than 50 micrograms per liter (• g/L)) in wells adjacent to the mature cottonwood tree (Figure 3). The median ratio of TCE to the degradation product *cis*-1,2-dichloroethene (DCE) was smallest adjacent to the mature tree (Figure 4), and vinyl chloride was detected in the shallow ground water (Science Applications International Corp., written commun., 1998) These findings indicate that degradation of TCE by reductive dechlorination occurs in the vicinity of the mature tree. On the basis of methane detected in well 511 near the mature tree, methanogenesis is the most probable reaction that is occurring.

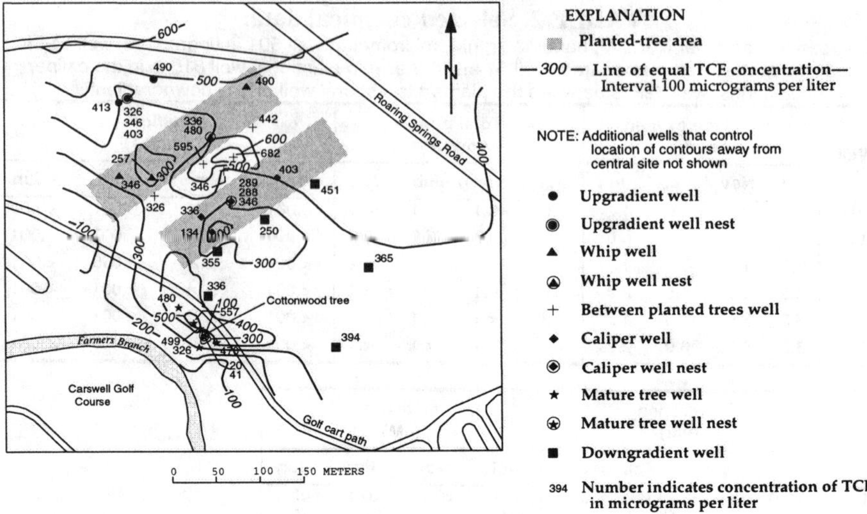

FIGURE 3. Distribution of TCE in ground water.

On the basis of isotopic analyses of • ^{13}C data from various locations away from the mature tree, TCE concentration is controlled by volatilization rather than reductive dechlorination. Laboratory experiments have demonstrated that reductive dechlorination of TCE strongly fractionates • ^{13}C in TCE, resulting in enrichment of greater than 10 per mil ($^0/_{00}$) in residual TCE, although dissolution and volatilization produce no significant fractionation (Slater et al., 1998). Isotopic data collected from five wells at the site showed • ^{13}C values that are identical (-26.9 to –27.3 $^0/_{00}$, within error of 0.5 $^0/_{00}$), which indicates that dechlorination is not a primary process at the site.

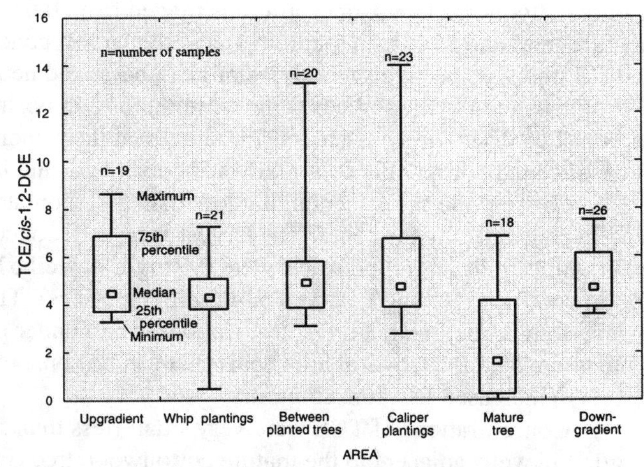

FIGURE 4. Boxplots showing TCE/*cis*-1,2-DCE concentrations for the six areas.

Isotopic data are not available from the zone of probable dechlorination activity near the mature cotton-wood because the concentration of TCE in ground water from well 511 was less than the method-detection limit (50 • g/L) necessary to obtain a • $^{13}C_{TCE}$ signature. If anaerobic conditions become more widespread at the site, future changes in • $^{13}C_{TCE}$ in conjunction with TCE concentrations could be an effective means to determine if reductive dechlorination is occurring.

CONCLUSIONS

Two years after planting the cottonwood trees at the phreatophyte demonstration site, ground-water chemistry in the terrace alluvial aquifer appears to be changing locally. DO concentrations show some evidence of decreasing at the southern end of the whip plantings, and total iron concentration is increasing. Ground-water chemistry near a mature cottonwood tree approximately 60 m from the caliper trees is different from that observed elsewhere at the demonstration site. Evidence from DO, dissolved total iron, molecular hydrogen, methane, and DOC analyses indicates development of anaerobic conditions near the mature cottonwood tree. A very small concentration of TCE in ground water, a small median ratio of TCE to the degradation product *cis*-1,2-DCE, and the presence of vinyl chloride support the conclusion that reductive dechlorination of TCE occurs in the aquifer near the mature tree. Decaying root tissue and root exudates containing hydrophilic acids are possible sources of labile organic carbon that can be used by microbes to produce anaerobic conditions in ground water conducive to reductive dechlorination. Decreases in DO observed in wells near the cottonwood plantings indicate that maturation of root systems at the demonstration site, similar to those of the mature cottonwood tree, might be affecting ground-water chemistry.

ACKNOWLEDGMENTS

This research project is funded by the Aeronautical Systems Center/Environmental Management Directorate at Wright-Patterson Air Force Base. Laboratory analyses for carbon isotopes on TCE were done at the University of Toronto with the assistance of Neil Arner.

SELECTED REFERENCES

Bouwer, Herman, and R. C. Rice. 1976. "A Slug Test for Determining Hydraulic Conductivity of Unconfined Aquifers With Completely or Partially Penetrating Wells." Water Resources Research. 12(3): 423–428.

Chapelle, F. H. 1996. "Identifying Redox Conditions That Favor the Natural Attenuation of Chlorinated Ethenes in Contaminated Ground-Water Systems." In *Symposium on Natural Attenuation of Chlorinated Organics in Ground Water, September 1996, Proceedings*, pp. 17–20. EPA/540/R–96/509.

Leenheer, J. A., and E. W. D. Huffman, Jr. 1979. *Analytical Method for Dissolved-Organic Carbon Fraction.* U.S. Geological Survey Water-Resources Investigations 79–4.

Rust Geotech. 1995. Air Force Plant 4, *Remedial Investigation and Preliminary Assessment/Site Inspection Report*. Vol. I. U.S. Air Force, Aeronautical Systems Center, Wright-Patterson Air Force Base, Ohio, Department of Energy Contract No. DE–AC04–861D12584, GJPO–WMP–7S.

Slater, G. F., H. D. Demptster, B. Sherwood Lollar, J. Spivack, M. Brennan, and P. Mackenzie. 1998. "Isotopic Tracers of Degradation of Dissolved Chlorinated Solvents." In First International Batelle Conference on Remediation of Chlorinated and Recalcitrant Compounds, Monterey, Calif., May 1998.

THE FATE OF CHLORINATED ORGANIC POLLUTANTS IN A REED-BED SYSTEM

C.J.A. MacLeod, B.J. Reid, ***K.T.Semple***
(Lancaster University, Lancaster, U.K.).

ABSTRACT: A laboratory-based microcosm study was carried out to investigate the fate of [UL-^{14}C]1,2-dichlororoethane and [UL-^{14}C]chlorobenzene in the presence and absence of reeds (*Phragmites* sp.), monitoring volatilization and mineralization of the pollutants. At the end of the incubation, the microcosms were destructively sampled to assess soil-bound and plant-associated fractions. After 47 d, 7.31%±1.30 for the reed system, and 6.20%±0.34 for the no-reed system of the added [UL-^{14}C]1,2-dichloroethane was mineralized. Of the added [UL-^{14}C]chlorobenzene, 26.99%±1.86 for the reed system and 15.91%±2.64 for the no-reed system was mineralized. However, the dominant loss process for both compounds was volatilization. After 20 d, 59.0%±9.6 and 79.7%±5.8 of the added [UL-^{14}C]1,2-dichloroethane was volatilized in the reed and no-reed systems respectively. [UL-^{14}C]Chlorobenzene was volatilized to a smaller extent, with 31.6%±1.8 for the reed system and 31.4%±3.1 in the no-reed system being volatilized. This study shows that reeds and the associated rhizosphere can enhance the degradation of semi-volatile compounds, such as chlorobenzene.

INTRODUCTION

In 1939, the Middleton Wood site (Lancashire, UK) was developed by industry to produce aviation fuel from gasoline. After the War, the main site was closed for three years, then reopened by Shell and ICI to refine crude oil and finally closed in 1980. From the Richards, Moorehead and Laing report (1997) and from chemical analyses carried out by the Environment Agency (pers comm), the soils on the site and the 'surface water' effluent being pumped from the site into Morecambe Bay contain a range of pollutants. Chlorinated organic compounds were found to be present and include trichloromethane (chloroform) 1,2-dichloroethane, 1,1,1-trichlorethane, tri- and tertachloroethene, chlorobenzene and chlorotoluene.

The site requires remediation for a number of reasons, these include its proposed development as a community woodland, as well as improving the quality of future effluent discharge to Morecambe Bay to comply with legislation introduced during the 1990s. A number of potential treatments strategies could be employed to deal with the chemical and physical problems/hazards associated with the site. However, determining factors would be the availability of landfill/special waste sites, cost of the treatment and the impact on the thriving wetland that has become established at the site and the nearby Middleton Village. For these reasons, biological remediation has been proposed.

Reed-bed systems (*Phragmites* sp.) and their associated rhizosphere microbial communities have been successfully used to bioremediate a range of organic pollutants in contaminated land and water in the UK, Europe and the USA (Urbancbercic, 1994; Revitt *et al.*, 1997). In previous studies, the common reed (*Phragmites communis*) has been shown to be particularly suitable (McEldowney, *et al.,* 1993). This plant combines rapid growth with a deep, extensive rhizome and root system, providing a large surface area for microbial growth and also improves soil porosity.

The aim of this laboratory-based study was to characterise the fate of chlorobenzene and 1,2-dichloroethane in the presence and absence of reeds (*Phragmites* sp.). And to assess the feasibility of reed systems and their associated rhizosphere microbial communities to reduce the concentrations of chlorobenzene and 1,2-dichloroethane in the surface water effluent.

MATERIALS AND METHODS

Chemicals. All scintillation fluids and solvents were obtained from Canberra-Packard (UK). Both [UL-^{14}C]chlorobenzene and [UL-^{14}C]1,2-dichloroethane were supplied by Sigma Chemical Co UK. Supelco ORBO solvent desorption devices were obtained from Sigma-Aldrich Co Ltd. UK. Potassium hydroxide used in CO_2 traps was obtained from Fissons, Ltd., UK

Samples. Soil was taken from the Middleton Wood site and sieved through a 2 mm sieve to remove stones and plant material. The reeds (*Phragmites* sp.) where also sampled from the Middleton Wood site. Effluent from the site was provided by the Environment Agency, UK.

Microcosms. Glass microcosms (3 l - 12 cm i.d.) were constructed which would hold effluent contaminated soil-plant systems. The total mass of soil and reeds was 1.2 kg and added to acetone rinsed microcosms within 2 d of sampling. The reeds were cut back to the rhizome. An equal amount of rhizome was added to each microcosm. The soil and rhizome was equilibrated for 7 d in a green house under artificial light to establish growth of new shoots (8-10). The microcosms were then transferred to a fume cupboard where the light intensity across the microcosms was assessed (using a porometer AP4 Delta T devices, Cambridge UK), 120-140 μmol m^{-2} s^{-1}. The light source was controlled by an automatic timer and was switched on from 8 am until 6 pm. The temperature of the system was 20°C ± 2°C.

Microcosms were run in triplicate under the following conditions: (i) reeds and soil plus [UL-^{14}C]1,2-dicloroethane-spiked effluent (10.2 kBq); (ii) soil only plus [UL-^{14}C]1,2-dicloroethane-spiked effluent (10.2 kBq); (iii) reeds and soil plus [UL-^{14}C]chlorobenzene-spiked effluent (6.5 kBq) and (iv) soil only plus [UL-^{14}C]chlorobenzene-spiked effluent (6.5 kBq). Effluent (100 ml) was added to each microcosm containing either [UL-^{14}C]1,2-dichloroethane or [UL-^{14}C]chlorobenzene at a concentration of 97.8 μg l^{-1} or 100 μg l^{-1}, respectively. Additionally, microcosms containing unspiked effluent and reeds were set up. After addition of the effluent to the microcosms, air was pumped through the microcosm at a rate of 0.2 l min^{-1} for 15 min every 2 h after having passed through 300 ml of KOH (1 M) solution to remove atmospheric CO_2.

Compound Mass Balance. The volatile fraction of the gas phase was trapped on Supelco ORBO solvent desorption sampling devices which contained activated coconut charcoal (600 mg). The ORBO tubes were attached to gas impermeable c-flex tubing using rip ties. At each sampling event, the ORBO tubes were eluted with four 10 ml washes of dichloromethane. The trapping efficiency of the ORBO tubes was found to be 90%±5%. The maximum sample volume was determined and not exceeded. To each elution Ultima Gold XR scintillation fluid (10 ml) was added, which were then analysed by liquid scintillation counting. Values were corrected for quenching using an external standard source.

The $^{14}CO_2$ was trapped in two 10 ml solutions of KOH (1 M) in series (phenol indicator to assess when traps were exhausted, a situation which never arose in the study). Samples of the KOH (3ml) were removed and added to 17 ml Ultima Gold scintillation fluid and analysed by liquid scintillation counting. The microcosms were sampled at 7, 14, 20, 27, 35 and 47 days. During sampling, the microcosms were sealed. Before commencement of the experiment, the trapping efficiency of the system was established and found to be 95% ± 5 %.

At the end of the microcosm incubation, soil samples (~1 g field wet soil), plant leaves (~1 g) and root material (~1 g) were sample oxidised (combusted). This provided a measure of the amount of ^{14}C-activity left in the soil and to account for any uptake/partitioning by the reeds.

RESULTS AND DISCUSSION

Volatilisation. Table 1 shows the volatilisation of both [UL-^{14}C]chlorobenzene and [UL-^{14}C]1,2-dichloroethane in the presence and absence of reeds. In all cases, there was an initial rapid loss of the compounds through volatilisation, followed by a much slower rate of loss. The initial rate (data not shown) and extent of loss through volatilisation were different for the two compounds. Chlorobenzene was volatilised more slowly and to a lesser extent than 1,2-dichlorethane. Of the chlorobenzene, 31.6%±1.8 and 31.4%±3.1 was volatilized after 20 d in the reed and no reed systems, respectively. However, 59.0%±9.6 and 79.7%±5.8 of the 1,2-dichlorethane was volatilized after 20 d in the reed and no reed systems, respectively. The vapour pressure of 1,2-dichlorethane (590 kPa at 20 °C) is over 2 orders of magnitude greater than that of chlorobenzene (1.5 kPa at 20 °C) (CRC Handbook of Physics and Chemistry, 1991), which may account for the greater losses through volatilization.

TABLE 1. Volatilization (%±SE) of [UL-^{14}C]chlorobenzene (CB) and [UL-^{14}C]1,2-dichloroethane (DCE) in the presence and absence of reeds.

Microcosm	Incubation Time (Days)					
	7	14	20	27	35	47
Reed- DCE	51.4±10.6	55.6±9.7	59.0±9.6	61.4±8.7	63.6±7.3	64.8±6.5
No Reed- DCE	66.1±2.0	77.1±5.1	79.7±5.8	81.7±5.3	81.7±5.3	82.0±5.2
Reed- CB	31.4±1.7	31.6±1.8	31.6±1.8	31.7±1.8	33.1±2.8	33.1±2.8
No Reed- CB	30.4±2.6	31.3±3.1	31.4±3.1	31.5±3.2	31.5±3.2	31.5±3.2

Mineralization. Table 2 shows that chlorobenzene was mineralized to a greater extent than 1,2-dichloroethane. After 47 d 26.99%±1.86 and 15.91%±2.64 of the chlorobenzene was mineralized in the reed and no reed systems, respectively. However, 7.31%±1.30 and 6.20%±0.34 of the 1,2-dichloroethane was mineralized in the reed and no reed systems, respectively.

TABLE 2. Mineralization (%±SE) of [UL-14C]chlorobenzene (CB) and [UL-14C]1,2-dichloroethane (DCE) in the presence and absence of reeds

Microcosm	Incubation Time (Days)					
	7	14	20	27	35	47
Reed- DCE	2.69±0.03	4.43±0.69	5.50±1.09	5.66±1.13	6.59±1.35	7.31±1.30
No Reed- DCE	1.49±0.16	3.59±0.22	4.56±0.32	4.80±0.33	5.50±0.34	6.20±0.34
Reed- CB	4.86±0.89	12.72±0.9	22.44±1.6	22.79±1.5	25.74±2.3	26.99±1.9
No Reed- CB	3.04±0.94	8.81±1.22	11.30±1.6	12.17±2.0	13.42±2.2	15.91±2.6

Mass Balance. Figure 1 shows the mass balances achieved in the microcosm incubations as a percentage of the total ^{14}C-activity added to each set of microcosms. The total amount of ^{14}C-activity accounted for at the end of the experiment was >72%, however for three of the 4 conditions, this value was >87%.

Chlorinated organic solvents with high vapour pressures, such as chlorobenzene and 1,2-dichloroethane will volatilize from the soil to the atmosphere at 20°C. Volatilization of a compound is temperature dependant, with greater losses expected to occur in the summer than in the winter.

Figure 1 shows that after 47 d, 5.13%±2.64 and 1.59%±0.52 of the ^{14}C-label associated with chlorobenzene and 1,2-dichloroethane was taken up or adsorbed by reeds, respectively. The results do not indicate whether the compound was taken up by reeds or adsorbed to the plant surface. But, since a smaller percentage of the added activity was found to be associated with the root material, adsorption to the leaf surfaces is the more likely to occur.

There was more ^{14}C-activity associated with the soil after 47 d in the chlorobenzene microcosms (26.29%±1.64 and 24.01%±3.58, with and without reeds, respectively) than in the 1,2-dichloroethane microcosms (11.94%±4.58 and 3.72%±0.94, with and without reeds, respectively). This can be explained by the higher vapour pressure of 1,2-dichloroethane (CRC handbook, 1991). The data shows that of the total ^{14}C-labelled pollutants left in the soil and the fraction that was mineralized (after 47 d), approx. 52% of the chlorobenzene and 40% of the

1,2-dicloroethane was mineralized in the reed systems. Before degradation of the added compounds can take place, several criteria need to be satisfied. These include, the proliferation of microorganisms with the relevant catabolic potential present in the environment containing the compound in an available form (Alexander, 1994).

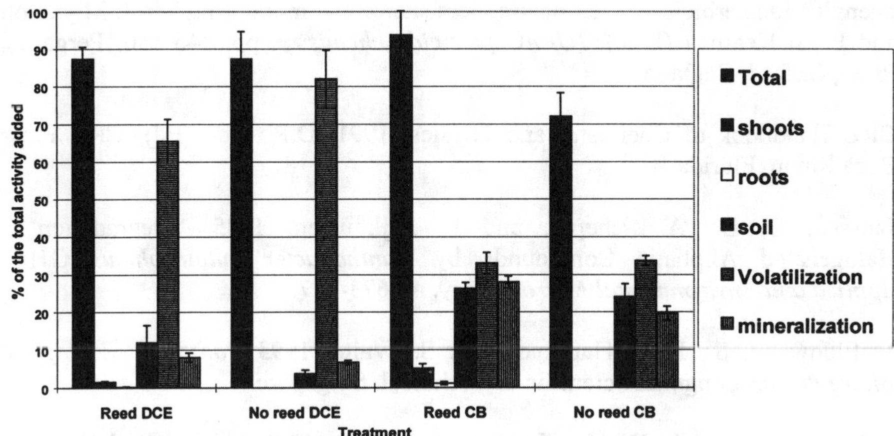

FIGURE 1. Mass balance of ^{14}C-label added to the microcosm as a percentage (%±SE) of the total activity added.

Soil microorganisms have been shown to degrade both chlorobenzene and 1,2-dichloroethane (Reineke *et al.*, 1984; Janssen *et al.*, 1985). Speitel *et al.* (1991) reported that the bioremediation of soils contaminated with 100 μg g^{-1} 1,2-dichloroethane would take several months due to the slow rate of degradation. For both compounds the amount of activity remaining associated with the soil was greater in the microcosms containing the reeds and their associated rhizosphere microbial communities, this could be due to the uptake of the compound by the microrganisms or the binding of a metabolite to the soil (Shen and Bartha, 1996; Bartha *et al.*, 1983).

CONCLUSION

This feasibility study shows that biodegradation of both compounds is occurring, but this may be controlled by (i) the physico-chemical properties of the chlorinated solvents (volatility, solubility); (ii) the physical nature of the environment (suspended solids, organic matter content); (iii) rooting of the reeds and rhizosphere colonisation by degradative microbes. Volatilisation was found to be the major fate of both compounds.

ACKNOWLEDGMENTS

This research was funded by English Partnerships. We would like to thank the Natural Environment Research Council (NERC, UK) for supporting this work.

REFERENCES

Alexander, M. 1994. *Biodegradation and Bioremediation*. Academic Press, Inc. San Diego, CA.

Bartha, R., I. S. You, and A. Saxena. 1983. "Humus-bound residues of phenylamide herbicides: their nature, persistance and monitoring." In J. Miyamoto and P. C. Kearney (Eds.), *IUPAC pesticide chemistry*, pp. 345-350. Pergamon Press, Oxford, England.

CRC Handbook of Chemistry and Physics. 1991. D.R. Lide (Ed). CRC Press, Boca Raton, Florida.

Janssen, D. B., A. Scheper, and L. Dijkhuizen. 1985. "Degradation of Halogenated Aliphatic Compounds by *Xanthobacter autotrophicus* GJ10." *Applied and Environmental Microbiology*. 49:673-677.

McEldowney, S., D. J. Hardman, and S. Waite. 1993 *Pollution: Ecology & Biotreatment*, Longman Scientific & Technical, Singapore.

Reineke, W., and H. J. Knackmuss. 1984. "Microbial metabolism of haloaromatics: isolation and properties of a chlorobenzene-degrading bacterium." *Applied and Environmental Microbiology*. 47:395-402.

Richards, Moorehead and Laing Ltd. 1997. "Middleton Wood scoping study for Lancaster City Council." Acc/1763.

Revitt, D. M., R. B. E., Shutes, N. R. Llewellyn, and P. Worrall. 1997. "Experimental reedbed systems for the treatment of airport runoff." *Water Science and Technology*. 36:385-390.

Shen, J., and R. Bartha. 1996. "Metabolic effeciency and turnover of soil microbial communities in biodegradation tests." *Applied and Environmental Microbiology*. 62: 2411-2415.

Urbancbercic, O. 1994. "Investigation into the use of constructed reedbeds for municipal waste dump leachate treatment." *Water Science and technology*. 29: 289-294.

AGRONOMIC MANAGEMENT FOR PHYTOREMEDIATION

Paul R. Thomas (Thomas Consultants, Inc., Cincinnati, OH, USA)
John K. Buck (Civil & Environmental Consultants, Inc., Pittsburgh, PA, USA)

ABSTRACT: A phytoremediation approach was accepted for use at a Superfund site in central North Carolina as a supplement to pump-and-treat technology. The site was characterized in order to allow design of an agronomic management program that would optimize the hydraulic function of the hybrid poplars proposed for use in the system. Agronomic and hydrologic parameters that could affect the function of the phytoremediation system were studied. A portion of the system was to be installed in or near thermally treated soils. The effect of thermal treatment on soil properties was considered. Based on preliminary characterization, a pilot planting was designed and installed in May, 1997. A full scale planting was performed in March, 1998 after treated soils had been placed over a portion of the site. Two hybrid poplars (*Populus spp.*) were used; *P. charcowiensis x P. incrassata* and *P. maximowiczii x P. trichocarpa*. Grasses were used as herbaceous ground cover. Results suggest that proper agronomic management can result in increased function of hybrid poplars by accelerating growth, decreasing mortality, protecting against drought, and postponing senescence.

INTRODUCTION

Background. The site is located in the Sand Hills region of central North Carolina Coastal Plain physiographic province. Local soils and groundwater were impacted by industrial activities associated with reformulating and packaging of chemical products. Phytoremediation was added to the Record of Decision (ROD) as the result of an Explanation of Significant Differences (ESD) and is now part of the groundwater remedy. The primary goal of the phytoremediation system is to help protect the quality of surface water bodies from impact by two separate groundwater plumes. This will be accomplished by operating phytoremediation systems that maximize evapotranspiration, resulting in a net loss of water from the saturated zone. A secondary goal is to encourage the movement of contaminated groundwater from the capillary fringe into a biologically active root zone in order to facilitate active degradation of the contaminants.

The local geology consists of interbedded sands and clays. The rolling topography (approximately 300 foot [90 meter] relief) and thick sand deposits have resulted in the "Sand Hills" appellation. The phytoremediation systems described here were installed in two discreet areas on sloping ground (5 to 10 percent slopes) adjacent to bodies of water. The movement of water in the subsurface is controlled by a complex sequence of interbedded unconsolidated sediments.

Objectives. The general objective of this work is to demonstrate the importance and function of agronomic management in phytoremediation. Proper agronomic management requires collection of information that is not typically included in the scope of standard remedial site investigations. Effective installation, operation, and management of phytoremediation systems require the detailed knowledge of soil chemistry, water retention, aeration, and fertility that result from thorough site characterization.

Function of these phytoremediation systems will hinge on the rate at which the introduced vegetation is able to move water from the soil. Optimization of this rate is the specific objective of the agronomic management methods assessed herein.

METHODS

Site Characterization. Phytoremediation systems were installed in both undisturbed native soil and highly disturbed, thermally treated soils. The County Soil Survey Report (USDA, 1995) mapped native soils in both phytoremediation operable units as part of the Vaucluse series. Site soils tended to contain less clay and silt and more sand than typical Vaucluse soils, but correlated well with descriptions of Vaucluse and related (e.g., Ailey) soil series. Samples up to 14 feet deep (4.2 meters) were collected to characterize soil physical, chemical, and moisture regime characteristics in the unsaturated zone. Native surface soils were typically acidic and sandy, with low organic matter and associated low plant-available water storage (PAW), with underlying soils having higher clay content (Tables 1 and 2).

TABLE 1: Texture and organic matter content of a typical soil profile.

ID	Sample Depth (m)	Soil Horizon	USDA Texture	<2.0 mm Fraction % Sand	% Silt	% Clay	% O.M.
MA	0-0.12	A*	sand	90.1	8.5	1.4	4.6
UMA	0.12-0.23	A	sand	89.0	7.0	3.9	0.8
E	0.23-0.52	E	sand	89.1	8.5	2.3	0.0
E2	0.52-0.79	E	coarse sand	88.3	6.6	5.1	0.0
B1	0.79-1.12	B	sandy loam (sl)	77.9	4.2	18.0	0.1
B2	1.12-2.75	B	sl /sandy clay loam	77.0	3.0	20.0	0.2
B3	2.75-3.40	C	sandy loam	81.3	2.6	16.1	0.2
B4	3.40-4.33	C	sandy loam	81.7	2.3	16.0	0.2

* After modification with manure, superabsorbent, fertilizer, limestone, and gypsum.

TABLE 2: PAW, pH, and conductivity of a typical soil profile.

ID	Sample Depth (m)	PAW %vol/vol	pH S.U.	2:1 Cond. (dS/cm)
MA	0-0.12	10.2	5.7	1.71
UMA	0.12-0.23	4.4	4.7	0.18
E	0.23-0.52	3.0	5.3	0.07
E2	0.52-0.79	0.8	5.0	0.06
B1	0.79-1.12	3.2	5.0	0.05
B2	1.12-2.75	4.6	4.8	0.07
B3	2.75-3.40	1.7	5.0	0.06
B4	3.40-4.33	4.9	5.0	0.08

To anticipate changes in soil properties resulting from thermal treatment of the most contaminated soils, physical and chemical tests were performed on samples before and after laboratory thermal treatment to 1800° F (982° C). As expected, heat treatments destroyed soil organic matter and PAW dropped substantially (Table 3). As the heat treatment ashed soil organic matter, the soil pH increased, presumably the result of released organically bound basic cations and destruction of organic acids.

TABLE 3: Effects of treatment on selected soil properties.

ID	Soil Horizon	PAW %vol/vol	Thermal PAW Decrease %	pH S.U.	% O.M.	2:1 Cond. dS/cm
1A	A	4.31		5.9	0.8	0.06
1AT	Heat-Treated A	3.65	15.26	7.0	0.0	0.10
1B	E	2.62		6.2	0.1	0.06
1BT	Heat-Treated E	2.52	3.74	7.2	0.0	0.08
2A	A	6.69		4.6	3.8	0.06
2AT2	Heat-Treated A	6.58	1.67	6.9	0.0	0.07
2B	E	2.63		5.5	0.1	0.05
2BT2	Heat-Treated E	1.40	46.81	6.4	0.0	0.07
5A	A	2.48		5.4	0.4	0.06
5AT2	Heat-Treated A	2.35	5.08	6.8	0.0	0.08
5B	E	3.08		5.7	0.4	0.05
5BT2	Heat-Treated E	1.62	47.27	6.4	0.0	0.08

In summary, the most significant properties of soils in the phytoremediation planting areas were sandy texture, low clay and organic matter content, and associated low plant available water (PAW) storage and narrow capillary fringe. The soil displayed moderately acidic pH, well within the pH tolerance envelope (pH 4.0 and higher) of hybrid poplar cultivars. High surface soil permeability was inferred from the sandy surface soil texture. Low subsoil permeability was inferred from higher subsoil clay content, and slow water percolation rates observed during the backfilling and watering stage of pilot planting. Low native fertility was noted relative to potential nutrient use of hybrid poplars. Any intrinsic microbiological communities in thermally treated soils were destroyed.

Soil Amendments and Planting Techniques. Soil amendments were designed for surface and planting backfill soils to improve PAW storage, moderate pH, address fertility limitations and to inoculate roots and soils with beneficial microorganisms. Surface soils were amended with superabsorbent polymers (435 lbs/acre) (484 kg/hectare) and composted wood chips/horse manure (20 tons/acre) (44 metric tons /hec) to improve PAW storage and soil aeration. Surface soils were limed per SMP buffer lime requirements (2 tons/acre) (4.5 mt/hec) and were amended with slow-release nitrogen (as sulfur coated urea to supply 80 lbs. N/acre) (89 kg N/hec), triple superphosphate (to supply 200 lbs. P_2O_5/acre) (222 kg P_2O_5/hec), and muriate of potash (to supply 200 lbs./acre K_2O)(222 kg/hec K_2O). Gypsum was applied at a rate of approximately 1,500 lbs./acre (1,668 kg/hec).

Backfill soils were amended using higher rates of superabsorbent and compost. Slow-release 20-10-5 (plus micronutrients) fertilizer pellets were added to the backfill just above plant roots at a rate to supply 34-17-8 grams N- P_2O_5-K_2O/tree. Lower fertilizer pellet rates were used experimentally in pilot plantings performed in 1997.

Shallow planting (root collars set 0.3 to 1.2 meters below ground surface, bgs) was done using a trenching machine. All planting rows were set 3.8 meters apart to maximize air circulation and to allow for follow-up maintenance. Trencher-planted trees were bare-root stock approximately 1.5 to 2.5 meters tall, and were spaced 1 meter apart within the row, using alternating cultivars (*P. charcowiensis x P. incrassata* and *P. maximowiczii x P. trichocarpa*). Deep planting (up to about 3 meters bgs) was done using a 20 cm or wider auger drill. Auger-planted trees were bare-root stock approximately 3 to 4.5 meters tall, spaced 3.8 meters apart. Only *P. maximowiczii x P. trichocarpa* was used in auger plantings. All tree roots were inoculated with a proprietary preparation of beneficial endo- and ecto-mycorrhizal fungi plus superabsorbents and other growth stimulants.

Backfilling was done by hand using the amendments described above. Backfill soils were liberally watered to saturate the compost and superabsorbents. Most trees were planted in the capillary fringe or within 2 meters above the capillary fringe. No post-planting irrigation was used with the 3,500 trees planted in March/April 1998.

All disturbed surface soils were sown with a mixture of annual ryegrass and bermudagrass. Grass was kept mowed to a 30 cm or shorter height.

RESULTS AND DISCUSSION

Plant Response. Approximately 185 dormant trees were planted in a pilot study in May of 1997 (about 2 months later than an ideal planting date), and about 3,500 trees in the full scale planting in March/April 1998. Plants sprouted leaves within a few days of removal from cold storage and planting. First year survival of both plantings exceeded 99 percent. First-year height was typically up to 10 ft (3 m) for the pilot planting and higher, up to about 16 feet (4.8 m) for the full scale planting. First year growth and survival of the 1998 planting was remarkable considering that there were several unusually hot dry months and that no irrigation was used.

Insect predation by both cottonwood leaf beetle and poplar tent-maker caterpillar was noted and successfully controlled using appropriate commercial formulations of the natural microorganism *Bacillus thuringiensis*.

Some fungal disease was noted (isolated cankers near the leader tip), as was leaf rust, but plants performed well despite these pathogens. Some water stress was noted in October 1998, with evidence of plants having dropped many of the lower and innermost leaves, especially among the most deeply planted (and widely spaced) trees. Presumably water stress will be reduced as roots of deeply planted trees grow down and contact the capillary fringe – a perennial source of water.

Visual symptoms of salt stress (blotchy necrotic areas within leaves, and necrosis of leaf tips and margins) was noted in an isolated portion of an area planted into thermally treated soil. Follow-up leaf tissue and soil analyses indicated that sodium (Na) toxicity was the most likely cause, with direct Na toxicity indicated (such sandy soils do not have the potential for Na-induced clay dispersion that can cause secondary Na stress from poor drainage/aeration). Gypsum has since been applied to this area to improve Ca/Na ratios in the soil solution.

Plant tissue analysis was done on the pilot planting in September 1997 and in both the pilot and full-scale plantings in early October 1998. 1997 plant tissue levels did not vary as expected among trees given varying levels of fertilizer pellets – this effect was probably obscured by the fact that surface soils were amended prior to planting, and fertilizer-rich soils comprised much of the backfill. 1998 plant tissue data must be interpreted with some caution, because trees appeared to be entering into early senescence, and some nutrients may have already been mobilized from the leaves. Despite these limitations the data suggest potential nutrient deficiencies in some stands, particularly with regard to copper and nitrogen. Premature senescence may have been aggravated by a combination of drought and nutrient deficiency.

Transpiration Rates. The average 24-hour sap flow rate from October 3 through October 7, 1998 was observed to be about 3.9 gallons per minute per acre (gpm/acre) (37 liters per minute per hectare - lpm/hec) of phytoremediation planting. Scaling up for increased leaf area of larger trees, it is estimated that (for the same time period) flow rates in other portions of the planted areas ranged from 10.6 to 15.4 gpm/acre (100 to 146 lpm/hec). Using the measured rate of 3.9 gpm/acre, annualized water usage (accounting for a six month growing season) would exceed 37 acre-inches. The average 24-hour flow rates were observed in first season transplants over a wide range of average daily photosynthetically active solar radiation (PAR) near the end of the growing season. Collection of xylem sap-flow data from both hybrids is planned for spring, midsummer, and late summer during the 1999-growing season.

Conclusions. Use of aggressive agronomic management techniques allowed the introduction of hybrid poplars into droughty, low fertility, and thermally treated (virtually sterile) soils. First season mortality was less than one percent. First season growth rates were very impressive due to high permeability soils and nutrient availability. Measured sap flow rates in first year transplants exceeded expectations.

Thorough characterization of the site with respect to the collection of agronomic parameters allowed for design of an effective soil amendment program. It is possible to install effective phytoremediation systems in less than ideal conditions when proper agronomic management methods are used.

REFERENCES

Kozlowski, Theodore T., and S. G. Pallardy. 1997. *Physiology of Woody Plants,* Second Edition. Academic Press.

Kramer, Paul J. 1983. *The Water Relations of Plants.* Academic Press.

US Department of Agriculture, Natural Resources Conservation Service. 1995. Soil Survey of Moore County.

CONJUGATE FORMATION DURING THE METABOLISM OF 2,4,6-TRINITROTOLUENE IN PLANT ROOTS

Rajiv Bhadra, Darcey G. Wayment, Joseph B. Hughes, and
Jacqueline V. Shanks (Rice University, Houston, TX)

ABSTRACT: This paper presents evidence for the occurrence of conjugation as a fate process during TNT metabolism by plants. Hairy root cultures of the terrestrial plant *Catharanthus roseus* were used as a axenic or microbe free model root system. A number of TNT-metabolites were isolated and characterized from the root cultures and their temporal profiles were determined. Analytical evidence from NMR spectroscopy, mass spectroscopy, HPLC-UV, and hydrolysis established their identities as conjugated derivatives of 2-amino-4,6-dinitrotoluene or 4-amino-2,6-dinitrotoluene. Mass balance analysis in 75-hour studies indicates significant extractable fraction of these conjugates: 20-26% of initial TNT. In the same time-frame, TNT disappeared, and its aminated derivatives accounted for only 4% of initial TNT, while bound residues increased to 29%. More than one conjugated derivative of each monoamino-dinitrotoluene (2-amino-4,6-dinitrotoluene or 4-amino-2,6-dinitrotoluene) was isolated, emphasizing numerous possible outcomes of conjugation processes during TNT metabolism. Two of the conjugates were also detected at significant levels in the biomass phase of the aquatic plant *Myriophyllum aquaticum* (Parrot feather) that was exposed to TNT. The occurrence of the same conjugate in a diversity of plant species (aquatic vs. terrestrial) reiterates the importance of conjugation as a fate process in TNT metabolism.

INTRODUCTION

Contamination of soil and groundwater at explosives manufacturing, loading, and packaging sites by 2,4,6-trinitrotoluene (TNT) poses serious environmental risk. Phytoremediation may be a viable strategy for in situ, cost-effective clean-up. For this technology to be evaluated objectively, the fate of the TNT molecule in contact with plants needs to be determined. In most studies of plant exposure to TNT, appearance of the monoamine derivatives 2-amino-4,6-dinitrotoluene (2A46DNT) and 4-amino-2,6-dinitrotoluene(4A26DNT) (Palazzo and Leggett 1986; Harvey et al. 1990; Hughes et al. 1997; Vanderford et al. 1997; Thompson et al. 1998) follows the disappearance of TNT. The product profile of TNT biocatalysis by plants beyond these early stages remains largely unknown. Our previous studies with aquatic systems of *Myriophyllum aquaticum* (Parrot feather), and axenic root cultures of the terrestrial flowering plant *Catharanthus roseus* (Vinca) have demonstrated the ability of plant metabolism to sequester TNT-metabolites into biomass (Hughes et al. 1997; Vanderford et al. 1997). These findings suggest the formation of conjugates (Sandermann 1994), a vital component of xenobiotic metabolism in plants, that leads to lowered toxicity and biological half-life of the toxicant to the plant (Coleman et al. 1997).

This research was directed at the immediate goal of determining transformation pathways by which conjugates formed during TNT metabolism by plants. In this study, "hairy root" cultures of *C. roseus* provided a model root system without metabolic interference from microbes or other pathogens. TNT-derived conjugate-metabolites were isolated from root cultures exposed to TNT in aqueous media. Temporal analyses of product profiles and ^{14}C distribution were also determined. More detailed results of these studies are in press (Bhadra et. al. , 1999; Wayment et al., submitted). Representative 'natural' aquatic systems of *M. aquaticum* were also probed for a similar product profile.

MATERIALS AND METHODS

Four-to-five week old, stationary phase, "hairy root" cultures of *C. roseus* (Bhadra and Shanks 1995) were employed to study nitroaromatic transformation. Hairy roots were grown to stationary phase in 50 mL of Gamborg's B5/2 medium (Bhadra and Shanks 1995). Two types of studies were conducted: temporal studies of TNT transformation with mass (^{14}C) balance analysis, and product analysis of nitroaromatic metabolites on exposure to either TNT (initial, 25-30 mg/L), 2A46DNT or 4A26DNT (initial 27 mg/L).

In transient studies of TNT transformation, levels of extracellular nitroaromatics and ^{14}C were determined by sampling the aqueous medium of root cultures. Intracellular and mass balance analysis conducted by sacrificing cultures. Mass balance on the root systems was performed by determining ^{14}C in the following compartments – extracellular ^{14}C, intracellular extractable ^{14}C, and intracellular bound ^{14}C. HPLC analysis of monomeric TNT derivatives was conducted by C_8 reverse phase HPLC (Hughes et al. 1997), while azoxy dimers were analyzed by C_{18} reverse phase HPLC with UV/VIS detection and an isocratic mobile phase of 60/40 (% v/v) CH_3CN/H_2O at 1.2 mL/min.

Products of plant transformation of TNT or monoamino-dinitrotoluenes were separated and purified from extracts of root tissues by a series of chromatography steps: solid phase extraction (SPE) on silica with hexane/ethyl acetate, 60/40 (% v/v), followed by SPE on C_8-bonded phase with a gradient of water/2-propanol. Purified metabolites were analyzed by NMR spectroscopy (Bruker AC-250 MHz), LC-MS (HP1100 with API-ES and AP-CI), direct probe MS-CI (Finnigan MAT 95), and enzymatic hydrolysis (ß-glucuronidase from Helix pomatia; and ß-glucosidase from almonds).

RESULTS AND DISCUSSION

Four TNT-related metabolites of plant biocatalysis were isolated from nitroaromatic-amended axenic root cultures of *C. roseus*. Two metabolites –TNT-6.7 and TNT-5.5 – were isolated from TNT-amendment studies. Another two – 4A-6.6 and 2A-5.6 – were isolated from amendment of the same system with 4A26DNT and 2A46DNT, respectively. The characterization of all four compounds is summarized in Table 1. The analytical evidence – NMR (^1H chemical shifts and long-range coupling), MS spectra, and UV/VIS spectra – indicate the structural similarity of these compounds to a monoamine derivatives of TNT, i.e. either 4A26DNT or 2A46DNT. However, the analytical evidence also indicates a greater molecular mass than TNT, 4A26DNT or 2A46DNT. Enzymatic (with ß-glucuronidase and ß-glucosidase) and acid hydrolysis (at pH=5) of extracts of exposed roots containing one or more of these compounds, resulted in the disappearance of the transformant, and the appearance of the corresponding monoamine derivative of TNT. Thus, the TNT-related plant-metabolites reported here are conjugates formed via monoamino-dinitrotoluene intermediates. For example, the ion with M+1, 360 (Table 1), could be that of a monoaminodinitrotoluene-hexose conjugate ($M_w = 359$).

The temporal profiles of TNT and transformants in TNT-amended root cultures are presented in Figure 1. In the aqueous phase, TNT decreased to below detection limit, with the appearance of the reduction derivatives 2A46DNT and 4A26DNT and the conjugates TNT-6.7, 4A6.6, and 2A5.6 (Figure 1A). The accumulation of TNT in the plant root matrix was minimal, as were 2A46DNT and 4A26DNT (Figure 1B). However, the conjugates TNT-6.7, TNT-5.5, and 4A-6.6 accumulated to high levels over the 80-hour study duration. Thus in the same system, a number of pathways exist for conjugation via a common intermediate (2A46DNT or 4A26DNT), occurring at differing net kinetic rates. This diversity serves to illustrate the numerous possible outcomes for the plant-based conjugation of 2A46DNT or 4A26DNT, perhaps due to the variety of molecules (sugars, amino acids and organic acids) available for substitution at reactive sites, in this case -NH_2. In comparing the two sets of profiles (intracellular vs. extracellular), it also appears that conjugate formation is an intracellular process. In studies where *C. roseus* root cultures were amended with either 2A46DNT or 4A26DNT, only 2A-5.6 or 4A-6.6, respectively, accumulated. Furthermore, the activity of a conjugation pathway possibly could be dependent on the flux of the precursor: neither 2A46DNT nor 4A26DNT accumulated to levels higher than 2 mg/L in TNT-exposure studies, whereas in the 2A46DNT- and 4A26DNT-exposure studies, the initial levels were 27 mg/L.

TABLE 1: Summary of chemical analysis of TNT-related plant-metabolites isolated from axenic (microbe-free) roots of *C. roseus*.

Analyte	Aromatic (^{1}H-NMR)	M_w>197[(1)]	M_w>227[(2)]	Negative ion AP-ES[(3)]	Positive ion AP-CI[(4)]	UV/VIS Spectra	Type
TNT-6.7	√	√	√	488, 472, 374	360, 395	229 nm, ~360 nm	4A26DNT conjugate
4A-6.6	√	√	√	472, 238	360, 323	229 nm, ~360 nm	4A26DNT conjugate
TNT-5.5	√	√	√			224 nm, 368 nm	2A46DNT conjugate
2A-5.6	√	√	√	472, 238	360, 323	224 nm, 368 nm	2A46DNT conjugate

[(1)] molecular mass of 2,4,6-TNT; [(2)] molecular mass of either 4-A-2,6-DNT or 2-A-4,6-DNT; [(3)] AP-ES atmospheric pressure electrospray MS; [(4)] AP-CI atmospheric pressure chemical ionization (M+1)

A breakdown of the transient ^{14}C balances is presented in Figure 2 comprising three fractions: extracellular ^{14}C, intracellular extractable ^{14}C, and intracellular bound ^{14}C. Total ^{14}C decreased in the extracellular aqueous medium in concert with TNT (Figure 1A). Within 47 hours however, extracellular ^{14}C decreased to ~20% of initial (Figure 2A), while extracellular TNT decreased to below detection limit. During the same period, the intracellular profile of extractable ^{14}C mirrored this trend, increasing to ~60% of initial. The conjugate-metabolites accounted for 15-26% of initial TNT (Figure 2B), and the total level of conjugate-metabolites was comparable to or greater than that of TNT, 2A46DNT and 4A26DNT, beyond 7 hours from TNT-amendment. Significant binding processes were observed within the initial 75-hour period, with the bound fraction of TNT-derived ^{14}C increasing to 29% of initial. Since the conjugates are likely "gateways" to bound residues, the latter possibly represents conjugates integrated into plant biomass.

The identification of conjugates in this study fits the metabolic model of xenobiotic transformation in plants known as the 'green liver': transformation, conjugation and sequestration (Sandermann 1994). The first and final stages of this metabolic model have been previously established for TNT (Palazzo and Leggett 1986; Harvey et al. 1990; Hughes et al. 1997; Thompson et al. 1998) i.e. the initial formation of 2A46DNT and 4A26DNT and the ultimate formation of bound residues. This study provides direct evidence for the intermediate stage of conjugate

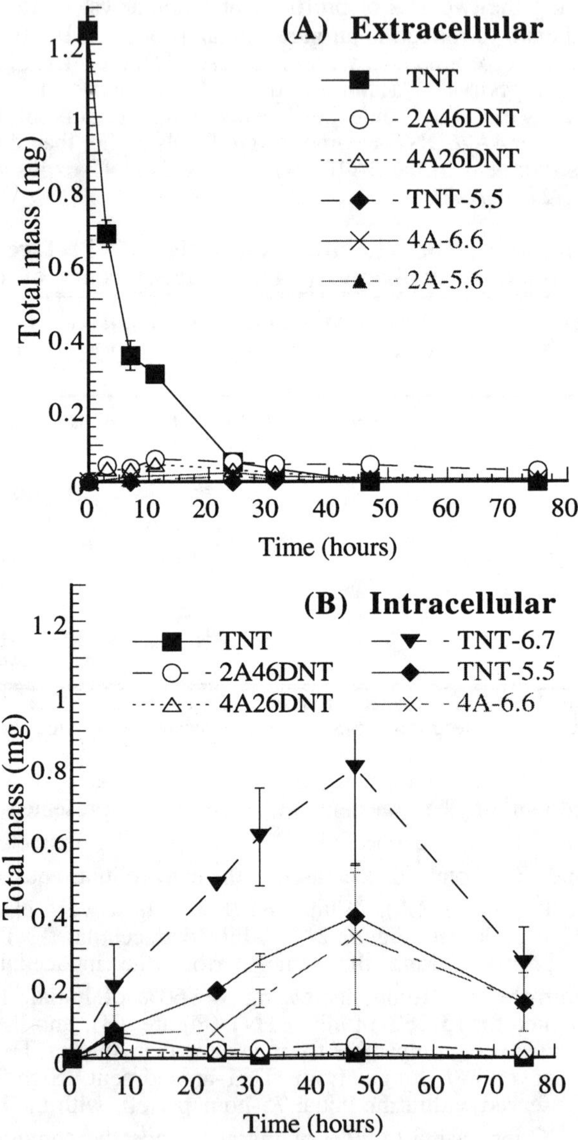

Figure 1. The temporal profiles of TNT, 2A46DNT, 4A26DNT, and conjugates (TNT-6.7, 4A6.6, TNT-5.5, and 2A5.6) in TNT-amended root cultures of *C. roseus*: (A) extracellular, aqueous medium; and (B) intracellular or biomass. Each data point represents an average of at least triplicate cultures; error bars are standard deviations.

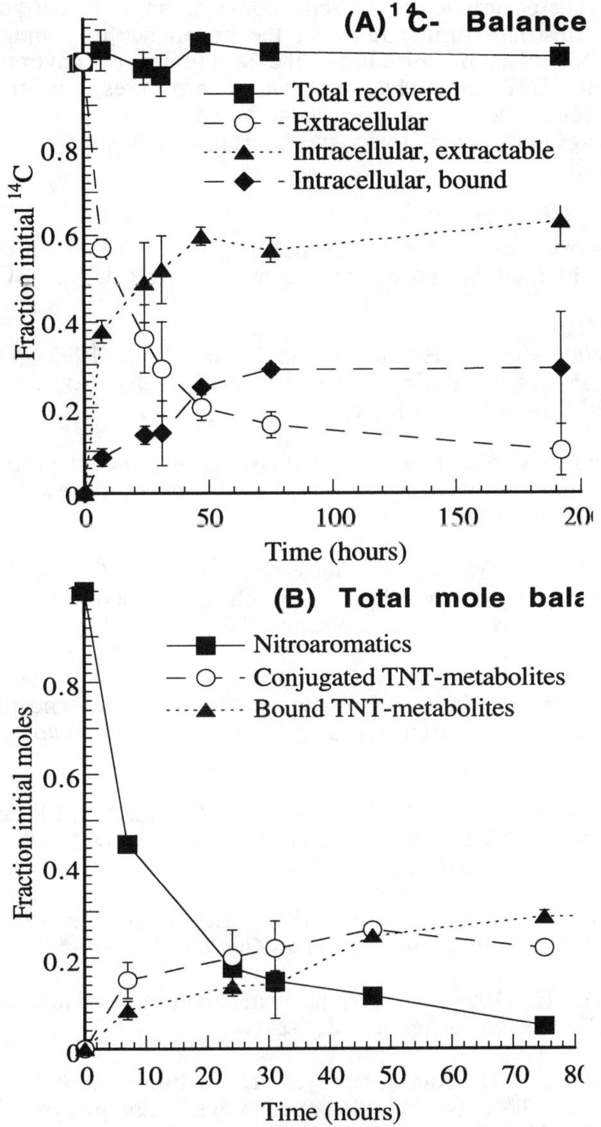

FIGURE 2. Balance on (A) ^{14}C and (B) moles, in hairy root cultures of *C. roseus* over a 8-day period. Definitions: total, intracellular plus extracellular; known nitroaromatics, TNT, 2A46DNT and 4A26DNT; and conjugate metabolites, TNT-6.7, TNT-5.5, 4A6.6, and 2A5.6. Each data point represents an average of at least triplicate cultures; error bars are standard deviations.

formation. Further, of importance is that the conjugates, TNT-6.7 and TNT-5.5, were also detected at significant levels in the biomass phase of the aquatic plant

Myriophyllum aquaticum (Parrot feather) that was exposed to TNT at levels from 15-60 mg/L. The occurrence of the same conjugate in a diversity of plant species (aquatic vs. terrestrial), further reiterates the importance of conjugation as a fate process in TNT metabolism. From the standpoint of environmental impact assessment of TNT and phytoremediation processes, understanding plant conjugation, could facilitate a knowledge-based approach to assessing 'true' toxicity, mutagenicity, and toxic mechanism(s) beyond the stage of TNT 'disappearance.'

ACKNOWLEDGMENTS

This work was funded by DSWA project # 01-97-1-0020, Texas ATP (003604-045), by the NSF Young Investigator Award to J.V.S. (BCS-9257983).

REFERENCES

Bhadra, R., Wayment, D., Hughes, J. and Shanks, J.V., (1999) "Confirmation of Conjugation Processes during TNT Metabolism by Axenic Plant Roots" *Environmental Sci. and Tech.,* in press.

Bhadra, R. and J. V. Shanks (1995). "Statistical design of the effect of inoculum conditions on growth of hairy roots of *Catharanthus roseus.*" *Biotechnology Techniques* 9(9): 681-686.

Coleman, J. O. D., M. M. A. Blake-Kalff and T. G. E. Davies (1997). "Detoxification of xenobiotics by plants: chemical modification and vacuolar compartmentation." *Trends in Plant Science* 2(4): 1144-151.

Harvey, S., R. J. Fellows, D. A. Cataldo and R. M. Bean (1990). "Analysis of 2,4,6-trinitrotoluene and its transformation products in soils and plant tissues by hiugh performance liquid chromatography." *Journal of Chromatography* 518: 361-374.

Hughes, J. B., J. V. Shanks, M. Vanderford, J. Lauritzen and R. Bhadra (1997). "Transformation of TNT by aquatic plants and plant tissue cultures." *Environmental Science and Technology* 31(1): 266-271.

Palazzo, A. J. and D. C. Leggett (1986). "Effect and disposition of TNT in terrestrial plants." *Journal of Environmental Quality* 15(1): 49-52.

Sandermann, J., H. (1994). "Higher plant metabolism of xenobiotics: the 'green liver' concept." *Pharmacogenetics* 4: 225-241.

Thompson, P. L., L. Ramer and J. L. Schnoor (1998). "Uptake and transformation of TNT by hybrid poplar trees." *Environmental Science and Technology* 32: 975-980.

Vanderford, M., J. V. Shanks and J. B. Hughes (1997). "Phytotransformation of trinitrotoluene (TNT) and distribution of metabolic products in *Myriophyllum aquaticum.*" *Biotechnology Letters* 19(3): 277-280.

Wayment, D.G., Bhadra, R., Lauritzen, J., Hughes, J. and Shanks, J.V., "A Transient Study of Formation of Conjugates During TNT Metabolism by Axenic Plant Roots", submitted to *Phytoremediation.*

LABORATORY STUDIES ON PLANT UPTAKE OF TCE

Brady J. Orchard, William J. Doucette, Julie K. Chard, and Bruce Bugbee
(Utah State University, Logan, Utah)

ABSTRACT: Contradictory reports of plant uptake and phytovolatilization make it difficult to assess the importance of these mechanisms in the phytoremediation of trichloroethylene (TCE)-contaminated sites. Experimental artifacts associated with unnatural plant growth environments that occur when using low flow or semi-static laboratory systems may account for part of the discrepancy. Exposure concentration and duration may be equally important. Poor mass recoveries and inadequate controls may also contribute to the inconsistent findings. Four continuous-high-flow plant growth chamber systems designed to provide high mass recoveries, a natural plant environment, and complete separation of foliar and root uptake were used to determine the uptake of [^{14}C]TCE by hydroponically-grown hybrid poplar as a function of root-zone oxygen status, exposure concentration, and exposure duration. Mass recoveries ranged from 92 to 101%. No volatilization of [^{14}C]TCE from the leaves was detected in eight out of nine chambers. Shoot concentrations (2 to 350 mg/kg) were dependent on water transpiration and exposure concentration, but not on root-zone oxygen status. Measured transpiration stream concentration factors (TSCFs) appear to be independent of exposure concentration and duration. These values (0.09 ± 0.06) are much less than those reported previously or predicted from empirical expressions.

INTRODUCTION

Understanding plant uptake, translocation, and transformation of organic chemicals is critical in assessing the effectiveness of phytoremediation and evaluating the potential for food chain contamination. Much of the experimental plant uptake data and related theory has been generated for commercial pesticides; much less information is available for other industrial chemicals. The fate of volatile organic solvents, such as trichloroethylene (TCE), in planted systems is of particular importance because of their widespread occurrence as groundwater contaminants.

Plant uptake of TCE has been recently examined in several laboratory studies using a variety of experimental designs (Anderson and Walton, 1995; Narayanan et al., 1995; Gordon et al., 1997; Newman et al., 1997; Schnabel et al., 1997; Burken and Schnoor, 1998). Conclusions regarding the magnitude of plant uptake and translocation are inconsistent, but collectively they have led to the widespread belief that plant uptake and transpiration of TCE are significant fate processes. Experimental details are often insufficient to explain the inconsistencies among studies. Low mass recoveries can lead to inaccurate conclusions. Leaks from the root to the foliar chamber can mistakenly infer root uptake and translocation. Static or low flow rates in the upper chamber can result in elevated humidity and unrealistic atmospheric contaminant levels. Plant stress due to the use of co-solvents, anaerobic root zones, or high root-zone TCE concentrations may increase the permeability of the plasma membrane that protects the root cells. Exposure duration and plant age may also be important.

The uptake and translocation of TCE by hydroponically-grown hybrid poplar trees was evaluated using a dual-vacuum, continuous-high-flow plant growth chamber system specifically designed to provide a natural plant environment, complete root/shoot separation, the ability to quantify transpired TCE, and high mass recovery. Transpiration stream concentration factors (TSCFs) are used to estimate the magnitude of plant uptake in phytoremediation applications. Variables that may impact plant uptake predictions are discussed.

MATERIALS AND METHODS

Four identical systems, each consisting of separate foliar and root chambers with independent airflow control and large volume liquid traps (0.5 to 2 L) for collecting CO_2 and volatile organics from the exiting air stream of both the foliar and root chambers were constructed (Figure 1). Each system is constructed mainly of glass, stainless steel, and Teflon® (PTFE) for maximum inertness. Airflow through the foliar chamber (5 to 10 L/min) minimizes humidity and the potential build-up of volatilized contaminant, if transpired. Flow rates through the root zone are maintained between 20 and 50 mL/min to ensure mixing and control root-zone oxygen levels. The foliar and root chambers are isolated by both a physical seal (glass and rope caulk) and a pressure differential (-10" H_2O foliar-root). The combination of a physical seal with a pressure differential minimizes the potential for leaks of TCE from the lower to the upper chamber. Conversely, any contaminant transpired is quickly swept from the foliar chamber by the high airflow and is not drawn into the root zone.

Plant Growth Conditions. Imperial Carolina hybrid poplar trees (*Populus deltoides x nigra*, DN34) were rooted hydroponically in a greenhouse, selected for uniformity, and transplanted into the chamber system prior to dosing. A complete and appropriately dilute nutrient solution was used. The four chamber systems were assembled and enclosed within a walk-in plant growth room that provided temperature (29/24 ± 0.5°C day/night) and photoperiod control (16-h). Cool white fluorescent (CWF) lamps provided a photosynthetic photon flux (PPF) of approximately 600 $\mu mol/m^2$-s at the tops of the chambers.

System Dosing and Trapping Solutions. The system was dosed via the inlet air stream with a mixture of [^{14}C]TCE and certified ACS Grade non-labeled TCE using a multi-syringe pump. The root-zone solution was monitored daily to ensure constant solution concentrations. A solution of 80% ethylene glycol monomethyl ether (EGME)/20% ethylene glycol monobutyl ether (EGBE) was used in the volatile organic traps. The CO_2 traps contained 2.0 N KOH.

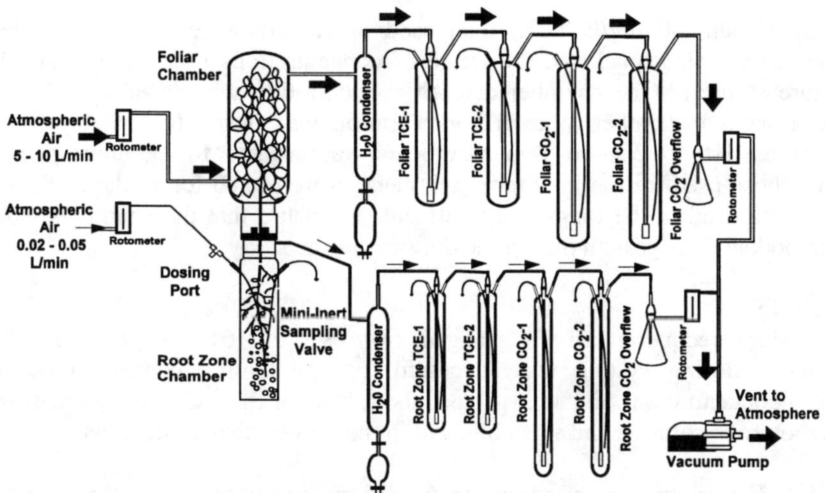

FIGURE 1. Schematic of the dual-vacuum, continuous-high-flow chamber system used to determine the uptake of TCE by hybrid poplar.

Trap Sampling and Analysis. All samples were taken in triplicate using gas-tight syringes. Organic traps were sampled at 3-d intervals and analyzed directly by liquid scintillation counting (LSC). CO_2 traps were sampled at the end of each experiment, precipitated with barium chloride, the CO_2 re-evolved and trapped in a solution of 50% Ready Gel®/40% methanol/10% monoethanolamine, and subsequently analyzed by LSC.

Plant Tissue Analysis. At harvest, plant tissue was separated into leaves, stems, and roots. Sub-samples (1-2 g fresh weight) were combusted using a biological oxidizer to determine total ^{14}C in each tissue fraction. Remaining tissue was extracted and analyzed for the TCE metabolites (trichloroacetic acid (TCAA), dichloroacetic acid (DCAA), and trichloroethanol (TCEt)) using a gas chromatograph with electron capture detector (Doucette et al., 1998).

Experimental Design. Three studies have been conducted using this system. In the first study, hybrid poplars in three separate chambers were exposed to an aerobic root-zone solution containing an average concentration of 0.6 mg/L TCE for 13 days. This concentration was used because it was representative of a TCE-contaminated groundwater site previously investigated (Doucette et al., 1998). The fourth chamber was used as a non-dosed, planted control.

In the second study, the effect of oxygen stress was examined by exposing plants to an aqueous solution of 0.9 mg/L TCE under either aerobic or oxygen-reduced conditions for 11 days. Two unplanted controls, one aerobic and one oxygen-reduced, were run at the same time. The unplanted controls used a stainless steel rod in place of the plant. Oxygen-reduced conditions were obtained by pulling nitrogen gas through the root zone instead of atmospheric air.

The impact of exposure concentration and duration was evaluated in the third uptake study. Two exposure concentrations approximately 10 and 100 times greater (10 and 70 mg/L TCE) than used previously were evaluated. These concentrations also bracketed the range of concentrations used in several other literature studies. One chamber at each concentration was dosed for 12 days, while a second chamber at each concentration was dosed for 26 days. The cuttings used for the short-term treatments were rooted for 35 days prior to dosing; those used for the long-term treatments were rooted for 12 days. Smaller plants were used in the long-term treatments to ensure that the chambers would accommodate their growth during the duration of the study.

RESULTS

Mass recoveries of ^{14}C ranged from 92 to 101% (Table 1). The continuous dosing system was successful in maintaining constant root-zone solution concentrations and as expected, about 90% of the ^{14}C directly volatilized from root-zone solution and was collected in the root-zone organic traps.

TABLE 1. Summary of exposure conditions, tissue concentrations, and calculated transpiration stream concentration factors (TSCFs).

Treatment	Root Zone Conc.[a] (mg/L)	Exposure Duration (d)	^{14}C Mass Recovery	Root Conc.[b] (mg/kg)	Shoot Conc.[b] (mg/kg)	Total Shoot Dry Mass (g)	Total Water Uptake (L)	TSCF[c]
Study 1								
Aerobic	0.6	13	93%	58 ± 19	2.2 ± 0.3	16.8	1.7	0.04
	0.6	13	92%	49 ± 30	3.6 ± 0.5	14.7	1.3	0.07
Aerobic (non-dosed)	-	13	-	ND	ND	10.2	1.3	-
Aerobic	0.6	13	94%	40 ± 40	3.7 ± 0.4	11.4	1.5	0.05
Study 2								
Aerobic	0.9	11	98%	26 ± 15	9.7 ± 3.9	10.9	1.3	0.10
Reduced	0.9	11	93%	14 ± 0	8.1 ± 2.0	11.3	1.1	0.09
Aerobic (unplanted)	1.2	11	93%	-	-	-	-	-
Reduced (unplanted)	0.9	11	101%	-	-	-	-	-
Study 3								
Aerobic	12.1	26	92%	3077 ± 1106	350 ± 33	4.0	0.9	0.18
Aerobic	9.2	12	96%	855 ± 241	168 ± 21	10.1	1.2	0.19
Aerobic	67.5	12	94%	336 ± 97	41 ± 5.7	11.8	1.1	0.03
Aerobic	74.0	26	93%	208 ± 79	154 ± 26	1.8	0.2	0.02

a- Time-weighted average solution concentration.

b- ^{14}C converted to TCE mass equivalents.

c- $$TSCF = \frac{(\text{Shoot conc., mg/kg})*(\text{Shoot dry mass, kg}) + (\text{TCE tranpired, mg}) + (^{14}CO_2 \text{ transpired, mg})}{(\text{Water uptake, L})*(\text{Solution concentration, mg/L})}$$

No transpiration of [^{14}C]TCE from the leaves was detected in eight out of nine chambers. Transpiration of [^{14}C]TCE (1.64 mg) was observed only in the short-term (12-d), high concentration (70 mg/L) treatment. This may be due to a reduction in the plant's ability to metabolize TCE at higher concentrations. Phytotoxicity, as evidenced by reduced plant size and water uptake, was observed in the long-term (26-d), high concentration (70 mg/L) treatment. Efflux of $^{14}CO_2$

from the shoots was observed (0.4 mg TCE equivalents) in both of the 10 mg/L treatments and may be due to increased microbial and/or plant metabolism. The interaction between plant size, phytotoxicity, uptake, translocation, and metabolism warrants further investigation.

Average shoot concentrations ranged from 3 to 10 mg/kg (expressed as TCE mass equivalents) for Studies 1 and 2; no significant differences in shoot concentrations were observed between aerobic and oxygen-reduced root-zone environments. Root concentrations in Studies 1 and 2 ranged from 14 to 49 mg/kg. No metabolites were identified above detection limits (20 μg/kg TCEt; 43 μg/kg TCAA; 145 μg/kg DCAA) in any of the plant tissue. TSCFs, defined as the ratio of TCE in the transpiration stream (tissue + transpired) normalized for water transpiration and root-zone solution concentration, ranged from 0.04 to 0.10 for the first two studies (Table 1). These TSCFs are much less than reported in other studies (Burken and Schnoor, 1998) or predicted from empirical expressions based on octanol/water partition coefficients (Briggs et al., 1982).

As shown in Table 1, shoot concentrations for Study 3 ranged from 41 to 350 mg/kg, with concentrations two to four times higher in the 10 mg/L treatments than in the 70 mg/L treatments. Toxicity may have limited uptake in the high concentration (70 mg/L) studies. The distribution of ^{14}C in the shoot tissue (stem concentrations higher than leaf concentrations in the short-term treatments and leaf concentrations higher than stem concentrations in the long-term treatments) suggests increased translocation of TCE and/or its metabolites over time. Root concentrations were significantly greater than in Studies 1 and 2 (208-3080 mg/kg). TCEt, TCAA, and DCAA were identified in the root and shoot tissue at concentrations ranging from 0.2 to 5.3 mg/kg, primarily DCAA. Metabolite concentrations were greatest in the roots, suggesting that degradation may be occurring in or on the root surface. Measured TSCFs (Table 1), assuming all ^{14}C was TCE, confirm that plant uptake was limited in the highest exposure treatments. Exposure duration did not significantly influence the measured TSCFs.

TSCFs describe the efficiency of movement of a compound from roots to shoots and are used along with groundwater concentration, canopy transpiration rates, and the fraction of groundwater use to provide estimates of contaminant mass removed by plant uptake [Plant uptake = (groundwater concentration) * (TSCF) * (transpiration) * (fraction of groundwater use)]. A TSCF of 1.0 indicates unrestricted passive uptake (i.e., the plant takes up water and the contaminant at the same rate). TSCFs less than 1.0 indicate that the plant is excluding the contaminant and TSCFs greater than 1.0 represent active uptake (typical of N, P, and K ions). TSCFs are generally assumed to be constant. Data from this study suggest that the TSCF for TCE (0.09 ± 0.06) is constant over a wide range of concentrations.

Predicting the importance of plant uptake in field-scale phytoremediation efforts also requires better information on canopy transpiration rates and the fraction of groundwater utilized by the plants. Transpiration rates in the field vary widely (200 to 1800 L/m^2-yr) depending on the soil water availability and evaporative demand. The fraction of groundwater utilized by the plants is not often considered in field-scale uptake predictions and depends on the depth to groundwater, oxygen availability in the capillary fringe, precipitation patterns, and plant species. Until more data is available, a range of groundwater use fractions from 0.1 to 0.5 is probably appropriate. A better understanding of the impact of environmental variables on TSCFs, canopy transpiration rates, and the fraction of groundwater used by plants is needed to improve uptake predictions.

REFERENCES

Anderson, T. A., and B. T. Walton. 1995. "Comparative Fate of ^{14}C Trichloroethylene in the Root Zone of Plants from a Former Solvent Disposal Site." *Appl. Environ. Microbiol.* 56(14): 1012-1026.

Briggs, G. G., R. H. Bromilow, and A. A. Evans. 1982. "Relationships Between Lipophilicity and Root Uptake and Translocation of Non-ionised Chemicals by Barley." *Pest. Sci.* 13:495-504.

Burken, J. G., and J. L. Schnoor. 1998. "Predictive Relationships for Uptake of Organic Contaminants by Hybrid Poplar Trees." *Environ. Sci. Technol.* 32(21): 3379-3385.

Doucette, W. J., B. Bugbee, S. Hayhurst, W. A. Plaehn, D. C. Downey, S. A. Taffinder, and R. Edwards. 1998. "Phytoremediation of Dissolved-Phase Trichloroethylene Using Mature Vegetation." In G. B. Wickramanyake and R. E. Hinchee (Eds.), Bioremediation and Phytoremediation: Chlorinated and Recalcitrant Compounds, pp. 251-256. Battelle Press, Columbus, OH.

Gordon, M., N. Choe, J. Duffy, G. Ekuan, P. Heilman, I. Muiznieks, L. Newman, M. Ruszaj, B.B. Shurtleff, S. Strand, and J. Wilmoth. 1997. "Phytoremediation of Trichloroethylene with Hybrid Poplars." In E. L. Kruger, T. A. Anderson, and J. R. Coats (Eds.), *Phytoremediation of Soil and Water Contaminants*, pp. 177-185. American Chemical Society Symposium Series 664, Washington, DC.

Narayanan, M., L. C. Davis, and L. E. Erickson. 1995. "Fate of Volatile Chlorinated Organic Compounds in a Laboratory Chamber with Alfalfa Plant." *Environ. Sci. Technol.* 29(9): 2437-2444.

Newman, L. A., S. E. Strand, N. Choe, J. Duffy, G. Ekuan, M. Ruszaj, B. B. Shurtleff, J. Wilmoth, and M. P. Gordon. 1997. "Uptake and Biotransformation of Trichloroethylene by Hybrid Poplars." *Environ. Sci. Technol.* 31(4): 1062-1067.

Schnabel, W. E., A. C. Dietz, J. G. Burken, J. L. Schnoor, P. J. Alvarez. 1997. "Uptake and Transformation of Trichloroethylene by Edible Garden Plants." *Water Res.* 31(4): 816-824.

BIODEGRADATION OF CARBON TETRACHLORIDE BY POPLAR TREES: RESULTS FROM CELL CULTURE AND FIELD EXPERIMENTS

Xiaoping Wang, Lee A. Newman, Milton P. Gordon, and Stuart E. Strand
(University of Washington, Seattle, Washington)

ABSTRACT: Cell culture and field experiments were conducted to determine the oxidative metabolites of carbon tetrachloride (CT) degradation by poplar trees. The results from the cell culture experiment showed that poplar cells could completely oxidize CT to carbon dioxide. Some of the carbon from CT was fixed in poplar cells. The field experiment indicated that chlorine from the dechlorination of CT was not retained in poplar trees but was released as chloride ion into the soil. CT or its chlorinated intermediates did not significantly accumulate in poplar tree tissues. These results suggest that poplar trees could break down CT into harmless components.

INTRODUCTION

Carbon tetrachloride (CT) is an important groundwater pollutant at many sites in the United States (Datskou and North, 1996). In the groundwater environment the degradation of CT by bacteria occurs only under anaerobic conditions (Vogel et al., 1987). CT has been reported to be degraded to carbon dioxide and chloroform in a variety of anaerobic bacterial cultures, but further degradation of hazardous chloroform is slow (Krone et al., 1989). Recently it has been found that under field conditions poplar trees can take up and metabolize more than 95% of the dosed CT (Strand et al., 1998), however, the metabolic pathway of CT transformation by poplar trees remains unknown.

It is well known that hepatic transformation of CT goes through an oxidative metabolic pathway in mammalian systems. Carbon dioxide has been found as a major oxidative metabolite of CT with covalent binding of some reactive intermediates to cell proteins and lipids (Paul and Rubinstein, 1963; Diaz Gomez et al., 1973). Although there have been no reports of studies of CT transformation by plants, oxidative transformations of several organic pollutants, including trichloroethylene (Newman et al., 1997) and polychlorinated biphenyls (Fletcher et al., 1987), have been demonstrated in studies with various plant cell cultures. Recent results from our laboratory indicated that trichloroethylene can be completely mineralized by poplar trees with the release of chloride ions into soil (Wang et al., 1997). Therefore, it is possible that transformation of CT by poplar trees follows an oxidative pathway. The objective of this study was to investigate the metabolism of CT degradation by poplar tissue and trees.

The data presented in this paper were obtained from the trials involving both poplar cell culture and whole poplar trees grown under field conditions. The ability of poplar cells to oxidize CT to carbon dioxide was determined using a radiolabelled cell culture experiment. The production of carbon dioxide involves a complete dechlorination of CT, which could result in an accumulation of chloride ions. The accumulation of chloride ion after uptake and degradation of CT by poplar trees was examined using a field experiment. The results from these experiments demonstrate that poplar trees are capable of degrading CT into harmless components and support the feasibility of using poplar trees to clean up soil and groundwater contaminated with CT.

MATERIALS AND METHODS

Cell Culture Experiment. A radiolabelled cell culture experiment was carried out to determine if poplar cells were capable of oxidizing CT to carbon dioxide in the absence of soil-plant-microbe interaction with CT metabolism. Axenic tumor cells derived from the hybrid poplar *Populus trichocarpa* x *P. deltoides* clone [H11-11] were used for the experiment because they grow rapidly and are handled easily. The cells were grown in the Murashige and Skoog basal salt medium (MS medium) without hormones (Murashige and Skoog, 1962). Cell culture suspension (20 ml) was placed in a 125 ml two-necked flask with a screw cap containing a lined Teflon septum on the side neck and a septum valve mounted in a screw cap on the top neck (Mininert). Killed controls were prepared by autoclaving the cell suspension for 20 minutes. The flasks were then sealed and 2 μl of radiolabelled CT was added through the septum valve. The cells were incubated for 24 hours with shaking. At the end of the incubation period, a tube containing activated carbon was connected to the septum valve of the flask and an impinger containing 20 ml of 1 N NaOH was positioned between the activated carbon tube and a vacuum pump. Glass and Teflon tubes were used to make all connections in the system. Vacuum was supplied for two hours and gas flowed from the flask through the activated carbon tube into the impinger. Non-transformed CT was captured on activated carbon and carbon dioxide was trapped in 1N NaOH solution. Any reactive intermediates of CT bound to cell lipids and proteins remained in the cells. After gassing of the cell cultures, the cells were separated from the MS medium by filtration using Ahlstrom filter paper. The amount of radioactivity in NaOH and MS medium was determined by direct counting in liquid scintillation cocktail (Ultima Gold) using a Beckman LS 7000 scintillation counter. The amount of radioactivity associated with the cells was determined by combusting cell samples with a sample oxidizer (Packard), collecting $^{14}CO_2$ in the scintillation cocktail and quantifying ^{14}C radioactivity in the scintillation counter.

Field experiment. A field experiment was conducted to investigate the uptake and metabolism of CT by hybrid poplar trees under field conditions. In this study, the accumulation of chloride ions after uptake and metabolism of CT by poplar trees were examined. This controlled field experiment with artificial aquifers was set up in Fife, Washington in 1995. Double walled cells were constructed of heavy gauge polyethylene liner. Each cell was 1.2 m deep, 3.7 m wide and 6.1 m long. The cells contained about 0.2 m of coarse sand overlaid with 1 m of Sultan silt loam soil. In order to ensure a uniform flow of influent water, a "T" shaped influent pipe was placed at the bottom and the bottom was oriented at a 1/40 slope toward an effluent well at the opposite end of the cell. The cells were planted with 15 hybrid poplar trees (H11-11) and continuously dosed with water containing 15 mg/L CT during the growing seasons (May to October) of 1996 and 1997. Trees receiving water containing no CT were used as controls. Soil samples at 5 depths and tree tissue samples (leaves, stems and roots) were taken in August 1997, after two years of CT dosing. The mercury (ll) thiocyanate method (Adriano and Doner 1982) was used to analyze the concentration of chloride ion in soil and plant samples. The method was modified when used to analyze chloride in plant tissues. Activated carbon powder was added to remove the color produced during the aqueous extraction of chloride from plant tissues. HNO_3 (1:1) was added to remove the color formed after addition of the $Fe(NO_3)_3$ reagent. The chloride recovery rate of the method after modification ranged between 90 to 110% as determined from plant samples spiked with different concentrations of chloride. Fixed organic halides in the plant tissues were analyzed by Analytical Resources, Inc., using EPA standard method for total organic halide (TOX). Prior to TOX analysis the samples were washed free of chloride using deionized water.

RESULTS AND DISCUSSION

Table 1 shows that CO_2 production by poplar H11-11 cells was significantly higher compared with the dead cell control, indicating mineralization of CT by poplar cells. About 3% of the dosed radioactivity was recovered in the cell fraction, which represented the binding of CT intermediates to cell proteins and lipids. This finding demonstrates that poplar cells were able to mineralize CT.

TABLE 1. Oxidative transformation of carbon tetrachloride by poplar cells

Distribution of dosed radioactivity	Dead cells (%)	Living cells (%)
Recovered as CO_2	0.3 ± 0.3	1.5 ± 0.2
Binding to cell tissue	1.1 ± 0.5	3.0 ± 0.1
Remaining in media	12.6 ± 5.5	7.1 ± 2.8
Trapped on carbon tube	87.7 ± 33.0	84.5 ± 9.8
Total recovery	101.6 ± 32.9	96.0 ± 12.7

The oxidation of CT to carbon dioxide causes a complete removal of chloride from CT, which should result in production of chloride ions. However, analysis of chloride changes in cell culture was hampered by the high level of chloride ion presented in the cell growth media. Therefore, the accumulation of chloride ions during the uptake and metabolism of CT by poplar trees was investigated in a field experiment in which poplar trees had been continuously dosed with CT for two growing seasons in 1996 and 1997. Previous results from this field experiment indicated that poplar trees could take up 95% of the dosed

CT from the system and that less that 5% of the total CT uptake was transpired (Strand et al., 1998). However, tissue analysis did not show an increase in chloride ions in the poplar trees dosed with CT (Figure 1a). The concentrations of total organic halides in poplar tree tissues (Figure 1b) were much lower than the concentrations of chloride ions, and there was no increase in the concentrations of total organic halides in the tissues of CT-dosed trees compared to the controls not exposed to CT. These results indicate that CT and chlorinated organics formed from CT transformation did not significantly accumulate in poplar tree tissues.

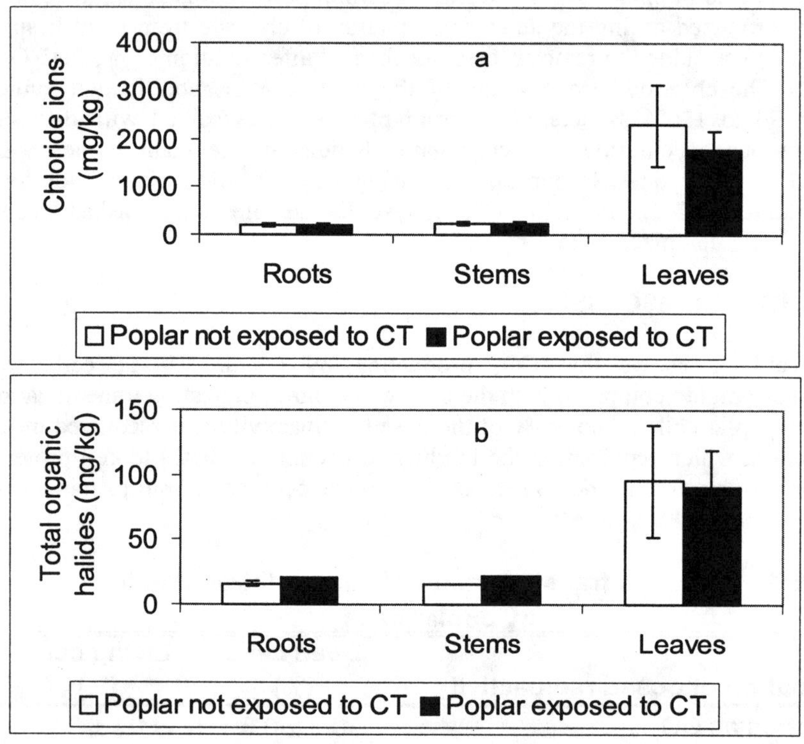

FIGURE 1. Concentrations of chloride ions (a) and total organic halides (b) in the tissues of poplar trees

While analysis of plant tissues did not show an increase in chloride ions or chlorinated metabolites by poplar trees exposed to CT, the analysis of soils indicated a higher concentration of chloride in cells treated with CT compared with the control soils (Fig 2). Statistical analysis (t-test) indicated that chloride concentrations were significantly elevated in the CT-dosed soils collected at 0.2 – 0.8 m depth. The cell culture work demonstrated the ability of poplar cells to dechlorinate CT, therefore, it is likely that the increase of chloride concentration in the soil of the cell treated with CT resulted at least in part from the dechlorination of CT by poplar trees. Chloride ions produced from CT dechlorination could be released into soil through root exudation. Mechanisms for transport of chloride into and out of plant root cells have been reported (Cram, 1983; Felle, 1994; Marten et al., 1991). Similar chloride ion accumulations have been observed in our previous field trails with phytoremediation of trichloroethylene (TCE) (Wang et al., 1997).

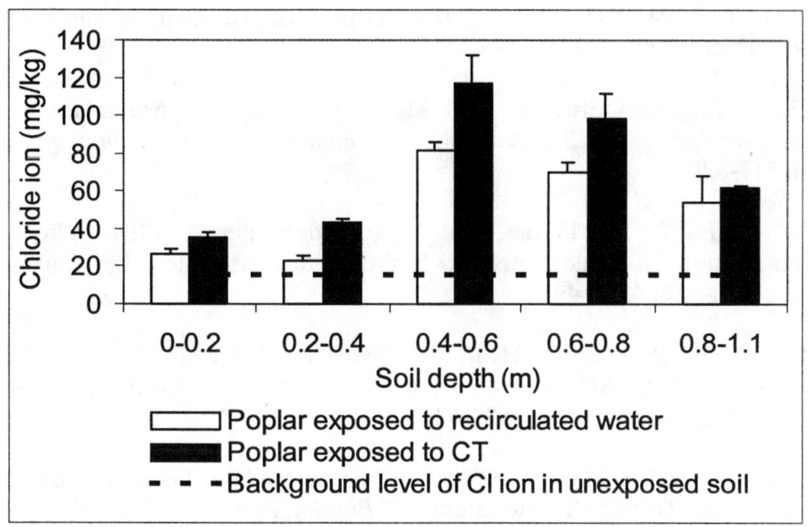

FIGURE 2. Concentrations of chloride ions in different depth of soils

In summary, the results presented suggest that planting of poplar trees has the potential to break down CT into harmless components. Poplar cells are capable of carrying out complete degradation of CT to carbon dioxide. Chloride ions from the dechlorination of CT are released into the soil. There is no significant accumulation of total organic halides in the tissues of poplar trees exposed to CT treatment. The findings from this investigation demonstrate the feasibility of using poplar trees to remediate CT-contaminated sites.

REFERENCES

Adriano, D. C., and H. E Doner. 1982. "Bromine, Chlorine, and Fluorine." In A. L. Page, R. H. Miller and D. R. Keeney (Eds). *Methods of Soil Analysis part 2*, pp. 461-462. Madison, WI.

Cram, W. J. 1983. "Chloride Accumulation in Plant Cells as a Homeostatic System: Energy Supply as a Dependent Variable." *J. Membr. Biol.* 74: 51-58.

Datskou, I., and K. North. 1996. "Risks due to Groundwater Contamination at a Plutonium Processing Facility." *Water Air Soil Pollution.* 90: 133-141.

Diaz Gomez, M. I., J. A. Castro, J. A. De Ferreyra, N. D'Acosta, and C. R. De Castro. 1973. "Irreversible Binding of ^{14}C from $^{14}CCl_4$ to Liver Microsomal Lipids and Proteins from Rats Pretreated with Compounds Altering Microsomal Mixed Function Oxygenase Activity." *Toxicol. Appl. Pharmacol.* 25: 534-541.

Felle, H. H. 1994. "The H^+/Cl^- Symporter in Root-Hair Cells of *Sinapis Alba.*" *Plant Physiol.* 106: 1131-1136.

Fletcher, J., A. Groeger, J. McCrady, and J. McFarlane. 1987. "Polychlorobiphenyl (PCB) Metabolism by Plant Cells." *Biotechnology Letters.* 9(11): 817-820.

Krone, U. E., R. K. Thauer, and H. P. Hogenkamp. 1989. "Reductive Dehalogenation of Chlorinated Cl-Hydrocarbons Mediated by Corrinoids." *Biochemistry.* 28: 4908-4914.

Marten, I., G. Lohse, and R. Hedrich. 1991. "Plant-Growth Hormones Control Voltage-Dependent Activity of Anion Channels in Plasma-Membrane of Guard-Cells." *Nature.* 353: 758-762.

Murashige R., and F. Skoog. 1962. "A Revised Medium for Rapid Growth and Bioassays with Tobacco Tissue Cultures." *Physiol. Plant.* 15: 473-497.

Newman L. A., S. S. Strand, N. Choe, J. Duffy, G. Ekuan, M. Ruszaj, B. B. Shurtleff, J. Wilmoth, P. Heilman, and M. P. Gordon. 1997. "Uptake and Biotransformation of Trichloroethylene by Hybird Poplars." *Enviorn. Sci. Technol.* 31: 1062-1067.

Paul, B. B., and D. Rubinstein. 1963. "Metabolism of CT and Chloroform by the Rat." *J. Pharmac. Exp. Ther.* 141: 141-148.

Strand S. E., L. A. Newman, X. Wang, D. Martin, N. Choe, B. Shurtleff, J. Wilmoth, I. Muiznieks, G. Ekuan, J. Duffy, J. Massman, P. Heilman and M. P. Gordon. 1998. "Phytoremediation of Chlorinated Solvents: Laboratory and Pilot-Scale Results." *Proceedings of the International Conference and Special Seminars on Groundwater Quality: Remediation and Protection*, Tubingen, Germany.

Vogel, T. M., C. S. Criddle, and P. L. McCarty. 1987. "Transformation of Halogenated Compounds." *Environ. Sci. Technol.* 21: 722-736.

Wang, X., S. E. Strand, L. A. Newman, D. M. Martin, Q. T. Shang, and M. P. Gordon. 1997. "The Fate of Chlorinated Organic Compounds after Degradation and Transformation by Poplar Trees." *Agronomy Abstracts of 1997 Annual Meeting of American Society of Agronomy, Crop Science Society of America and Soil Science Society of America.* pp. 38. Anaheim, CA.

BIOAUGMENTATION OF POPLAR ROOTS WITH *AMYCOLATA* SP. CB1190 TO ENHANCE PHYTOREMEDIATION OF 1,4-DIOXANE

Sara L. Kelley (University of Iowa, Iowa City, Iowa)
Eric W. Aitchison (Ecolotree, Iowa City, IA), Jerald L. Schnoor and
Pedro J.J. Alvarez (University of Iowa, Iowa City, Iowa)

ABSTRACT: Bioaugmentation experiments were conducted with the dioxane-degrading actinomycete, *Amycolata* sp. CB1190. Reactors planted with hybrid poplar trees removed more dioxane within 26 days than in unplanted reactors, regardless of whether CB1190 was added. Addition of CB1190 enhanced mineralization of dioxane in all experiments. CB1190 increased the percentage of dioxane mineralized in unplanted soil from 9% to 26%. Experiments were also conducted with excised tree reactors that offer a root zone (exuding potentially stimulatory substrates) but do not remove dioxane through plant evapotranspiration. More dioxane was mineralized in the excised-tree reactor with CB1190 (35%) than in the one without CB1190 (7%). Although CB1190 also enhanced dioxane mineralization in planted soil, this enhancement was not statistically significant because plant uptake reduced the availability of dioxane for microbial mineralization. Rapid uptake of 1,4-dioxane by hybrid poplar trees coupled with CB1190 degradation makes phytoremediation an attractive alternative for dioxane-contaminated sites.

INTRODUCTION

1,4-dioxane, a suspected carcinogen, is a widely used industrial solvent in paints, varnishes, lacquers, cosmetics, deodorants, fumigants, and detergents (Howard, 1990). It is also produced as an undesirable by-product of polyester synthesis. The log K_{ow} of 1,4-dioxane is -0.27, which indicates a very low partitioning into the organic carbon fraction of soil. Thus, it is expected to be highly mobile in soil (Howard, 1990). Dioxane is one of the most recalcitrant toxic contaminants in subsurface environments. Its persistence and high mobility represents a major challenge to site remediation.

Phytoremediation, the use of plants to remove environmental pollutants from contaminated sites, shows great promise as an approach to hazardous waste management. Plants can enhance the removal of xenobiotics by at least two mechanisms: (1) direct uptake and, in some cases, in-plant transformation to less toxic metabolites, and (2) stimulation of microbial activity and biochemical transformations in the root zone through the release of exudates and enzymes (Schnoor et al., 1995). This latter mechanism, however, is not very effective for removing recalcitrant xenobiotics, such as dioxane. Therefore, bioaugmentation of the rhizosphere with specialized microorganisms might enhance the biofiltration capabilities of the root zone (Aitchison et al., 1997; Alvey and Crowley, 1996; Crowley et al., 1996).

The objectives of this project were to determine

(1) if poplar trees can enhance the removal and mineralization of dioxane from contaminated soil, and

(2) if bioaugmentation with CB1190 can enhance the cleanup process in both planted and unplanted soil.

MATERIALS AND METHODS

Amycolata sp. CB1190 is the dioxane-degrading microorganism that was used for bioaugmentation. This actinomycete can grow aerobically on dioxane as the sole carbon and energy source (Parales et al., 1995). CB1190 was obtained from Rebecca Parales, Department of Microbiology, University of Iowa.

^{14}C-labeled dioxane was used as a tracer to determine whether adding CB1190 to the poplar rhizosphere or to unplanted soil could enhance dioxane mineralization and removal kinetics. Experiments were also conducted with one excised-tree reactor that offered a root zone (exuding potentially stimulatory substrates) but did not remove dioxane through plant evapotranspiration.

Hybrid poplar cuttings (*Populus deltoides* × *nigra,* DN34, Imperial Carolina) were planted individually in 280-mL flasks with uncontaminated soil (f_{oc} = 8.4%), watered with an inorganic nutrient solution, and the flasks were sealed. A modified 1-L Erlenmeyer flask was placed over the aerial portion of each cutting (Figure 1). The top flasks were sealed to the bottom flasks with Parafilm to enclose the aerial compartment. Activated carbon traps were affixed to the top flasks to capture dioxane volatilized from the leaves, and to the bottom flasks to capture volatilized dioxane. NaOH solution was used to trap $^{14}CO_2$. Each reactor contained 10 mg dioxane per kg of air-dried soil. The primary mechanisms influencing the fate of the added ^{14}C-dioxane (i.e., plant uptake and in-plant accumulation, volatilization through evapotranspiration, microbial degradation, and adsorption onto soil or bioreactor components) were studied interactively. The bioaugmented reactors contained CB1190 at a concentration of 2×10^7 cells/g-soil. All reactors were prepared in duplicate, except for a single excised-tree reactor, and were incubated at 24 °C for 26 days.

Cold dioxane was analyzed using a Hewlett-Packard 5890A gas chromatograph with a flame ionization detector and a Hewlett-Packard autosampler. Separate plant components and soil were oxidized with a RJ Harvey OX-600 Biological Oxidizer at the end of the experiment to convert ^{14}C-dioxane to $^{14}CO_2$ and determine where ^{14}C dioxane resides. A Beckman 6000IC liquid scintillation counter was used to analyze $^{14}CO_2$ and to calculate total dioxane mass in soil and plant tissue.

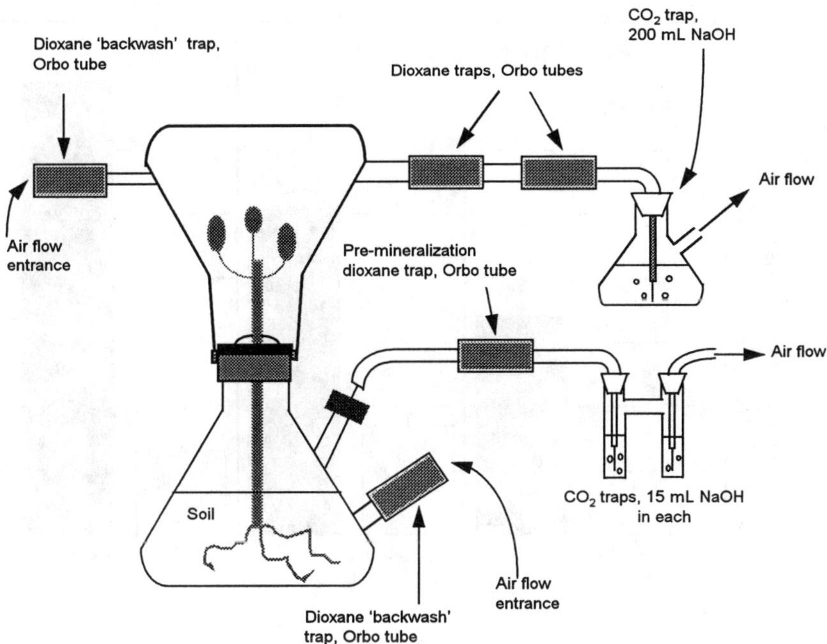

FIGURE 1. Planted Bioreactor Setup. Bottom compartment is sealed to separate microbial mineralization in the rhizosphere from evapotranspiration from leaves. Air flows through bottom compartment into two 15-mL NaOH traps in series, and through top compartment into one 250-mL NaOH trap to capture $^{14}CO_2$. Activated carbon Orbo tubes capture ^{14}C 1,4-dioxane released through evapotranspiration or volatilization from soil.

RESULTS AND DISCUSSION

Whether bioaugmentation with CB1190 enhances dioxane mineralization was investigated using both planted and unplanted soil (Figure 2). Bioaugmentation significantly enhanced dioxane mineralization in unplanted soil ($p<0.05$). Reactors amended with CB1190 mineralized 26 ± 5 % of the added ^{14}C-labeled dioxane, compared to 9 ± 1 % for no-treatment controls. Excised-tree reactors (which cannot evapotranspire dioxane) were used to evaluate the "rhizosphere effect," on biodegradation. Bioaugmentation of excised-tree reactors significantly enhanced dioxane mineralization, from 7 ± 0.4% to 35%. In the absence of evapotranspiration, roots also enhanced the performance of CB1190, which mineralized more dioxane in the excised-tree reactor (35%) than in unplanted soil (26 ± 5%). However, the beneficial effect of CB1190 was not statistically significant for planted soil. This was attributed to the fact that plant uptake reduces dioxane availability for microbial mineralization.

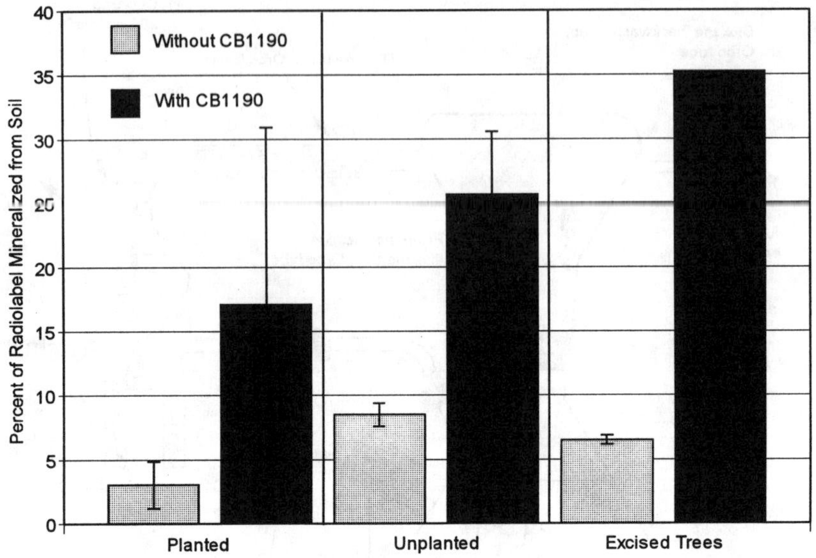

FIGURE 2. Percentage of Radiolabel Mineralized from Soil. Each reactor initially contained 10 mg/kg 1,4-dioxane, and was incubated for 18-26 days.

Although CB1190 significantly enhanced dioxane mineralization in unplanted and excised-tree reactors, it did not significantly reduce the residual dioxane concentration in those soils (Figure 3). The lack of enhancement of dioxane removal from soil was more pronounced in planted reactors, where more dioxane remained in bioaugmented soil (11 ± 6%) than in soil without CB1190 (4 ± 1%). These effects are not statistically significant, and are possibly due to the high variability at low dioxane concentrations that remained in the soil. These effects may also be partly attributed to an experimental artifact. Reactors without CB1190 had a higher percentage of dioxane removed by air suction (13% vs. 1% for excised-tree reactors, 40% vs. 30% for unplanted reactors, and 8% vs. 6% for planted reactors).

This experiment verified that poplars can take up and transpire dioxane from contaminated soil, as shown by the recovery of radiolabel (23 ± 14%) from Orbo tube traps connected to the top compartment (Figure 1). Trees enhanced dioxane removal from soil. Figure 3 shows that the percent dioxane remaining in the soil was significantly lower ($p<0.05$) in the planted reactors without CB1190 (4 ± 1%) than in the unplanted controls (17 ± 4%).

The percentage of dioxane remaining in unplanted reactors was lower than expected due to an experimental artifact: removal by air suction. The experimental design incorporated a constant airflow to allow for the capture of ^{14}C dioxane and $^{14}CO_2$. Because the unplanted reactors were not watered during the course of the experiment, the soil dried out, allowing ^{14}C dioxane to volatilize from the soil. The percent removed by air suction was significantly higher (p<0.05) in the unplanted reactors (40 ± 4%) than in the planted reactors (8 ± 7%).

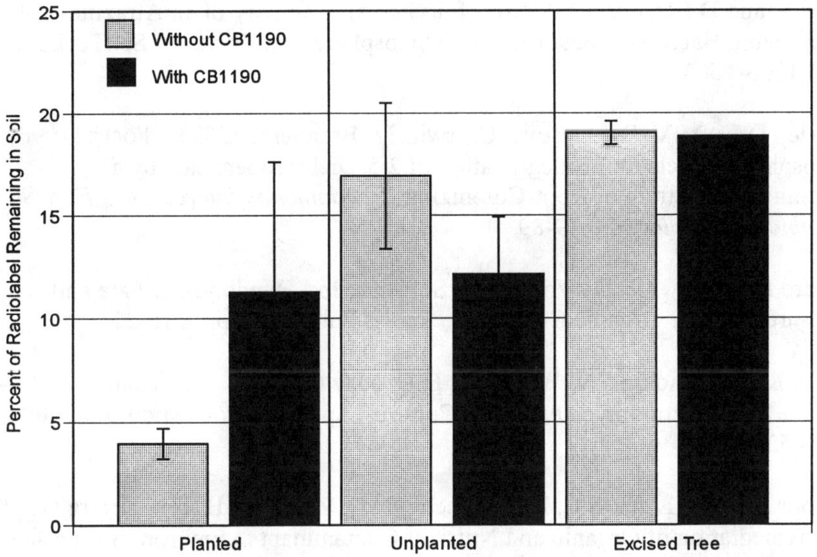

FIGURE 3. Percentage of radiolabel remaining in soil. Each reactor initially contained 10 mg/kg 1,4-dioxane, and was incubated for 18-26 days.

CONCLUSIONS

Bioaugmentation of planted or unplanted soil with CB1190 can enhance the remediation process. Rapid uptake and volatilization of 1,4-dioxane by hybrid poplar trees coupled with mineralization by *Amycolata* CB1190 in the rhizosphere makes phytoremediation an attractive alternative to remove dioxane from shallow contaminated sites.

ACKNOWLEDGEMENTS

We thank Rebecca Parales, who provided *Amycolata* CB1190, and Milind Deshpande for his assistance in growing large quantities of CB1190 at the Center for Biocatalyis and Bioprocessing of the University of Iowa. Hoechst-Celanese Corporation and the National Science Foundation funded this work.

REFERENCES

Aitchison EA, S.L. Kelley, J.L. Schnoor, and P.J.J. Alvarez (1997). Phytoremediation of 1,4-Dioxane by Hybrid Poplars, *Proc. WEF 70th Annual Meeting and Exposition.* North Chicago, IL, October 18-22, 1997.

Alvey S. and D.E. Crowley (1996). Survival and Activity of an Atrazine-Mineralizing Bacterial Consortium in Rhizosphere Soil. Environ. Sci.Technol. 30(5):1596-1603.

Crowley D.E., M.V. Brennerova, C. Irwin, V. Brenner, and D.D. Focht (1996). Rhizosphere Effects on Biodegradation of 2,5-Dichlorobenzoate by a Bioluminescent Strain of Root-Colonizing *Pseudomonas fluorescens. FEMS Microbiology Ecology* 20:79-89.

Howard P. II (ed). (1990) Volume II, Solvents, In: Handbook of Fate and Exposure Data for Organic Chemicals. Lewis Publishers, pp. 216-221.

Parales RE, J.E. Adams, N. White and H.D. May (1994). Degradation of 1,4-Dioxane by an Actinomycete in Pure Culture. Appl. Environ. Microbiol. 60: 4527-4530.

Schnoor JL, L.A. Licht, S.C. McCutcheon, N.L. Wolfe, andL.H. Carriera (1995). Phytoremediation of Organic and Nutrient Contaminants. Environ. Sci.Technol. 29: 318-323A.

ISOLATION, PURIFICATION AND PARTIAL CHARACTERIZATION OF PLANT DEHALOGENASE-LIKE ACTIVITY FROM WATERWEED (*Elodea canadensis*)

Om P. Dhankher (University of Georgia, Athens, GA)
Jason L. Tucker, Valentine A. Nzengung, University of Georgia, Athens, GA and
N. Lee Wolfe (U.S. EPA, Athens, GA)

ABSTRACT: Plants have been shown to effectively degrade halogenated contaminants. This is the first report of a dehalogenase-like activity isolated and purified from plants. This activity is suspected to be cofactor of a dehalogenase. The putative cofactor is heat stable, has broad pH stability, and is protease resistant. It is soluble in methanol, active both in aerobic and anaerobic conditions and has a broad substrate specificity with a pH optima of pH 8.5. The activity is completely inhibited by EDTA and requires ascorbate as the electron donor. An analytical methodology has been developed to isolate and purify the cofactor as well as to quantify the cofactor dehalogenase-like activity.

INTRODUCTION

Many plants can directely or indirectly degrade organic pollutants or accumulate and sequester inorganic contaminants. For organic pollutants, it is possible to use selected plants to transform toxicants to environmentally acceptable compounds. Several field studies indicate that a wide range of organic compounds can be degraded or removed from contaminated media (Ferro et al., 1994).

Plant enzymes that have been shown to be responsible for the transformation of organic pollutants are dehalogenase(s), nitroreductase(s), laccase(s), nitrilase(s), peroxidase(s) and phosphatase(s) (Schnoor et al., 1995). Dehalogenases constitute a multi-enzyme family in bacteria and its members are involved in reductive dechlorination of chlorinated solvents. A group of enzymes such as heloalkane , heloacid and heloperoxidase dehalogenases have been isolated from many bacterial species and employed for the bioremediation of diverse chlorinated compounds (Slater et al., 1995). A number of plants appear to have dehalogenase activity and the isolation of a purified enzyme is the priority of several research groups. Hybrid poplar plants have been shown to detoxify atrazine possibly by the action of dehalogenases and laccases to several non toxic compounds including short chain metabolites (Black, Dec. 1995). Studies with chlorinated methanes, ethanes and ethenes in hydroponic systems using selected aquatic plants have demonstrated that dehalogenation can occur rapidly (Nzengung et al., 1998). Recent vapor phase studies with terrestrial plants have shown that methyl bromide, hexachloroethane and tetrabromoethane are dehalogenated as well (Jeffers and Wolfe, 1998). The present research is focused on a plant-based dehalogenase system that has the potential to breakdown halogenated compounds in sediments, soils and natural waters. This is the first report on the isolation, purification and characterization of a plant derived component that dehalogenates selected organics.

MATERIALS AND METHODS

Plant Materials and Buffer Solutions: Plant samples of Elodea (*Elodea canadensis)*were collected from a local lake near Athens, GA and washed thoroughly in running water, then were frozen in liquid nitrogen and ground in a mortar pestle to fine powder. To study the effect of elevated temperature/autoclaving on the stability of dehalogenase-like activity, plants were autoclaved for 15 minutes and used to prepare extract for activity analysis. The extraction and assay buffer was 25mM Tris, pH 8.5 + L-ascorbic acid (final concentration of 1mM). All the substrate (HCA, PCE, and TCE,) stocks were prepared in methanol to a stock concentration of 1mg/ml (1000ppm). The final concentrations, of all the substrates used in assays were 5ug/ml (5ppm).

Isolation and Purification: The fresh Elodea plants were freeze dried in liquid nitrogen, ground to fine powder in mortar and pastle and stored at –70°C. Further steps of the isolation and purification were carried out at room temperature. A 100g sample of frozen Elodea (fine powder) was homogenized to a slurry in 200 ml of buffer using a polytron. This fine slurry was filtered through 8 layers of cheesecloth and the filtrate centrifuged at 10,000rpm for 30 min. The supernatant was first purified by pH precipitation. The pH was lowered to pH 2.0 with HCl to allow the precipitates to flocculate for about 4hrs at 4°C and centrifuged as above. The pH of the supernatant was raised to 12.0 with NaOH and after another 4hrs at 4°C, it was again centrifuged for 30 min. The supernatant was adjusted to pH 8.5 and was concentrated 5X to 6X by boiling in a microwave. After cooling and centrifugation as above, the supernatant was filtered through an Amicon 3K centricon filter and the filtrate was further concentrated, by boiling, to about 10X of starting volume. The resulting filtrate, a partially purified extract was used for most of the experiments. For further purification, this <3K filtrate was freeze dried and resuspended in methanol, thereby removing the water soluble components. The clarified methanol solution was again lyophilized and the powder resuspended in a small volume of the 25mM Tris, pH 8.5 buffer. This was then further purified on a Pharmacia Biotech Fast Protein Liquid Chromatography (FPLC) equipped with a dual wavelength UV detector. Conditions for FPLC purification are as follows: a Superdex peptide PE 7.5/300 column; 100 ul injection volume; 25 mM tris-HCl buffer with 150 mM NaCl, pH 8.5; 0.5 ml fractional volume; 0.28 ml/min flow rate and a UV detector at 280 nm.

HCA assays - batch studies: Crude and purified fractions of the elodea extracts were assayed for activity with HCA. 2 mL of the extracts were placed in 20 mL batch centrifuge tubes and then 18 mL of 25 mM Tris buffer (pH 8.5) + 0.2 mL ascorbate was added. Each vial was spiked with a stock solution of HCA to obtain a final concentration of 5PPM and sealed immediately. Head space was minimized in all batches. The controls contained only the buffer solution, ascorbate and HCA and were prepared and handled in parallel with all samples. These experiments were performed in triplicate. The samples and their controls were incubated for different time periods ranging from 2hrs to 40hrs before analysis. At predetermined time intervals, the sample and control batch tubes were sacrificed for analysis. All samples were extracted (1:1 ratio) into the hexane phase before analysis by GC/ECD.

GC Analysis: A Hewlett Packard 5890 series-II gas chromatograph fitted with an electron capture detector and a DB-1 capillary column was used for analysis of volatile halogenated compound HCA. Column dimensions were 20mm x 0.18 mm x 0.4 um; Column phase: 100% methyl siloxane. All injections were made with an auto-injector (HP 7673) (0.5 ul volume of sample), and the solvent for all injected samples was hexane. Column temperature was initially 30°C for 2.00 min., then ramped to 175°C at 6.00°C/min and then ramped to 275°C at 10.0 °C/min. and held for 2.00 min.

RESULTS AND DISCUSSION

Earlier studies in our laboratory had shown that plant mediated transformation processes can make a significant contribution to the overall processes of attenuation of organic contaminants in environmental systems. To provide further insight into the role of plants in transforming halogenated organics, we have isolated and characterized a compound having dehalogenase-like activity from Elodea that can degrade a variety of chlorinated methanes and ethanes.

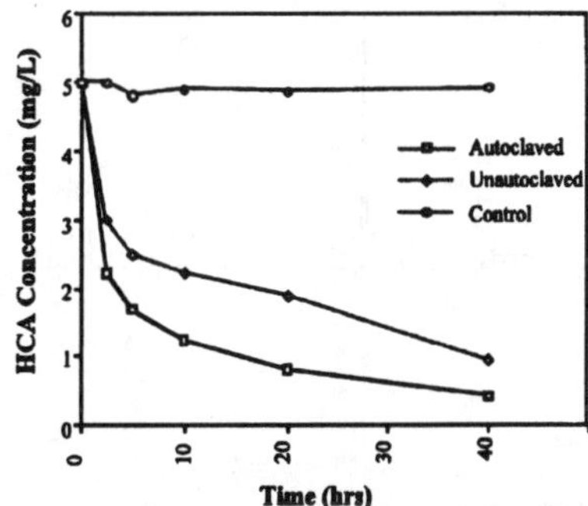

FIGURE 1. HCA degradation by extractsfrom autoclaved and non-autoclaved Elodea plant.

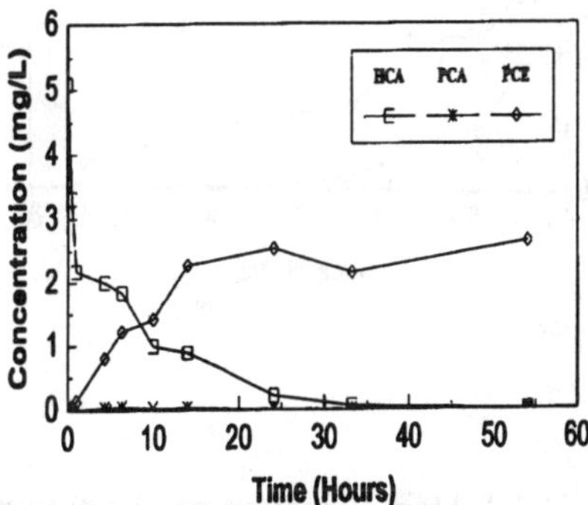

FIGURE 2. HCA degradation and formation of breakdown products by crude extract.

Homogeneous Kinetics: To assess dehalogenase-like activity, HCA was selected as a substrate. It was chosen because it has high water solubility and good analytical sensitivity and forms intermediates that are readily analyzed by GC/ECD. Previous in-vivo research suggested that the dehalogenase-like activity could be due to bacterial contamination. In order to eliminate this possibility, we decided to surface sterilize the plants before extraction. There was no significant difference in dehalogenase-like activity between extracts from surface sterlized plants with 10% bleach for 4 min and non-sterlized plants (data not shown). Also, the activity was maintained in extracts from autoclaved plants (Figure 1). This indicated that the extractable dehalogenase-like activity discussed below was derived from plants and not bacteria. The HCA transformation products were confirmed by GC/MS. Figure 2 is a plot of the concentration versus time for the disappearance of HCA and the formation of PCE and TCE with the crude plant extract. The rapid dehalogenation of HCA was observed in all experiments. The rate of transformation of HCA was faster with the purified fraction than with crude fraction (Figure 3). The control experiments showed that a buffered solution of ascorbate without the extract were not responsible for the transformation of HCA in these experiments. A first-order decay model described the Kinetic data for the plant extract and solute concentration, used in this study. The half-life was approximately 5.5 and 16 hours for the purified fraction and crude extract, respectively.

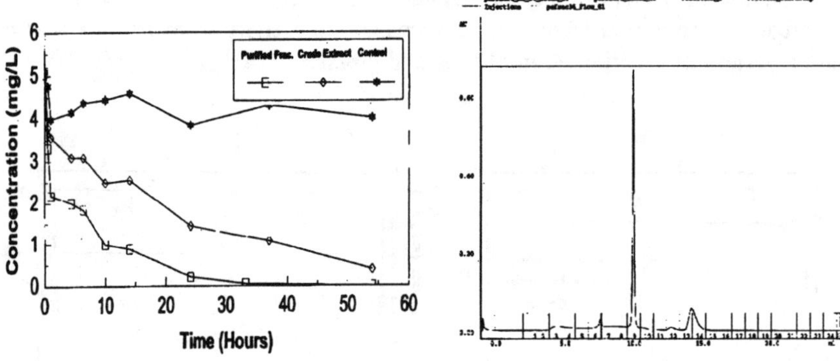

FIGURE 3. Homogenous transformation kinetic of HCA using purified fraction and crude extract.

FIGURE 4. Purification of dehalogenase cofactor to a single Peak resolution by gel filtration.

Purification: While standardizing the pH optima, it was observed that on lowering or raising the pH from the extraction pH of 8.5, there was precipitation without any loss of the dehalogenase-like activity. As the activity was heat stable, boiling was used as a concentration step after the pH purification step. Further purification was carried out by gel filtration on FPLC. This active dehalogenating compound did not bind to ion exchange and hydrophobic interaction columns but the column was later found to be useful to remove some impurities from the partially purified fraction. After desalting, the fraction was purified on a Superdex Peptide column and new fractions were collected. As shown in Figure 4, a single peak elutes from the Superdex peptide column. This fraction was shown to have dehalogenase-like activity. For these studies and for further characterization, it was assumed that this fraction was pure. Calibration of the Superdex peptide column gave a molecular weight for this compound of approximately 1.2 kDa. The UV-Vis for this compound showed absorption maxima at 216 and 270 nm. To determine substrate specificity, the purified fraction was tested against a series of compounds such as PCE, TCE, PCP, TCB, DCB, CTC, toxaphene and DDT. Results suggested that several of these compounds can be degraded (Table 1). The dehalogenase-like activity was also completely inhibited by 2mM EDTA (results not shown), suggesting that this dehalogenating component may contain one or more metals. It is believed that this component is the co-factor from the plant dehalogenase and carrying out the reductive dechlorination of the substrates. Further characterization of this dehalogenating fraction is in progress and results will be discussed somewhere else.

TABLE 1. Substrate specificity for dehalogenase-like activity extracted from Elodea plants in a 24 hrs assay.

Substrate	% Degradation
HCA (Hexachloroethane)	100
PCE (Tetrachloroethylene)	59
TCE (Trichloroethylene)	49
PCP (Pentachlorophenol)	0
TCB (Trichlorobenzene)	30
DCB (Dichlorobenzene)	16
CTC (Carbon tetrachloride)	100
Toxaphene	0
DDT	97

IMPLICATIONS AND USES

Further work on the characterization of the dehalogenating component is needed to provide a better understanding of the role of this compound in plants. Identification of it's function in plant biochemistry, the electron source for the reductive transformations, and the detailed pathways for the transformations are needed. These details will provide the basis for the development of predictive models for fate and transport as well as exposure assessments for this class of compounds.

These preliminary findings will lead to a better understanding of plant mediated transformation processes and the role that plants can play in the phytoremediation of chlorinated contaminants in soils and waters. It may well be possible to obtain the component from plants and use it directly for remediation. Also when the holoenzyme is identified, it could provide an opportunity for transgenics to and the development of plants with enhance dehalogenase activity.

REFERENCES

Ferro, A.M., R.C. Sims, and B. Bugbee. 1994. Hycrest Crested wheatgrass Accelerates the Degradation of Pentachlorophenol in Soil. *J. Environ. Qual.* 23: 272-279.

Harvey Black. 1995. Absorbing Possibilities: Phytoremediation. *Environmental Health Perspective.* 103(12): 1106-1108.

Jeffers, P. M. and N. Lee Wolfe. 1998. Green plants: A terrestrial Sink for Atmospheric Methyl bromide. Geophys. Res. Lett. 25:43-46.

Nzengung, V.A., N.L. Wolfe, D.E. Rennels and S.C. McCutcheon. 1998. Algae- and Aquatic Plant-mediated Transformation of Halogenated Organic Compounds. Submitted to *J. Phytoremediation.*

Schnoor, J.L., L.A. Licht, S.C. McCutcheon, N.L. Wolfe and L.H. Carreira. 1995. Phytoremediation of Organic and Nutrient Contaminants. *Environ. Sci. Technol.* 29: 318- 323.

Slater, J.H., A.T. Bull, and D.J. Hardman. 1995. Microbial Dehalogenation. Biodegradation. 6; 181-189.

Wolfe, N.L., T. Ou, L. Carreira and D. Gunnison. 1994. *Alternative Methods for Biological Destruction of TNT: A Preliminary Feasibility Assessment of Enzymatic Degradation.* Prepared for U.S. Army Corps of Engineers, Waterways Experiment Station, AD-285-645.

PHYTOREMEDIATION OF ORGANOPHOSPHOROUS (OP) COMPOUNDS USING AXENIC PLANT TISSUE CULTURES AND ENZYME EXTRACTS

Jianping Gao, A. Wayne Garrison, Chris Mazur, N. Lee Wolfe (U.S. EPA, Athens, GA)
Chris Hoehamer (University of Georgia, Athens, GA)

ABSTRACT: The bioremediation of OP compounds (malathion, demeton-s-methyl, ruelene) was investigated *in vitro* using axenic plant tissue cultures of parrot feather, duckweed, and elodea. The decay profile in all these cases followed first-order kinetics. However, extents and rates of biotransformation were different, depending on both physico-chemical properties of the OP compounds and the nature of the plant species. Malathion exhibited a similar disappearance pattern in all three plants, with 29-48% degradation. The most effective transformation was observed for demeton-s-methyl: less than 1% was recovered in parrot feather and elodea, and about 17% in duckweed. No significant biotransformation of ruelene occurred in elodea, while 17-24% degraded in the other plants. The results using enzyme extracts derived from duckweed provided strong evidence for a direct degradation relationship between organophosphorous hydrolase (OPH, EC 3.1.8.1) or multiple enzyme systems and OP compounds. This study showed that axenic tissue cultures of several aquatic plants have the enzymatic potential to metabolize OP compounds.

INTRODUCTION

Aquatic plants have a great potential to function as *in-situ*, on-site biosinks and biofilters of aquatic pollutants because of their abundance and limited mobility. Plant tissue/cell cultures have been used in a number of pesticide metabolism studies to evaluate their suitability as model systems for bioremediation by whole plants (Hughes et al., 1997). Organophosphorus (OP) compounds have contaminated environmental compartments in many countries. Most studies of OP compound bioremediation have been limited to microorganisms or plant-associated microflora (Zayed et al., 1998), so the ability of plants to transform OP compounds without the participation of associated microbes remained arguable (Hughes et al., 1997). In this study, axenic tissue cultures of the aquatic plants parrot feather (*Myriophyllum aquaticum*), duckweed (*Spirodela oligorrhiza L.*), and elodea (*Elodea canadensis*) were investigated *in vitro* for their ability to transform three OP compounds: malathion, demeton-s-methyl, and ruelene.

Little is known about the enzyme-based mechanisms that are responsible for degradation of OP compounds. The organophosphorous hydrolase (OPH, EC 3.1.8.1) is a family of enzymes involved in the transformation of OP compounds with P-O, P-S, P-CN, and P-F bonds; they are present in aquatic plants and have been shown to have variable substrate specificities (Morita et al., 1996). A direct link between OPH isolated from the tested plant and biodegradation of OP compounds would provide strong evidence for the metabolic pathways involved in phytoremediation. In order to characterize such plant/cell-lines more fully, the hydrolysis of the three OP compounds was also studied using duckweed enzyme extracts.

EXPERIMENTAL SECTION

Reagents and Axenic Plant Tissue Culture. Analytical grade (>99% purity) malathion, demeton-s-methyl, and ruelene were used. Axenic parrot feather was supplied by the Department of Biochemistry, University of Georgia. These plants were propagated vegetatively in culture boxes on NH_4^+-free Murashige Skoog supplemented with agar. Duckweed was taken from ponds and wetlands in the Athens, GA vicinity and cultured in 20-gal plastic containers with half-strength Hoagland's culture solution at pH 7.0. After cultivation for 3 weeks, the duckweed was rinsed with flowing tap water, then transferred into sterilized water containing 1% sodium hypochlorite. This solution was stirred with a magnetic bar for 15 min, the plants were then removed and rinsed with sterilized water. Elodea plants were collected from Lake Herrick in Athens, GA. The procedures of culture and sterilization were similar to that for duckweed as described above.

Phytoremediation Studies. Two g (wet wt.) of axenic plants were placed in each of three 50-ml Erlenmeyer flask, containing 20 ml of sterile Hoagland nutrient solution and spiked with 1 ppm malathion, 10 ppm demedon-s-methyl, or 10 ppm ruelene. These mixtures were incubated in a rotating incubator at 22 °C with fluorescent lights. 0.5 ml of the culture media was sampled at time intervals of 0, 4, 8, 16, and 24h, and then one time each day for 8 days; samples were extracted with 0.5 ml n-hexane by shaking in a small vial and analyzed by GC-ECD (HP 5890 series II) using a RTS-5 column (30 m, 0.53 mm ID, 0.5 μm df). Column temperature was programmed: 120 °C for 2 min, then 20 °C/min to 220 °C, followed by a 1 min hold, finally at 4 °C/min to 240 °C with a 10 min hold. The detector was at 300 °C. Helium was the carrier gas at 30 cm/sec, nitrogen was the make-up gas at 30 ml/min. The injector was at 250 °C in the split mode at 1:40.

Mass Recovery Studies. At the end of the incubation (8th day), the plants were rinsed thoroughly with water and ground in liquid nitrogen using a motor and pestle. Then the plant materials were sonicated and cleaned up by passing through a silica gel and then a C-18 column. The eluate from the C-18 was gently evaporated with nitrogen and residues were redissolved in 1 ml hexane. Mass recovery was analyzed by GC-ECD.

Degradation Studies Using Isolated Enzyme. Plant culture, enzyme extraction and enzyme purification were modified according to Morita et al. (1996). To assay the degradation of OP compounds, 1 ml of plant enzyme extract was mixed with 10 μl of 3.03 μM malathion, 38.71 μM demeton-s, or 34.28 μM ruelene. Samples were incubated in a rotating incubator at 22 °C, with fluorescent lights. At time intervals of 0, 4, 8, 16, 24, and 48h, 0.5 ml of the aqueous culture media was taken out for measuring the degradation rate by GC-ECD. All experiments were replicated at least 3 times.

Data Analysis. A first-order one-compartment model ($C_t/C_0=e^{-kt}$) was used to estimate the kinetic parameters of OP compound degradation, where C_t is as the concentration of the pollutant at time t, C_0 as its initial concentration, and k is the first-order rate constant. The half-life time ($t_{1/2}$) is given by $t_{1/2}=\ln(2)/k$. Mean values and standard deviations were calculated for each test compound based on the values obtained for each kinetic run. The bioconcentration factor (BCF) was calculated as OP compound concentration in plant tissue (mg/kg)/ initial OP concentration in the media (mg/L).

RESULTS AND DISCUSSION

Phytotransformation with Parrot Feather. About 8.6% of malathion disappeared from the viable parrot feather culture medium within 16h of incubation. After that, the concentration in the culture medium decreased gradually; at the end of incubation (8d) a total of 83% had disappeared (Figure 1). An initial decrease in malathion concentration was also observed with autoclaved plants, this was attributed to sorption on the plant surface. Thereafter, the concentration remained constant. Similar trends for demeton-s-methyl and ruelene were observed, but to a lesser extent (2.1% and 1.8% at 16h; 78% and 58% at 8d, respectively, Figure 1). After 8d, the ruelene concentration continued to decline slowly and about 77% had disappeared at the end of the experiment (14d).

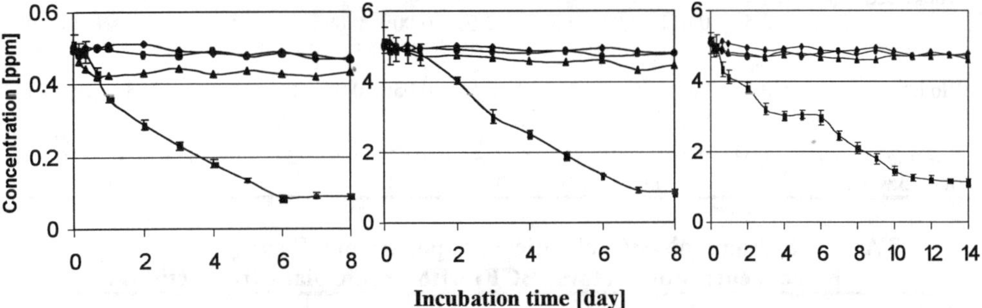

FIGURE 1. Disappearance of malathion (a), demeton-s-methyl (b), and ruelene (c) from media exposed to axenic Parrot Feather under controlled conditions. ◆, control; ●, culture medium; ▲, autoclaved parrot feather; ■, parrot feather.

Before undergoing any transformation, OP compounds must be taken up by the plants, which is a complex process that may involve an active process specified by compound, and/or a passive process. It is not clear whether the transformation occurs on the surface or during transport into the plant tissue (Hughes et al., 1997). At any rate, metabolism of the OP compounds is mediated by enzymes after they reach the plant cells. This transformation is dependent on three factors: 1) physico-chemical properties of the compounds, 2) plant species, and 3) environmental factors. In the case of parrot feather, the physical and chemical characteristics of the OP compounds are apparently the predominant factors. Unlike malathion, demeton-s-methyl has a P-O bond and two diethyl groups, while ruelene has a P-O bond and a chlorophenyl moiety. These differences in molecular structure may increase persistence under field conditions; metabolic pathways may be mediated by different or multiple enzymes (Tarrant et al., 1992).

The concentration of malathion in the parrot feather culture medium followed a logarithmic decay, with a first-order decay constant of 0.012 hr^{-1}. This 58-hour half-life was achieved with a mass-ratio of plant to medium of about 0.1 (20 ml medium to 2 g of parrot feather). The half-lives ($t_{1/2}$) and disappearance rate constants (k) for all three OP compounds are given in Table 1. Mass recovery studies after 8d showed that about 29%, 82%, and 17% of malathion, demeton-s-methyl, and ruelene, respectively were degraded and/or bound in a non-extractable manner with plant material (Figure 2). The BCF shows that demeton-s-methyl accumulates in parrot feather much less than malathion and ruelene (Table 2). These results suggest that axenic parrot feather contains the enzymes necessary to metabolize OP compounds.

TABLE 1. Disappearance rate constants (k) and half-lives ($t_{1/2}$) of OP compounds exposed to axenic plant tissue cultures and duckweed enzyme extracts

Plants	Malathion				Demeton-s-methyl				Ruelene			
	C_0 [ppm]	k [h^{-1}]	r^2	$t_{1/2}$ [h]	C_0 [ppm]	k [h^{-1}]	r^2	$t_{1/2}$ [h]	C_0 [ppm]	k [h^{-1}]	r^2	$t_{1/2}$ [h]
Parrot Feather	0.5	0.012	0.91	58	5.0	0.011	0.93	69	5.0	0.005	0.86	139
Duckweed	0.5	0.011	0.92	63	5.0	0.009	0.89	77	5.0	0.007	0.90	99
Elodea	0.5	0.006	0.88	116	5.0	0.056	0.87	12	5.0	0.003	0.91	2302
Enzyme extracts of duckweed	3.03*	0.019 * [μM]	0.94	36	38.71*	0.013 * [μM]	0.92	53	34.28*	0.009 * μM]	0.90	78

TABLE 2. Some physico-chemical properties of OP compounds and their bioconcentration factors (BCF) with axenic plant tissue cultures

Compound	Water solubility [ppm]	Vapor pressure [mm Hg]	K_{ow} [ml/ml]	K_{oc} [ml/g]	BCF [L/kg] Parrot Feather	Duckweed	Elodea
Malathion	145	4×10^{-5}	280	NA	3	23	1.2
Demeton-s-methyl	3300	4.8×10^{-4}	20.9	66	0.023	13.73	0.37
Ruelene	<1	NA	NA	NA	2.67	2.59	0.046

Phytotransformation in Duckweed. Compared to parrot feather the results with duckweed (data not shown) show some differences, as follows: 1) Malathion and demeton-s-methyl disappear from the medium more rapidly from 1d-3d, and reach non-detectable levels after 5d and 8d, respectively. After reaching equilibrium, the growth medium concentration of malathion and demeton-s-methyl is lower in duckweed (2.7%, 1.2%) than in parrot feather (18%, 17%). 2) The initial time of concentration reduction from the medium is delayed for demeton-s-methyl and ruelene, by 2d and 3d after the incubation, respectively. 3) The equilibrium time for ruelene is reached much early in duckweed (7d) than in parrot feather (11d). 4) Comparing the kinetic parameters shown in Table 1, duckweed demonstrates a similar potential to parrot feather to transform the

OP compounds. However, it generally accumulates much more OP compounds than does parrot feather, except for ruelene (Table 2). Mass recovery data show that slightly higher percentages of OP compounds were degraded by duckweed (Figure 2). Generally, Duckweed provides a rapid process for OP compound transformation and uptake.

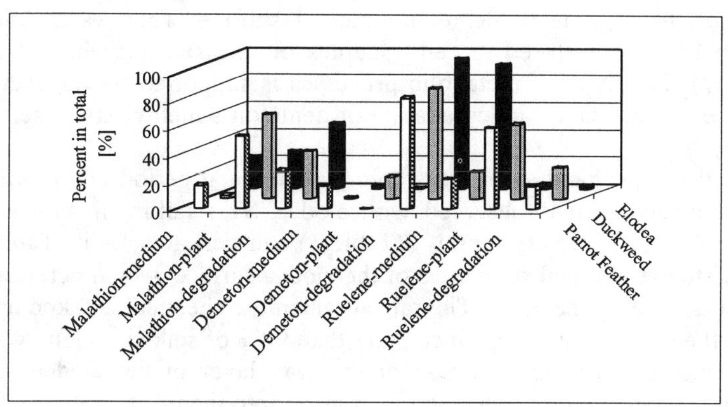

FIGURE 2. Biodegradation and mass recovery of OP compounds from the axenic plant tissue cultures.

The duckweed enzyme extracts had transformed about 25% of malathion by 24h of incubation, with maximum degradation occuring before 8h; no further transformation occurred before the end of incubation (48h). The demeton-s-methyl concentration exhibited no significant change after 24h; however, it was reduced about 10% by the end of incubation (48h). Similarly, the ruelene concentration decreased about 20% at 16h after the incubation, no further change was observed. This delay of the initial biodegradation of demeton-s-methyl and ruelene is in agreement with that in the axenic tissue culture of duckweed. These results indicate that in addition to uptake potential, there exist other factors that determine the transformation of the two compounds.

Compared to the plant tissue culture of duckweed, enzyme extracts provide more effective and rapid degradation of OP compounds (Table 1). This tissue extract is known to contain high activity of OPH (Morita et al., 1996). These data support the observation that there is a broad-spectrum of OPH capable of hydrolyzing a variety of OP compounds with P-O as well as P-S bonds. Although in some case this enzymatic hydrolysis can occur at several locations in a given OP pesticide, the most common reaction involves cleavage at a phosphate ester linkage as a result of base-catalysis. In addition to OPH, other enzymes may also mediate degradation of OP compounds. These involve mixed function oxidase, flavin-containing monoxygenase, glutathione S-transferase, and carboxylesterase (Edwards and Onen, 1998).

Phytotransformation in Elodea. Unlike with parrot feather and duckweed, malathion concentration in elodea doesn't decrease until 2 days after incubation, but similar reduction is observed after that (data not shown). The reason for this delayed transformation is not clear. Again, plant uptake may play an important role. Demeton-s-methyl exhibits a different behavior with an initial fast disappearance from the medium, followed by a slower reduction; more than 90% is lost within 48h. Its short $t_{1/2}$ (Table 1) and low BCF (Table 2) demonstrate that elodea may be an effective plant for remediation of this OP compound in the environment. This high transformation rate (>90%) of demeton-s-methyl may be due to its metabolism in plants. Generally, demeton-s-methyl first undergoes sulfoxidation to produce demeton-s-methyl sulfoxide which in less stable and continues to degrade to demeton-s-methyl sulfone. This water soluble, nontoxic product can be further stored in cell vacuoles or released into the extracellular space (WHO, 1997). This type of metabolic processes is supported by the fact that a similar decay profile and efficient biodegradation of demeton-s-methyl are observed in all three tested plants.

On the other hand, over the whole experimental period no significant reduction in ruelene concentration is observed with elodea. This failure in biotransformation of ruelene is indicated by a very long $t_{1/2}$ (Table 1) and very low BCF (Table 2). This may be attributed to the physical properties of the elodea surface, which determine the amount of ruelene adsorbed by the plant. The rate at which the chemical is taken up into the plant is probably the rate-limiting step since more than 90% of spiked ruelene was recovered at the end of incubation. The thickness of the wax layer in the elodea subsurface may influence the absorption of ruelene and its transfer into the plant.

This study indicated that some OP compounds can be taken up by aquatic plants and subsequently transformed as a function of time. This plant-mediated degradation of OP compounds can be attributed to enzymatic reactions. The intrinsic ability of plant enzymes to transform OP pesticides was confirmed by using the tissue enzyme extracts derived from duckweed.

REFERENCES

Edwards, R., and W.J. Onen. 1998. "Regulation of Glutathione S-Transferases of Zea Mays in Plants and Cultures." *Planta*. *175*:99-106.

Hughes, J. B., J. Shanks, M. Vanderford, J. Lauritzen, and R. Bhadra. 1997. "Transformation of TNT by Aquatic Plants and Plant Tissue Culture." *Environ. Sci. Technol. 31*:266-271.

Morita, N., H. Nakazato, H. Okuyama, Y. Kim, and G. A. Thompson Jr. 1996. "Evidence for a Glycosklinositospholipid-anchored Akaline Phosphatase in the Aquatic Plant Spirodela oligorrhiza." *Biochem. Biophy. Acta. 31*:53-62.

Tarrant, K. A., H. M. Thompson, and A. R. Hardy. 1992. "Biochemical and Histological Effects of the Aphicide Demeton-s-methyl on House Sparrows (Passer Domesticus) under Field Condition." *Bull. Environ. Contam. Toxicol. 48*:360-366.

WHO (Eds.). 1997. *Demeton-s-methyl*. Vol. 197, pp. 1-19. Wissenschaftliche Verlagsgesellschaft GmbH, Stuttgart, Germany.

Zayed, A., S. Gowthaman, and N. Terry. 1998. "Phytoaccumulation of Trace Elements by Wetland Plants: I. Duckweed." *J. Environ. Qual. 27*:715-721.

ENHANCEMENT OF RECALCITRANT COMPOUNDS BIODEGRADATION BY FENTON'S OXIDATION TREATMENT

Luca Di Palma, Carlo Merli, Elisabetta Petrucci
(Università di Roma "La Sapienza" via Eudossiana 18, 00184 Roma, Italy)

ABSTRACT: As an alternative to the traditional anaerobic digestion systems an advanced treatment of toxic and recalcitrant compounds has been investigated by combining chemical and biological oxidation process. The chemical treatment has been carried out by Fenton oxidation system which generates hydroxyl radicals by the reactions of hydrogen peroxide and ferrous iron, at acid pH values.
In optimizing the Fenton's pre-treatment the iron and hydrogen peroxide concentration ratio was found to be crucial.
After the chemical treatment the samples were flocculated, decantated and the supernatant was fed to a post-activated sludge treatment.
The data obtained, in applying the proposed treatment system to the tannery and pulp mill wastewaters, confirm that the hydroxyl radicals provide an effective enhancement of the recalcitrant compounds.

INTRODUCTION

In the treatment of industrial wastewater it often proves necessary to use processes of chemical oxidation in order to break down polluting substances into harmless end products or intermediates of a more easily biodegradable nature.

The oxidation processes that use the generation of hydroxyl radicals or other active free radicals to bring about reactions are known as *Advanced Oxidation Processes (AOP)*.

The *hydroxyl radical* in particular is a highly unstable species but one endowed with extraordinary oxidizing potential, second only to that of fluorine:

$$HO^{\bullet} + H^{+} + e^{-} \Leftrightarrow H_2O \qquad E^0 = +2.8\ V$$

The category of oxidants acting through the generation of hydroxyl radicals also includes a system composed of a mixture of hydrogen peroxide and Fe (II) known as Fenton's reagent, whose reaction mechanism is as follows:

$$Fe^{2+} + H_2O_2 \rightarrow Fe^{3+} + HO^{\bullet} + HO^{-} \qquad (1)$$

$$HO^{\bullet} + RH \rightarrow H_2O + R^{\bullet} \qquad (2)$$

$$R^{\bullet} + Fe^{3+} \rightarrow R^{+} + Fe^{2+} \qquad (3)$$

$$R^{+} + H_2O \rightarrow ROH + H^{+} \qquad (4)$$

Alongside the reactions reported above, attention should be drawn to a number of undesirable "termination" reactions. Their effect can be minimized through suitable calibration of reaction conditions.

Objectives. The objective of this study is to verify the effect of the oxidation by Fenton's reagent on the biodegradability enhancement of two tipical recalcitrant industrial wastes: the tannery pretreatments wastewater and the condensate of the black liquor concentration in the pulp mill industry.

MATERIALS AND METHODS

Hydrogen peroxide and iron in the form of heptahydrated ferrous sulfate ($FeSO_4 \cdot 7\ H_2O$) were added in conditions of constant agitation and at room temperature to the sample, previously acidified to pH 3 by means of hydrochloric acid. The additions were made using different concentrations of reagents in order to determine the conditions offering maximum yield. On completion of the oxidation reaction, the sample was alkalized to pH 8.5 by means of powdered calcium hydroxide and subjected to clarifying flocculation. Floxan 9924, a polyacrylamide-based anionic flocculant produced by Misan Chimica, was used for this purpose. Cellulose filters (diameter 47 mm, porosity 0.45 μm) produced by Micro Filtration Systems were used to filter the supernatant liquor, upon which the necessary analytical measurements were then carried out.

RESULTS AND DISCUSSION

The application to liming and fleshing waste. The purification of tannery wastewater involves the treatment of liquid with a high organic content. While a variety of treatments are possible from the technical standpoint, the range is sharply reduced when the aim is not only to reduce the level of pollution but also to set up a process making it possible to recover raw materials prior to discharge while keeping treatment costs comparatively low.

Wastewater from the liming and fleshing processes presents a very high level of pollution, accounting by itself for 60% of the total quantity of organic substances contained in tannery wastewater.

The wastewater discharged from the liming process presents a high level of alkalinity (pH normally over 12) and has a high content of organic substances, sulfides and nitrified material of a proteinic nature.

All this makes liming wastewater inherently difficult to treat. The difficulty is further increased by the considerable difference between the relevant values of BOD_5 and COD, as a result of which the biocompatibility ratio (i.e. the ratio of BOD_5 to COD) presents very low values of below 0.1.

A suitable treatment system must involve a compromise between the need to break down the various components and the need to limit the use of chemical reagents in order to contain costs.

In the conventional treatment the water discharged, after acidification and flocculation for sulphur and proteins recovery, is normally subjected to biological treatment. Because of the considerable level of organic and nitrified compounds present in such wastewater, such treatment often proves difficult and the yields are consequently low.

The substantial modification with respect to this traditional system lies in subjecting the wastewater, subsequent to acidification and the resulting separation of the protein component, to chemical oxidation in acidic conditions by means of Fenton's reagent.

This makes it possible to obtain a substantial reduction in the level of dissolved organic substances which, together with the subsequent process of clarifying flocculation, leads to an improvement in the quality of the wastewater to be subjected to final biological treatment.

Two series of laboratory tests were carried out with a view to assessing the effectiveness of oxidation with Fenton's reagent as regards the biocompatibility of the wastewater produced in comparison with the conventional method of treatment. The tests carried out on the treatment of wastewater from the liming process without Fenton's reagent involved the following phases:

- acidification of the wastewater in conditions of constant agitation
- precipitation of proteinic material
- mixing with wastewater from the fleshing process
- flocculation in an alkaline environment

A sample of wastewater from the liming process was first subjected to acidification to pH 4.5 with 12 g/l of sulfuric acid in a closed recipient with a volume of 1 litre. During this phase the recipient was subjected to mechanical agitation. The transformation of sulfides into hydrogen sulfide was observed during this phase. The recipient was subjected to a light stream of nitrogen (37 ml/minute) through a porous membrane in order to obtain the stripping of the hydrogen sulfide formed during the process of acidification.
The observed duration of the conversion reaction was about 30 minutes.

At the same time as the gas was formed, the proteinic substances were separated inside the recipient, being precipitated through insolubilization from the liquid phase in the acid conditions created.

After removal of the product from the bottom, the supernatant liquor was mixed with a sample of wastewater from the fleshing process subjected to no form of pre-treatment, and then subjected to a flocculation treatment after neutralization with calcium hydroxide at pH = 8.5.

The two liquids were mixed in such a way as to obtain a sample in which the liming wastewater accounted for 65% of total volume. The average composition of the samples tested is shown in table 1.

TABLE 1. Average composition of mixture subjected to testing

Parameter	Unit of measurement	Value
pH		4.5
Conductance	μS/cm	8500
COD	mg/l	45000
BOD_5 vs COD ratio		0.17
Total suspended solids	mg/l	1500
Sulfides	mg/l	40
Amino nitrogen	mg/l	500

In the second series of tests, the sample was subjected, prior to flocculation, to a phase of oxidation with Fenton's reagent ($FeSO_4$ 7 H_2O + H_2O_2), after further acidification to pH 2.7 with hydrochloric acid. Under conditions of constant agitation and at ambient temperature, oxygenated water and bivalent ferrous salt were added in quick succession. After about twenty minutes, the sample was subjected to clarifying flocculation as above described.

Several concentration of bivalent iron and hydrogen peroxide were tested (see table 2).

The supernatant liquor was then tested to determine respirometric COD and BOD_5 (on 100 ml of sample) in accordance with the Standard Methods. The ratio of BOD_5 to COD was used as the basic parameter to determine to optimal concentration of reagents to be used in oxidation with Fenton's reagent.

TABLE 2. Experimental results

Iron (mg/l)	Peroxide (mg/l)	BOD_5/COD	COD effluent
0	0	0.17	10500
300	600	0.27	7650
400	800	0.28	7320
400	1000	0.32	7170
500	1000	0.32	7100
600	1000	0.36	7000
600	1200	0.37	6750
600	1500	0.30	6400

Treatment of wastewater from paper mills. A second series of experimental tests were carried on the condensate obtained through the concentration of black liquor, a by-product of chemical paper processing (sulfite pulp) in a multiple-effect plant.

The condensate, whose percent composition is reported in table 3, cannot be subjected directly to biological treatment as it proves somewhat refractory to bio-oxidation. The considerable difference between the values of BOD_5 and COD means a low level of biodegradability: the average COD measured was 4600, while BOD_5 was about 1000.

TABLE 3. Average composition of condensate

Parameter	Value %
Methanol	12
Acetic acid	71
Formaldehyde	6
Formic acid	1
Furfural	10

In order to increase the BOD_5/COD ratio, a process was thus devised involving combined chemical and biological treatment. This includes chemical pre-treatment with Fenton's reagent in order to lower the COD of the condensate and increase its BOD_5, thus making it possible to obtain a product with a higher degree of biocompatibility suitable for treatment by means of an activated sludge process. One advantage of applying Fenton's reagent to the condensate is the fact that the latter is already in itself extremely acidic (pH = 3) and can thus be subjected to the action of Fenton's reagent in its original state with no need for the preliminary acidification required in the treatment of other forms of industrial wastewater.

The experiment involved the initial preparation of 100 ml samples of condensate, which were then subjected to the test with Fenton's reagent. The tests differed with regard to the concentrations of bivalent Fe, introduced in the form of a salt ($FeSO_4$-$7H_2O$), and hydrogen peroxide. Experimental results are shown in table 4.

TABLE 4. Experimental results for the tests on pulp mill condensate

Fe^{++} (mg/l)	H_2O_2 (mg/l)	COD Removal (%)	BOD_5/COD
0	0	---	0.22
150	350	9.6	0.23
300	600	14.6	0.37
600	1000	15.9	0.35
800	1400	21.3	0.31
1000	1650	28.9	0.30
1200	1800	32.4	0.31

Conclusions. The experimental results, for tannery pretreatments wastewater and pulp mill wastewater, confirm that the hydroxyl radicals provide an effective enhancement of the recalcitrant compounds. In each case it can be verified an optimum of Fenton's reagent concentration to reach the maximum enhancement in the biodegradability. In the treatment of tannery wastes, though large amounts of reagent concentration leads to an higher COD removal and a correspondent higher BOD_5/COD ratio, in the tests carried out with concentration of peroxide higher than 1000 mg/L, the improvement does not justify further adding of reagents. Concentration of 600 mg/l of bivalent iron and 1000 mg/l of hydrogen peroxide was found optimal.

For the treatment of pulp mill condensate an optimal concentration value was found at 600 mg/l of peroxide and 300 mg/l of bivalent iron: for larger amount of reagents, while the COD removal increased, the BOD_5/COD ratio showed a slow decrease.

The BOD_5/COD ratio reached in both cases was next to the correspondent common value for the municipal wastewater, therefore a chemical pretreatment by Fenton's oxidation can be successfully realised before adding industrial wastewater to municipal treatment biological plants.

REFERENCES

Bae, J.H., S.K. Kim, and H.S. Chang. 1997. "Treatment of Landfill Leachates Ammonia Removal via Nitrification and Denitrification and further COD Reduction via Fenton's Treatment Followed by Activated Sludge." - *Wat. Sci. Tech.* 36 (12): 341-348.

Bower Carberry, J., and S. Yao Yang. 1994. "Enhancement of PCB Congener Biodegradation by Pre-Oxidation with Fenton's Reagent." *Wat. Sci. Tech.* 30 (7): 105-113.

Cecen, F., W. Urban and R. Haberl. 1992. "Biological and Advanced Treatment of Sulfate Pulp Bleaching Effluents." *Wat. Sci. Tech.* 26 (1-2): 435-444.

Davila, B., F. Kawahara, and J. Ireland. 1994. "Combining Biodegradation and Fenton's Reagent to Treat Crerosote-contaminated Soil." *U.S. EPA, Cincinnati, OH 45221, Rock S. and Vesper S.J., Department of Civil and Environmental Engineering, University of Cincinnati.*

Koyama, O., Y. Kamagata, and K. Nakamura. 1994. "Degradation of Chlorinated Aromatics by Fenton Oxidation and Methanogenic Digester Sludge." *Water Research.* 28 (4): 895-899.

Talinli, I. 1994. "Pretreatment of Tannery Wastewaters." *Wat. Sci. Tech.* 29 (9): 175-178.

Tünay, O., D. Orhon and I. Kabdasli. 1994. "Pretreatment Requirements for Leather Tanning Industry Wastewaters." *Wat. Sci. Tech.* 29 (9): 121-128.

Wang, T., C.K. Waite and W.J. Cooper. 1994. "Oxidant Reduction and Biodegradability Improvement of Paper Mill Effluent by Irradiation." *Water Research.* 28 (1): 237-241.

FATE OF PATHOGENIC MICROORGANISMS IN SOIL

Sandip Chattopadhyay (NRC, Kerr Environmental Research Center, Ada, OK)
Robert W. Puls (U.S. EPA, Ada, Oklahoma)

ABSTRACT: In order to forecast the effect of viruses contaminating the ground water supply, sorption of pathogens on soil and subsurface materials was studied. Considering that change in free energy for the process (ΔG) is directly proportional to the degree of sorption, a model has been developed using thermodynamic principles to predict sorption and to elicit information about the forces involved in the process.

INTRODUCTION

Artificial recharging of groundwater with wastewater seems to be attractive as the demand for groundwater increases. However, it is essential to ensure the safety of water quality by preventing contamination with disease-causing microorganisms. Therefore, a knowledge of the fate of pathogens in soil and aquatic environments is required to assess the necessity of disinfecting the groundwater to balance the benefits and the hazardous effects of disinfection. The present model is based on a mechanistic study of sorption, which is one of the controlling processes that lead to colonization of microorganisms at solid-water interfaces, and therefore, dictates the fate of pathogens in environment.

Thermodynamic principles have been used to derive information about the "non-specific" forces that lead to sorption of microorganisms on soil. The total force leading to sorption is assumed to be the summation of hydrophobic (ΔG^H) and electrostatic (ΔG^{EL}) forces.

$$\Delta G = \Delta G^{EL} + \Delta G^{H} \quad (1)$$

The relative importance of these forces is dictated by the nature of sorbent and sorbate at a particular set of environmental conditions. The sign and magnitude of ΔG also predict the feasibility of the process and estimate the relative degree of sorption between two or more sorbate-sorbent combinations.

Presently, we have investigated the forces that lead to the interactions between bacteriophages and clay particles, which are negatively charged colloidal particles. Colloids are particles that range in size from 1 nm to 10 μm typically. The study demonstrates that ΔG is dictated by three sets of variables, which are water contact angles (θ) of the sorbent and sorbate, their respective sizes and zeta potentials (ζ). Determination of these properties will enable us to predict ΔG.

MATERIALS AND METHODS

Three bacteriophages were selected as surrogate viruses, and their properties are shown in Table 1. Three phyllosilicates and a clay fraction of a landfill site (Norman, Oklahoma) were used as model sorbents. The phyllosilicates were kaolinite (KGa-2), hectorite (SHCa-1), and saponite (SapCa-1), and they were obtained from the Source and Special Clays Repository of the Clay Minerals Society (MI). Size fractionation of the landfill soil samples was conducted to collect the clay fraction, which contained an array of phyllosilicates and organic matter. X-ray diffraction measurements of landfill clays identified the presence of smectites, illite, kaolinite, mica, and quartz. The total organic carbon content, as determined with elemental pyrolysis, was found to be 0.18±0.06%. All clay samples were suspended in Millipore® water and a background electrolyte concentration of 0.01 M NaCl was maintained for all samples. The hectorite and saponite samples were Na-exchanged (Chattopadhyay, 1997).

TABLE 1. Properties of the selected bacteriophages.

	T-2	MS-2	ϕX-174
Bacteriophages			
Host bacterium	*E. Coli* 11303-B2	*E. Coli* 15597-B1	*E. Coli* 13706-B1
Type	tailed, double-stranded DNA	single-stranded RNA	single-stranded DNA
Nucleic acid content (%)	48	31	26
Dimensions (nm)	ϕ 65 × 95	24 - 26	25 - 27
Isoelectric points	4.2 (Sharp et al., 1946)	3.9 (Dowd et al., 1998)	6.6 (Dowd et al., 1998)

TABLE 2. Properties of the clays.

Clays	CEC (cmol kg^{-1})	BET Surface Area (m^2 g^{-1})
Hectorite (SHCa-1)	89.2	93.4
Saponite (SapCa-1)	80.4	34.6
Kaolinite (KGa-2)	3.3	14.11
Norman clay fraction	7.6	92.30

Electrophoretic mobilities (EM) of clay suspensions were measured for pH-values ranging from 3.5 to 11.5. The EM-values were determined with ZetaPlus (Brookhaven Instruments Corporation) which measured the velocity of charged, colloidal particles in liquids. The instruction manual of the instrument ZetaPlus was followed for the above experiments. The above instrument also calculated the ζ from solution conditions and particle mobilities.

The amounts of bacteriophages sorbed on clays were determined from the difference in the numbers of viable bacteriophages (expressed as numbers of plaque-forming units or PFU) present before and during the sorption experiments in the clay-bacteriophage suspensions. Assay of bacteriophages was conducted by following the procedure described by Adams (1959). Measurements were made over a period of 5 days at an interval of 24 hours. The initial amount of bacteriophages added to each clay suspension was 10^6 PFU, and the clay concentrations ranged between 6 to 10 mg mL^{-1} of suspension.

Particle size analyses of the clay samples, before and after the sorption experiments, were conducted by Autosizer 2C (Malvern Instruments, UK) using the principles of Photon Correlation Spectroscopy. Proper care was taken in sample preparation (dispersion, temperature) for particle size analyses to achieve higher percentage (> 85) of measured signal contained in the time span of the correlator.

Data Analyses. The magnitudes of ΔG^H and ΔG^{EL} were determined to obtain ΔG, with the assumption that ΔG is directly proportional to the degree of sorption. Surface hydrophobicities of bacteriophages and clays dictate the magnitude of

ΔG^H, which is a function of interfacial tensions (γ) existing in a system. A thermodynamic equation similar to that proposed by Chattopadhyay et al. (1995) has been developed to determine ΔG^H. Considering the different γ present before and after a bacteriophage sorbs on a clay particle, ΔG^H can be expressed as:

$$\Delta G^H = \gamma_{bc} - \gamma_{bw} - \gamma_{cw} \quad (2),$$

where "bc", "bw", and "cw" represent the interfaces between bacteriophage and clay, bacteriophage and water, and clay and water, respectively. Before sorption the sorbate and the sorbent particles were in contact with water only. After the bacteriophage attached on the clay particle, γ_{bw} and γ_{cw} were replaced by γ_{bc} at the point of contact. The above-mentioned γ-values can be determined with empirical equations, such as those proposed by Gerson (1982). Knowledge of θ_c (water contact angle on clays) and θ_b (water contact angle on bacteriophages) along with γ_{wv} (surface tension of water = 72.75 mJ m^{-2}) were necessary to obtain the solutions of the following equations (Gerson, 1982):

$$\frac{\gamma_{12} + \gamma_{13} - \gamma_{23}}{2\sqrt{\gamma_{12}\gamma_{13}}} - \exp\{\gamma_{23}(0.000065\gamma_{13} - 0.01)\} = 0 \quad (3),$$

$$\frac{\gamma_{12}(1 + \cos\theta)}{2\sqrt{\gamma_{12}\gamma_{13}}} - \exp\{(\gamma_{13} - \gamma_{12}\cos\theta)(0.000065\gamma_{13} - 0.01)\} = 0 \quad (4).$$

The above set of equations can be used to determine the γ-values between any two of the three phases (1, 2, and 3). Therefore, equations (2), (3), and (4) demonstrate that ΔG^H is a function of θ_c and θ_b only.

The magnitude of ΔG^{EL} depends on the surface charge of the interacting particles, and expressions for ΔG^{EL} can be obtained by solving the linearized Poisson-Boltzmann equation. In general, bacteriophages are shaped as spheres and clay particles behave as platelets. Furthermore, because of the size difference between the bacteriophages and the clay particles, the sorbates view the sorbents as flat plates. This justifies the use of the equation (shown below) proposed by Suresh and Walz (1996) for interaction between a sphere and a flat plate:

$$\Delta G^{EL} = \frac{16\varepsilon}{R}\left(\frac{kT}{e}\right)^2 \tanh\left(\frac{e\psi_{01}}{4kT}\right)\tanh\left(\frac{e\psi_{02}}{4kT}\right)e^{-\kappa h} \quad (5).$$

In the above equation, ε is permittivity of water, R is radius of bacteriophage, k is Boltzmann's constant, T is absolute temperature, e is charge of an electron, κ is inverse Debye length (Debye length is the thickness of double layer), h is the shortest distance between the interacting particles (assumed to be 1.5 Å), and ψ_{01} and ψ_{02} are the surface potentials on the interacting particles. Equation (5) has been adapted for the present study by substituting the values of the different constants (ε, k, T, e), and expressing ψ_0 as a function of ζ, as $\psi_0 = \zeta(1+z/\alpha)e^{z\kappa}$ where z is the distance between the surface of charged particle and slipping plane (assumed to be 5 Å), and α is particle Stokes radius. The modified expression is:

$$\Delta G^{EL} = \frac{69}{R}\tanh\left(11.66732\zeta_1\left(1+\frac{5}{r}\right)\right)\tanh\left(11.66732\zeta_2\left(1+\frac{5}{R}\right)\right) \quad (6).$$

Therefore, ΔG^{EL} is a function of R, r (radius of clay particles), ζ_1 (ζ on clay particles), and ζ_2 (ζ on bacteriophages) only. Combining the equations for ΔG^H and ΔG^{EL}, an analytical expression for ΔG can be obtained in terms of θ, ζ, and sizes of the sorbates and sorbents.

RESULTS AND DISCUSSION

Measurement of EM of the clay suspensions as a function of pH clearly demonstrated that under normal conditions (pH $\geqslant$ 6), the selected clays were negatively charged (Figure 1). Furthermore, in all cases, the negative charge on the clay particles increased with increase in pH of the suspension. At pH 6.0, the negative charge on saponite was highest, followed by kaolinite, hectorite, and Norman clay. Saponite has higher charge due to its large isomorphically substituted negative charge on tetrahedral layer. The low negative charge on the Norman clay particles can be due to its heterogeneity, or presence of organic matter.

Based on the IEPs of the selected bacteriophages (Table 1), it was evident that these bacteriophages are negatively charged under natural pH-conditions. Despite interaction between two negatively charged colloidal particles under normal pH-conditions, a high degree of sorption of the selected bacteriophages was observed on all the clays (Figures 2a and 2b). On comparing the degrees of sorption of bacteriophages on clays (Figure 2a), we observed that the percentage of the initial amount of bacteriophage sorbed on clays ranged from 97% to 88% for T-2, 94% to 85% for MS-2, and 93% to 84% for ϕX-174. Among the selected clays, maximum sorption was observed on hectorite, which might be due to its high CEC and BET surface area.

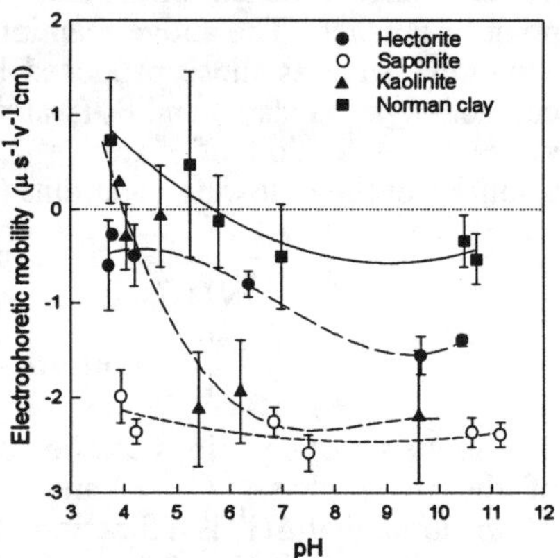

FIGURE 1. Change in EM of clay suspensions with pH.

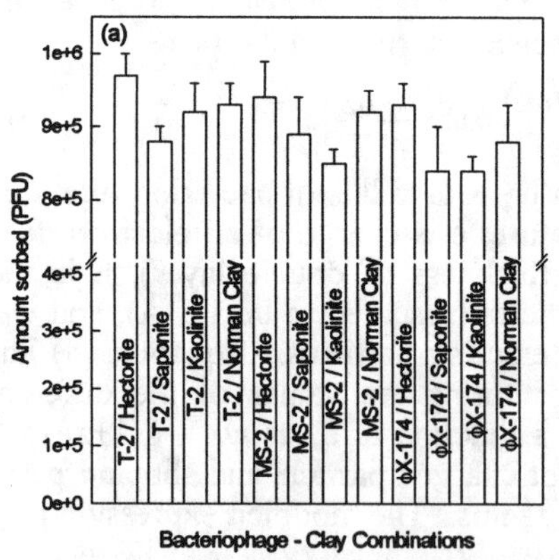

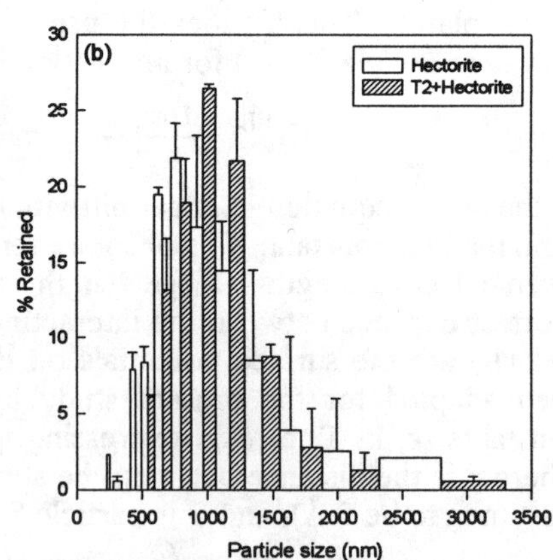

FIGURE 2. Evidence of sorption. Results obtained from (a) batch sorption experiments, and (b) particle size distribution.

However, despite the low CEC of the Norman clay fraction, large quantities of bacteriophages sorbed on Norman clay. Therefore, sorptive capacities of the sorbents were dependent primarily on their BET surface areas rather than on their ion exchange capacities. Among the selected bacteriophages, more T-2 sorbed on the selected clays, followed by MS-2 and ϕX-174. The high sorption of T-2 can be due to its size, shape, and surface hydrophobicities. T-2 is considerably larger than the other bacteriophages selected and this means that more sites are available for sorption. Also, the possibility of a sorbate coming under the influence of attractive forces increases with an increase in the surface area of the particles. The spikes and long fibers on T-2 also help the bacteriophage in attaching to the sorbent surface. Finally, the surface hydrophobicity of T-2 is the highest among the selected bacteriophages (data not presented here), and this is responsible for a higher degree of attractive hydrophobic forces in comparison with the repulsive electrostatic forces.

Particle-size analyses of all clay samples before and after the addition of the bacteriophage suspensions showed shifts towards higher mean sizes in the particle size distribution curve. Figure 2b shows a typical example of the particle size increase of hectorite resulting on interaction with the bacteriophage T-2. The increase in sorbent particle size might be due to sorption of bacteriophages on the clay particles, followed by aggregation of sorbed bacteriophages. The clay particles acted as templates for formation of aggregates of sorbed bacteriophages. It might also be possible that sorption of bacteriophages may lead to coagulation of the clay particles in the suspension. The particle size analyzer was also used in determining the average particle size (diameter) for the clay samples, which were found as follows: 840 nm for Na-hectorite, 1199 nm for Na-saponite, 1326 nm for kaolinite, and 410 nm for Norman clay.

Sensitivity tests were conducted to evaluate the effect of each of the six variables (r, R, ζ_c, ζ_b, θ_c, and θ_b) on ΔG. In each case, the change in free energy occurring over the entire range of the variable ($d(\Delta G)/dx$) was calculated, and this was followed by determination of ($(d(\Delta G)/dx)/\Delta G$ (Table 3).

TABLE 3. Sensitivity Analyses.

x	Δx		$d(\Delta G^1)/dx$	$(d(\Delta G^1)/dx)/\Delta G^1$ (%)		$(d(\Delta G)/dx)/\Delta G$ (%)	
	From	To		From	To	From	To
r	2000 Å	20000 Å	-2.20×10^{-9}	1.203×10^{-5}	1.205×10^{-5}	2.448×10^{-8}	2.448×10^{-8}
R	100 Å	1000 Å	-4.72×10^{-5}	-0.10	-1.05	5.261×10^{-4}	5.236×10^{-4}
ζ_c	30 mV	1 mV	-8.94×10^{-4}	3.33	95.97	9.878×10^{-3}	9.906×10^{-3}
	-1 mV	-60 mV	-8.02×10^{-4}	-86.09	-1.66	8.888×10^{-4}	8.935×10^{-4}
ζ_b	25 mV	1 mV	-7.31×10^{-4}	3.99	97.04	8.085×10^{-3}	8.101×10^{-3}
	-1 mV	-55 mV	-6.60×10^{-4}	-87.58	-1.81	7.313×10^{-3}	7.342×10^{-3}
θ_c	10°	100°	-0.28132	41.41	1.082	42.56	1.082
θ_b	10°	100°	-0.18374	37.45	1.079	38.90	1.080

(Note: Unless varied, the values of the variables (x) are as follows: r = 8000 Å, R = 250 Å, ζ_c=-20 mV, ζ_b = -25 mV, θ_b = 50°, and θ_c = 40°. $\Delta G^1 = \Delta G^H$ for x is θ and ΔG^{EL} for x is ζ and size.)

The following observations were made: (a) attractive hydrophobic forces ($-\Delta G^H$) increased with increase in surface hydrophobicities of clays (θ_c) and bacteriophages (θ_b); (b) considering sorption to be proportional to $(-)\Delta G$, increase in θ_c and θ_b led to increase in sorption; (c) repulsive electrostatic forces (ΔG^{EL}) decreased while sorption increased with increase in particles sizes of both clays and bacteriophages (r and R); (d) sorption decreased more sharply with decrease in positive charge on the particles than with increase in negative charge.

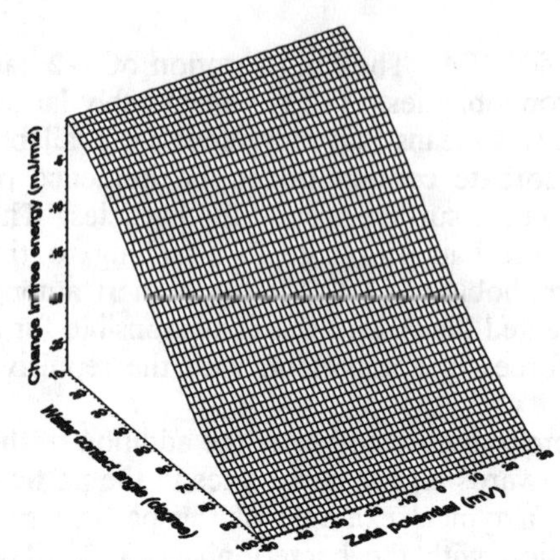

FIGURE 3. Effect of θ and ζ on ΔG.

The 3-dimensional plot of ΔG for a particular bacteriophage demonstrates that θ has a greater effect on ΔG for the selected conditions (R = 250 Å, r = 8000 Å, θ_b = 50°, and ζ_b = -25 mV). It is possible to increase the effect of ζ on ΔG by decreasing the surface hydrophobicities of the sorbate and the sorbent, and increasing the surface charge on the interacting particles.

Therefore, the above study shows that an equation can be constructed to predict the magnitude of ΔG. Such a mathematical correlation is useful in the development of models predicting the fate of pathogens in environment in conjunction with hydro-geologic data.

DISCLAIMER

Although the research described in this article has been funded wholly or in part by the US EPA, it has not been subjected to the Agency's peer and administrative review and therefore may not necessarily reflect the views of the Agency, and no official endorsement may be inferred.

REFERENCES

Adams, M.H. 1959. *Bacteriophages*. Interscience Publishers, New York.

Chattopadhyay, D., Rathman, J., Chalmers, J.J. 1995. "A thermodynamic approach to predict cell adhesion to air-medium interfaces." *Biotech. Bioeng. 48*:649-658.

Chattopadhyay, S. 1997. "A mechanistic study of sorption of ionic organic compounds on phyllosilicates." Ph.D. dissertation, The Ohio State University, Columbus, OH.

Dowd, S.E., Pillai, S.D., Wang, S., Corapcioglu, M.Y. 1998. "Delineating the specific influence of virus isoelectric point and size on virus adsorption and transport through sandy soils." *Appl. Environ. Microb. 64*(2):405-410.

Gerson, D.F. 1982. "An empirical equation-of-state for solid-fluid interfacial free energies." *Colloid Poly. Sci. 260*:539-544.

Sharp, D.G., Hook, A.E., Taylor, A.R., Beard, D, Beard, J.W. 1946. "Sedimentation characteristics and pH stability of the T-2 bacteriophage of *E. Coli*." *J. Biol. Chem. 165*:259-270.

Suresh, L.; Walz, J.Y. 1996. Effect of surface roughness on the interaction energy between a colloidal sphere and a flat plate. *J. Colloid Inter. Sci. 183*:199-213.

FIELD DEMONSTRATION OF A PERMEABLE BIOREACTIVE BARRIER FOR LEACHATE CONTROL

Robert M. Greenwald (HSI Geotrans, Freehold, New Jersey)
Peter A. Rich (HSI Geotrans, Sterling, Virginia)
L. Donald Ochs (Regenesis , Cinnaminson, New Jersey)

ABSTRACT: A biostimulation technique was selected for a pilot study to treat groundwater at a former New Jersey municipal landfill. The process will address a mixture of chlorinated compounds (e.g. vinyl chloride and chlorobenzene), hydrocarbons (e.g. benzene and methyl ethyl ketone) and heavy metals. The reactive barrier will be constructed using a series of injection points to implace a slurry of Oxygen Releasing Compound (ORC®) to enhance both biological and chemical oxidation of the chemicals of concern through the slow release of oxygen. ORC has been shown to be effective in the treatment of petroleum hydrocarbons (Koenigsberg et al., 1995) as well as some chlorinated hydrocarbons such as vinyl chloride (Dooley et al.,1998). Two mini-barriers, each 90 feet long, were installed at the site in January, 1999. Monitoring points were installed along the centerline of the barriers. Data was collected in two baseline sampling events prior to installation of the system. Data will be collected over an eight month period. Additionally, a companion study will be performed to investigate the viability of a reinjectable point system to reduce on-going operating costs. Data will be gathered to determine life expectancy for such a system.

INTRODUCTION

Contamination of groundwater from landfill leachate is a common problem that plagues many former landfill operations. Conventional treatment techniques include groundwater collection and treatment coupled with landfill capping to reduce further problems. This approach requires a large investment in the construction of the collection and treatment system, plus on-going operation and maintenance cost associated with the decades of system operation. Typical contaminants include volatile organics such as benzene and toluene, solvents such as methyl ethyl ketone and acetone, chlorinated hydrocarbons such as vinyl chloride, dichloroethylene and trichloroethylene , as well as soluble heavy metals. A new alternative treatment technology is the use of passive bioreactive barriers that promote accelerated *in-situ* bioremediation of the groundwater. To promote aerobic treatment slow release magnesium peroxide (ORC®) has been successfully used for a wide variety of compounds (Koenigsberg et al., 1995). Several small scale pilot tests have also demonstrated ORC®'s applicability for vinyl chloride (Dooley et al., 1998). This field demonstration was designed to confirm the application of this innovative technology for much larger projects such as a cut-off barrier for dissolved groundwater contamination at a landfill.

Objective. The objective of the field study is to quantify the concentration reductions for the chemicals of concern (VOC's and heavy metals) by the construction of two ORC bioreactive barriers. One is in an area of slightly elevated concentrations (the Western Flank), and the other is an area of higher concentrations (the Central Area). The field study will also confirm delivery of oxygen immediately downgradient of the ORC barriers. In addition, if the process is technically feasible, a detailed engineering cost estimate to refine ORC usage estimates (ORC mass and changeout interval) for a full-scale implementation will be developed, to determine comparative long term project costs compared to a conventional pump and treat system, potentially coupled with an upgraded landfill cap. ORC would serve as a permeable containment barrier at the edge of the landfill, with multiple treatment cycles required to maintain the effectiveness of the barrier. The pilot study will allow the Site-specific performance of ORC to be evaluated for incorporation into a Feasibility Study (FS).

Site Description. The site is a former municipal landfill that closed in 1986. It is located in New Jersey adjacent to a creek and associated wetlands. The landfill was capped with soil and clay as part of the closure process. The site is approximately one mile long and one half mile wide. A site map is depicted in Figure 1.

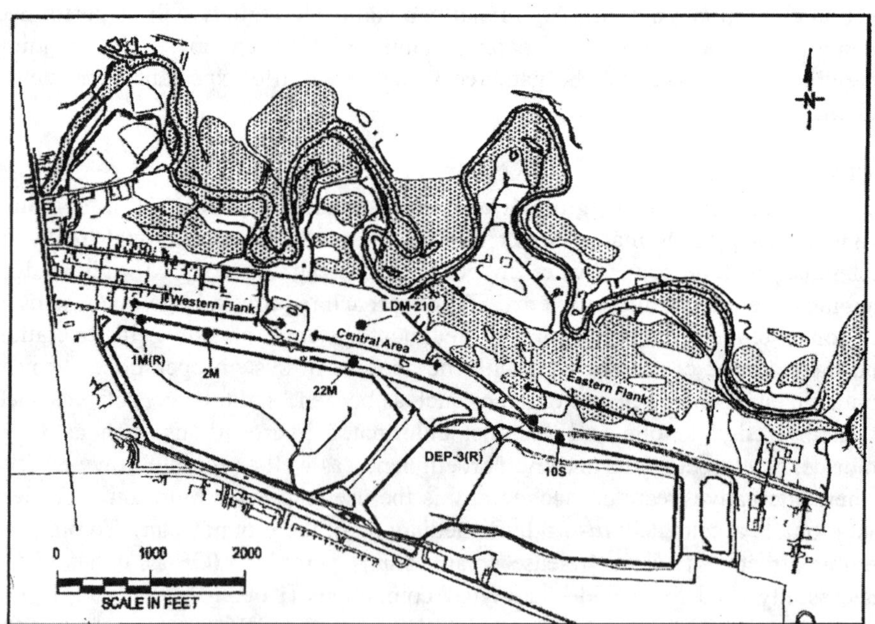

FIGURE 1. Locations of Western Flank and Central Area (Pilot Test Locations)

In the Western Flank Area, the chemicals of concern include benzene, chlorobenzene, ethylbenzene and arsenic. In the Central Area, the chemicals of concern include 1,2 DCA, MIBK, acetone, benzene, chlorobenzene, ethylbenzene, MEK, methylene chloride, vinyl chloride, xylene, arsenic and cadmium. Two aquifers are impacted. Both are located within 50 feet of the ground surface.

MATERIALS AND METHODS

The field demonstration consists of the installation of mini-barriers of ORC slurry in both the Western Flank and Central Areas. ORC is a proprietary formulation of magnesium peroxide (MgO_2) designed to release oxygen upon contact with water. ORC can be applied downgradient of landfills to replenish oxygen, which enhances in-situ biodegradation of organics and precipitation of metals (lowering concentrations of metals that are high due to anaerobic conditions). Due to its slow rate of oxygen release, ORC only needs to be re-introduced every six months to a year. Conceptually, an ORC barrier can be implemented to completely remediate groundwater at the barrier, or to reduce contaminant concentrations at the barrier such that acceptable concentrations are achieved at a compliance point downgradient (due to other attenuation mechanisms between the barrier and the compliance point). ORC can be placed in the ground using several techniques. One method is to place the ORC compound in replaceable socks that are hung within wells. Another method is to inject an ORC slurry into a geoprobe boring. The slurry injection method is more practical for a short-term pilot test. Concurrent with the pilot test, a promising alternative that employs a permanent injectable point will be evaluated for the full-scale, long-term system.

ORC has the potential to reduce concentrations of organics and metals in groundwater. By reducing metals concentrations in groundwater, potential impacts to wetlands from groundwater discharge downgradient of the landfill would also be reduced. The magnitude of concentration reduction in groundwater, due to ORC barriers at the landfill perimeter, will be an important factor in comparing ORC to other groundwater remedial alternatives in the FS. The Pilot Test will allow concentration reductions for a comprehensive set of organics and metals, due to ORC, to be quantified in a field setting.

Field Study Layout. To evaluate the effectiveness of ORC with respect to the objectives, the field studies were laid out as shown in Figures 2 and 3. One test is being performed in the Central Area, in the vicinity of well 22M, and one test is being performed in the Western Flank Area, in the vicinity of well 1M(R). A 90-foot ORC barrier is being evaluated in each area. These barriers are considerably smaller in magnitude than a full-scale implementation, but are long enough to provide a reasonable test with respect to field-scale heterogeneity. ORC slurry was injected into geoprobe boreholes beginning on January 18, 1999. In the Western Flank area, these injections are approximately 25-35 feet below ground surface. A 90-foot barrier was constructed with 13 injection points located 7 feet apart, with a total mass of ORC for the barrier of approximately 500 lbs. In the Central area, these injections are approximately 45-55 feet below ground surface. A 90-foot barrier was constructed with 13 injection points located 7 feet apart, with a total mass of ORC for the barrier of approximately 1,400 lbs.

For the pilot test, slurry injection will cost significantly less than drilling traditional 4-inch wells (for hanging ORC socks) or horizontal wells. Markers were left at ORC injection locations so that, if desired, ORC can be re-introduced at the same locations in the future. In the Central Area, a total of 6 wells are being monitored, consisting of existing well 22M, existing well LDM-210, and 4 new 2-inch monitoring wells. In the Western Flank Area, a total of 5 wells are being monitored, consisting of existing well 1M(R) and 4 new 2-inch monitoring wells. In each test, two of the monitor are clustered such that different 10-foot zones of the aquifer are monitored. This will allow patterns of contaminant concentration and oxygen delivery to be assessed with respect to depth.

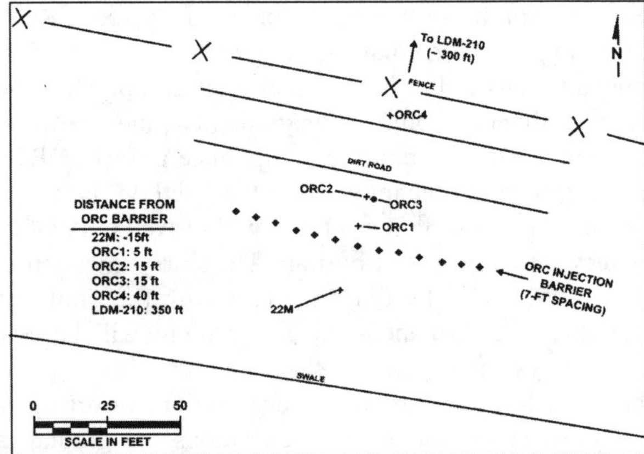

FIGURE 2. Pilot Test Layout: Central Area.

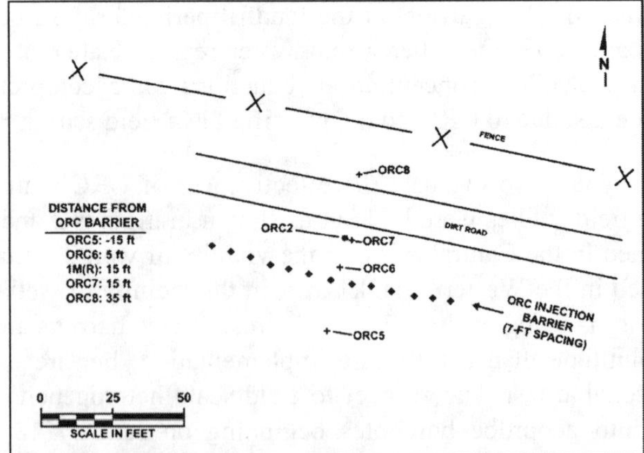

FIGURE 3. Pilot Test Layout: Western Flank.

Groundwater Sampling and Analysis. Table 1 provides a summary of analyses to be performed. Two baseline sampling events were conducted prior to introduction of ORC, during November and December of 1998.. Full rounds of sampling will be conducted in months 1, 4 and 7 (measured from when ORC is introduced). This will allow contaminant concentration reductions due to ORC to be assessed. For other months (months 2, 3, 5, 6) only field parameters (DO, pH, temperature, specific conductance, Nitrate, Sulfate, ORP) and COD will be measured. This will allow an assessment of oxygen delivery on a monthly basis. Water levels will also be measured each month, to allow for refined estimates of hydraulic gradient.

TABLE 1. Sampling parameters and frequency for 10 Pilot Study monitor wells.

	VOCs	TAL Metals	COD	BOD & TOC	DO, Temp, pH, ORP, spec. cond.	Nitrate, Sulfate	Water Levels
Baseline 1 (4 wks before ORC)	X	X	X	X	X	X	X
Baseline 2 (2 wks before ORC)	X	X	X	X	X	X	X
Month 1 after ORC	X	X	X	X	X	X	X
Month 2 after ORC			X		X		X
Month 3 after ORC			X		X		X
Month 4 after ORC	X	X	X	X	X	X	X
Month 5 after ORC			X		X		X
Month 6 after ORC			X		X		X
Month 7 after ORC	X	X	X	X	X	X	X

Project scale-up. Conceptualization of a full-scale ORC implementation in the Central Area is based on two ORC barriers. The first barrier would be near the landfill perimeter, and would reduce concentrations in groundwater such that a second barrier, located down-gradient, would remediate concentrations similar to those in the Flank Area. The Pilot Test incorporates only the near-landfill barrier in the Central Area, and is intended to demonstrate concentration reductions down-gradient of the barrier, but not complete remediation. If successful, the results will be used to design the location and ORC loading rates for a second (down-gradient) barrier. In the Western Flank area, this mass of ORC is intended to simulate a full-scale system.

RESULTS AND DISCUSSION

Because of delays in obtaining regulatory approval for the ORC injection the project was delayed from the early October, 1998 until mid January, 1999. Because of this delay, only the initial field readings are available at the time of publication of this presentation. Field observations indicated no dissolved oxygen in either test area prior to ORC injection. However, immediately after ORC injectionin the Flank Area, dissolved oxygen was detected in the nearest monitor well.

REFERENCES

Koenigsberg, S., C. A. Sandefur and R. D. Norris. 1995 "Management of Dissolved Phase Hydrocarbons with ORC® (Oxygen Releasing Compound)". *Proceedings of the Seventh IGT Symposium Gas, Oil and Environmental Biotechnology.* December 11-13.

Dooley, M, J. Johnson, and W. Murray. 1998. "In-situ Biodegradation of Chlorinated Solvents HRC™ Pilot Scale Results". AEHS Contaminated Soils Conference, Amherst, Massachusetts October 24-26.

POTENTIAL USE OF BIOPOLYMER GROUTS FOR LIQUEFACTION MITIGATION

Dawood Momeni (University of Southern California, Los Angeles, California)
Ramy Kamel, Geoffrey R. Martin, and Teh Fu Yen (University of Southern California, Los Angeles, California)

ABSTRACT: A pioneering research approach into the application of live species of bacteria, *Xanthomonas campestris,* to a loose silty soil to counter possible earthquake induced liquefaction is discussed in this paper. The subject bacterium produces the biopolymer xanthan gum, a well-known microbial polysaccharide of commercial significance. Both absorbtivity and turbidity measurements used to compare bacteria growth in different media indicated that YMTB (Yeast extract Malt Tryptone Broth) attained the highest turbidity among the four carefully selected growth media. The centrifuged cells amount from YMTB media was also the highest compared with the amount of cells separated from other media. The bacteria culture produced from laboratory fermentation of bacteria was added to bulk samples of a silty soil for subsequent tests on compacted samples. Permeability tests and cyclic simple shear tests were performed on the samples to assess any changes in both permeability and liquefaction resistance of silty soil samples.

INTRODUCTION

Liquefaction may occur when loose saturated deposits of sands or silts at shallow depths are subjected to earthquake ground shaking. Many existing light industrial and residential structures are sited on ground conditions where the potential for liquefaction at shallow depths is very high. Current methods of liquefaction mitigation commonly deployed include ground densification and insitu grouting techniques. Both densification and grouting methods are costly. Although chemical grouting is a viable option; chemical permeation grouts are very expensive and suffer from the inability to penetrate silty sands that are commonly encountered in practice.

To overcome the disadvantages of chemical grouts, this study is focusing on the feasibility of using nutrient enriched aqueous solutions of biopolymer forming *Xanthomonas campestris* injected into loose silty soils to increase their liquefaction resistance. *Xanthomonas campestris* is a gram-negative bacterium that grows at temperatures between 28°C and 35°C, is motile by means of a polar flagellum and produces colorless polysaccharide slime on glucose containing media. To establish the optimum growth conditions for the bacteria, several experimental techniques were employed to study the bacteria growth conditions in different culture media and to select the one with the highest amount of cells produced and with the least cost.

After *Xanthomonas campestris* was grown in YMTB solution (the media producing maximum growth), the resulting culture was added to bulk samples of a silty soil. Permeability data were obtained by the standard falling test method and cyclic simple shear tests were performed on both treated and untreated samples to evaluate the liquefaction resistance of the samples.

MATERIALS AND METHODS

Upon the delivery of *Xanthomonas campestris* NRRL B-1459 (ATCC#13951) from ATCC (American Type Culture Collection) it was kept at a freezer with temperature of –70°C at the Microbiology Department of USC for long time preservation and usage. A loop of that bacterium was then cultivated on a petri dish containing either ISP medium 2 (Yeast Extract 4 g/liter, Malt Extract 10 g/liter, Agar 20g/liter, Dextrose 4g/liter) or Nutrient Agar solid media (Pancreatic Digest of Gelatin 5.0 g/liter, Beef Extract 3.0 g/liter, Agar 15.0 g/liter). Those media contained both basic nutritional requirements and energy source for bacteria growth in addition to agar (a complex carbohydrate extracted from seaweed), an ideal solidifying agent for microbiological media. After incubating the plates for two days inside an incubator at the constant temperature of 37° C, each developed colony of bacteria was transferred with a loop to a sterilized tube of containing 10 ml of either YM broth (Yeast Extract 3g/liter, Malt Extract 3g/liter, Peptone 5g/liter, Dextrose 10g/liter) or Nutrient Broth (Pancreatic Digest of Gelatin 5.0 g/liter, Beef Extract 3.0 g/liter) for further cultivation of the bacteria. The broth media components were easily dissolved in water and contained basic essential ingredients for rapid multiplication of bacteria cells. After the tubes were incubated for 24 hrs at a constant temperature of 37° C, they were poured in the sterilized flasks with water solutions of different compositions of carbon sources, nitrogen sources, phosphate sources, various mineral nutrients, and a mixture of cofactors, such as vitamins. For each 100 ml of the solutions in the flasks, on tube of 10 ml broth was added. The flasks were then placed on a water bath shaker for two days at room temperature for optimum cultivation of bacteria in order to generate maximum amount of biopolymers. The choices for the growth media were based on their costs to provide the highest amount of growth with the least possible expenditure. Those choices are shown in Table 1. Culture dishes, test tubes, pipettes, transfer loops, and media were all sterilized at temperature of 121°C in an autoclave for 15 minutes for them to be free of viable microorganisms before they can be used for establishing pure cultures of *Xanthomonas campestris.*

TABLE 1. Selected Growth Media for *Xanthomonas campestris*

Choice 1:	YM-T Broth	
	D-glucose	1.2%
	Yeast extract	0.15%
	Malt extract	0.25%
	Peptone	0.25%
	$(NH_4)_2HPO_4$	0.15%
	K_2HPO_4	0.25%
	$MgSO_4$	0.005%
Choice 2:	Nutrient Broth	8 g/liters
Choice 3:	Trypticase Soy Broth	30 g/liters
Choice 4:	D-glucose	1.2% (weight per cent)

Turbidity Tests. Turbidity is an expression of the optical property that cause light to be scattered and absorbed rather than transmitted in straight lines through the media. Correlation of turbidity with the weight concentration of the suspended matter is difficult because the size, shape, and refractive index of the particulates also affect the light-scattering properties of the suspensions. To compare different growth media, however, turbidity is a good criterion to select the media with the highest concentration of suspended in the culture solution. A highly sensitive nephlometric turbidimeter, Model # 2100 N IS, with the range of 0~1000 NTU manufactured by Hach Instrument was employed to measure turbidity at different times for different media.

Absorbtivity Measurements. Concurrent with the turbidity measurements of different prepared culture media for the bacteria growth; a HP UV spectrometer was used to measure the absorptivity of samples taken from each media at different times.

Cell Mass Determination. Cell mass was selected as the index of *Xanthomonas campestris* growth in different media and was determined as follows: the contents of each flask after 14 days of growth were transferred to 35 mm polysulfone centrifuge tubes and the tubes were placed in an IEC model centrifuge for a period of one hour at the constant speed of 3000 rpm for the cells to be settled. The supernatant from each tube was then decanted; cells were washed with water and centrifuged again for 15 minutes. The added water was discarded and the tubes were then dried in an oven at 105°C to constant weight. The difference between that value and the weight of the empty tube was recorded as the nominal dry weight of bacterial cells.

Falling Head Permeablity Tests. Preliminary experiments were conducted utilizing a silty soil (Bonnie silt), a fine natural soil deposit found in Colorado, directly mixed with fresh culture media containing *Xanthomonas campestris* grown in YMTB media. Based on the Modified Proctor compaction test (ASTM D1557) for Bonnie silt, an optimum saturation content of 13.7% was targeted upon the mixing of bacteria solution with the soil. A blank control sample of Bonnie silt with no bacteria and five samples of mixed Bonnie silt with bacteria were used for permeability tests based on the falling head method (ASTM D5084/EPA 9100) and the results were recorded for a period of two weeks.

Cyclic Simple Shear Tests. One blank sample of Bonnie silt and one mixed with the bacteria culture media were used for undrained strain controlled cyclic simple shear tests to investigate the effects of bacteria on liquefaction resistance. Samples were 66.5 mm in diameter and 17 mm height and weighed 93 g with the final saturation of 13.7 per cent. A constant cyclic strain amplitude of 2.6% was used for testing. Tests were effectively undrained (constant height) and increases in pore-water pressure (vertical stress changes) were measured.

RESULTS AND DISCUSSION

Turbidity Tests. The results of turbidity measurements are shown in Figure 1. It was concluded that two selected media containing TSB (Trypticase Soy Broth) and YMTB (Yeast extract Malt Tryptone Broth) attained the highest turbidities among the four meticulously selected growth media.

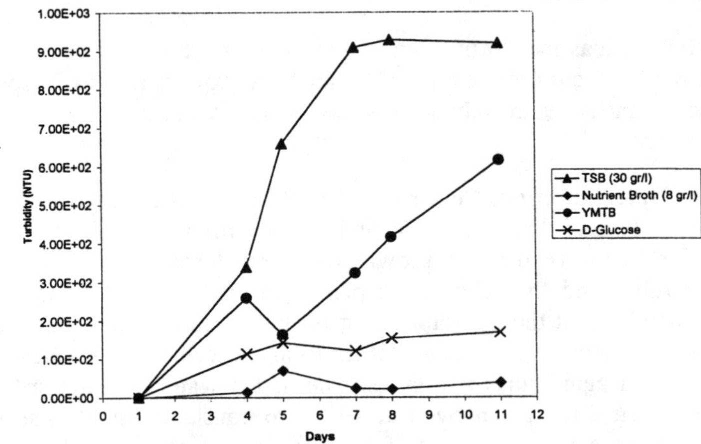

FIGURE 1. Media composition effects on *Xanthomonas campestris* growth.

Absorbtivity Measurements. Figure 2 is a record of the resulting UV analysis for the four selected culture media. It is evident that *Xanthomonas campestris* grew at a faster rate for the media composed of YMTB and for the one with high concentration of TSB content.

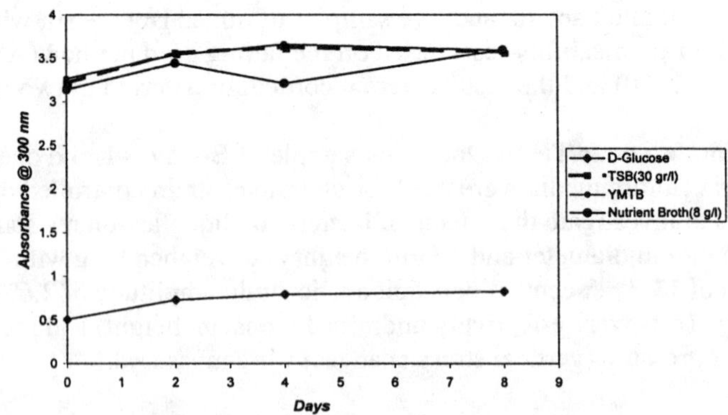

FIGURE 2. UV analysis data for different culture media.

Cell Mass Determination. The results for this part of study are presented in Figure 3. The highest amounts of cells resulted from the media containing YMTB and the one with 30 grams per liter of TSB.

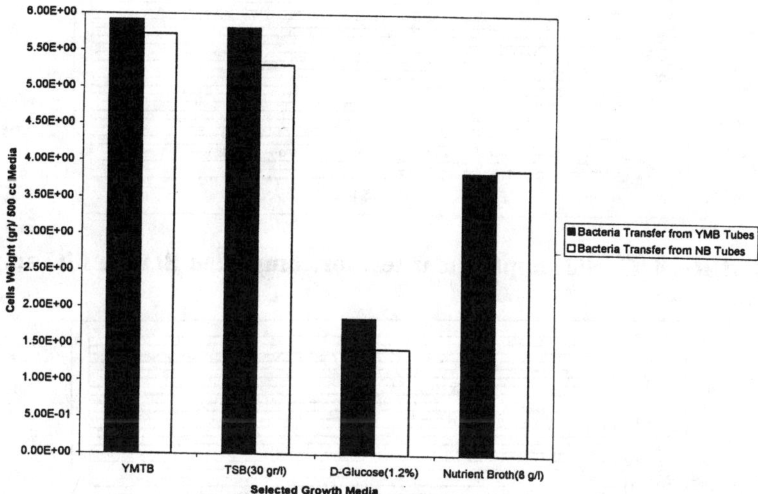

FIGURE 3. Centrifuged cell mass after two weeks of *Xanthomonas campestris* growth in selected media.

Falling Head Permeability Tests. The effects of bacterial plugging in porous media on physical soil properties have been investigated extensively in the past (Gruesback, 1982; Jack et al., 1989; Li et al., (1993)). Our tests were aimed at investigating the effects of the live bacteria solution on the permeability of compacted Bonnie silt. Previous work by Karimi (1997), using compacted samples of Bonnie silt mixed with powdered xanthan gum, indicated reduction in permeability in excess of one hundred fold. The preliminary results from this study showed that the permeability changed from 2.5×10^{-4} cm/sec for the blank control sample to 3×10^{-5} cm/sec for the sample mixed with bacteria solution, a factor of ten reduction. Further tests are needed to establish potential further reductions in permeability upon application of increased concentrations of *Xanthomonas campestris* to the soil.

Cyclic Simple Shear Tests. Preliminary results for this study are shown in Figures 4 and 5. These figures show increases in excess pore pressure ratios (pore pressure/initial vertical effective stress) with time (number of strain cycles).The blank sample liquefied (maximum pore pressure increase) after 5 strain cycles (60 seconds) while the sample with the bacteria liquefied after 7 cycles (85 seconds) This was only a relatively small increase in resistance. As for the permeability tests, further study is needed to examine possible increases of liquefaction resistance with increased bacteria concentration.

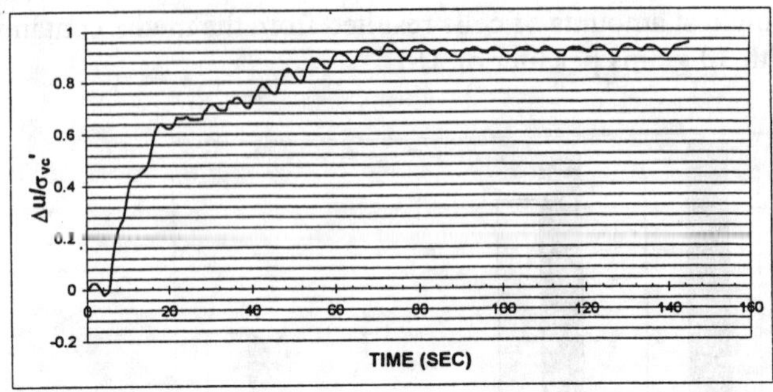

FIGURE 4. Cyclic simple shear test for compacted Bonnie silt sample.

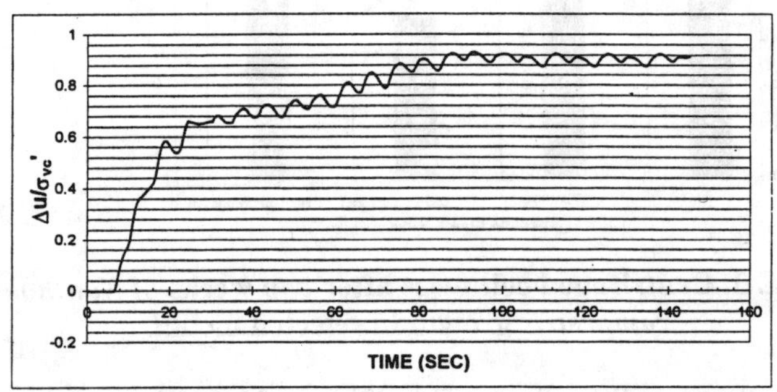

FIGURE 5. Cyclic simple shear test for compacted Bonnie silt mixed with *Xanthomonas campestris* culture media.

ACKNOWLEDGMENTS

The authors wish to acknowledge the financial support of the Pacific Earthquake Engineering Research Center. Teratest Labs (Irvine, CA) and UCLA (Los Angeles, CA) provided laboratory testing facilities for part of this research work.

REFERENCES

Gruesbeck, C., and R.E. Collins. 1982. "Entrainment and Deposition of Fine Particles in Porous Media," *Soc. Pet. Eng. J.*, *22*:847-856.

Jack, T.R., J. Shaw, N. Wardlow, and J.W. Costerton (Eds.). 1989. *Microbial Plugging in Enhancement Oil Recovery*, pp. 125-149. Elsevier Scientific Publishers, Amsterdam, The Netherlands.

Karimi, S. 1997. " A Study of Geotechnical Applications of Biopolymer Treated Soils with an Emphasis on Silt." Ph.D. Thesis, Civil Engineering Department, University of Southern California, Los Angeles, CA.

Li, Y., I.Y. Yang, K.Lee, and T.F. Yen. 1993. "Subsurface Application of *Alcaligenes eutrophus* for Plugging of Porous Media." In E.T. Premuzic and A. Woodhead (Eds.), *Proceedings of the 4th International Conference of Microbial Enhanced Oil Recovery*, pp. 65-77. Elsevier Scientific Publishers, Amsterdam, The Netherlands.

SOLID STATE BIOREACTOR DESIGN FOR LABORATORY SCALE LANDFILL STUDIES

Ann D. Christy, Michael J. Myers, Wendy R. B. Gagliano, Olli H. Tuovinen
The Ohio State University, Columbus, Ohio, USA.

ABSTRACT: This study investigates the use of *in situ* solid state landfill bioreactor design with leachate recirculation to biologically treat municipal solid waste at varying scales, including columns, laboratory bins and, eventually, pilot test cells. The objectives of the study are to evaluate the efficacy of sewage sludge co-disposal in anaerobic landfill bioreactors and to develop a bench-scale protocol to aid in the design and scale-up of landfill bioreactors. During the preliminary microbial acclimation phase described in this paper, temperatures range from 23.79°C to 28.87°C in the bench-scale bioreactor systems.

INTRODUCTION

One of the major challenges facing society today is the effective management of solid waste. The United States generates municipal solid waste (MSW) at an average rate of 176 kg/person/yr, resulting in approximately 40 million tons of waste produced per year (Daniel and Koerner, 1995). Landfilling accounts for over 70% of final solid waste disposal in the United States (Suflita et al., 1992). Many modern landfills are designed as dry vaults (Wall and Zeiss, 1992). They are equipped with top and bottom liners to minimize moisture entry into the system, and leachate and gas collection systems. This design reduces the volume of leachate generated, but also slows biodegradation due to low moisture levels. Converting landfills from dry vaults, which simply store waste, to solid-state bioreactors that promote biodegradation could reduce some of the environmental risks associated with waste disposal.

Recirculation of leachate in landfills can increase moisture content, which can lead to increased biodegradation (Warith and Sharma, 1998; Pohland, 1979). Several laboratory (Barlaz et al., 1989) and field studies (Townsend et al., 1994; Wall and Zeiss, 1992) have shown that leachate recirculation can effectively enhance biodegradation.

Sewage sludge addition as an inoculum can also be used to enhance biodegradation (Warith and Sharma, 1998). Several laboratory scale studies have used small amounts of sewage sludge to seed or accelerate biodegradation (Martin et al., 1997). The use of co-disposal of larger quantities of sewage sludge with solid waste to enhance biodegradation has been studied by several researchers (Reinhart and Townsend, 1997). These studies have shown that co-disposal of sewage sludge increased both the yield and quality of landfill gas produced.

Our research involves a small-scale column study done in conjunction with a larger scale bin study. The larger-scale bin study uses two 1.5 m^3 bins designed to simulate a landfill bioreactor. Both bins contain unshredded solid waste and are equipped with leachate recycle. One bin was amended with sewage sludge at a ratio of 1:5 by weight. The small-scale columns contain shredded solid waste and were also equipped with leachate recycle. The ratio of sewage sludge to solid waste was varied between columns to determine optimal ratios for biodegradation. Four treatments were analyzed, a control with no sludge addition, and sludge addition at a ratio of 1:10, 1:5, and 2:5.

MATERIALS AND METHODS

Laboratory scale columns: Glass columns, 4.5 cm in diameter and 40 cm in length, were packed with shredded solid waste (Figure 1). The solid waste was collected from the working face of a local landfill and fed though a wood chipper to reduce particle size. The shredded waste was mixed with the sludge and packed between top and bottom layers of glass wool and glass beads of two sizes (6 mm and 3 mm). Leachate was collected with 1.2 cm, O.D. butyl rubber tubing connected to the bottom of the column and stored in a 250 ml aspirator bottle. The top of the column was capped with a two-hole stopper. One hole was utilized for leachate recycling done using a 50 ml plastic syringe. The other hole served as a gas outlet using butyl rubber tubing. A collection port for gas sampling was inserted in the tubing, and the total volume of gas generated was collected. Total volume of leachate produced was measured and a 50 ml aliquot was recirculated weekly. Leachate parameters monitored weekly were pH, oxidation-reduction potential (ORP), biochemical oxygen demand, chemical oxygen demand, total organic carbon, total dissolved solids (TDS), and conductivity. Total gas production was monitored, and gas samples were analyzed for CH_4, CO_2, O_2 and H_2.

FIGURE 1. Landfill bioreactor column with leachate recirculation.

FIGURE 2. Landfill bioreactor bins.

Laboratory-scale bins: Each bin has a total volume of two cubic meters with the dimensions 91.4 cm x 182.9 cm x 91.4 cm (Figure 2). The bins were constructed of galvanized 14 gauge steel with a 60 x 78 cm Plexiglas window.

To measure preferential flow, the bottom of the bin was divided into eight sections, each section sloped towards the drain associated with that area. Above the sloped cement bin floor, a 1.52 mm high density polyethylene (HDPE) flexible membrane liner was installed, followed by a geotextile drainage layer. A protective layer composed of 15 cm of gravel was installed to protect the liner and to enhance leachate drainage. The cap design is the reverse of the bottom liner design with 10.2 cm of sand and gravel above the solid waste followed by a 1.52 mm HDPE flexible membrane liner.

Minimizing the heat transfer across the boundaries of the bin is essential to its use as a bioreactor and as a model of a full-size landfill which is insulated by the waste itself. Heat flow into and out of the bioreactor bin is controlled on all vertical sides and the bottom. The insulation consists of 15 cm of fiberglass board insulation with a removable section over the Plexiglas window.

Refuse placed in the bins was obtained directly from the working face of an existing landfill cell. After placement into the bin bioreactors, the refuse was compacted to 324 to 332 kg/m^3. Specific weight for normally compacted refuse in a landfill has been reported to range from 362 to 498 kg/m^3 with a typical value of 450 kg/m^3 (Tchobanoglous et al., 1993).

A gas extraction well was constructed for each bin using 10.2 cm O.D. schedule 40 PVC well screen and riser. The well screen was constructed of three 17.8 cm long sections. Each section represents 18 cm of the vertical profile. Each section is connected to the collection header by 1.6 cm O.D. clear PVC tubing. The gas extraction system has the capability to obtain samples from different depths and to quantify the total rate of gas production.

Leachate from each of the eight PVC drains flows into a 7.6 cm O.D. schedule 40 collection manifold. Leachate from the collection manifold then drains by gravity flow into a flow-through cell where probes are positioned to monitor pH, conductivity, ORP, dissolved oxygen, and TDS. Leachate then flows into a storage reservoir where sodium hydroxide is added as needed for buffering (Figure 3). The leachate is pumped from the reservoir up to an irrigation piping network similar to the herringbone design used in subsurface drainage systems. The network of 1.6 cm O.D. Perforated silicone tubing is located at the upper soil-waste interface.

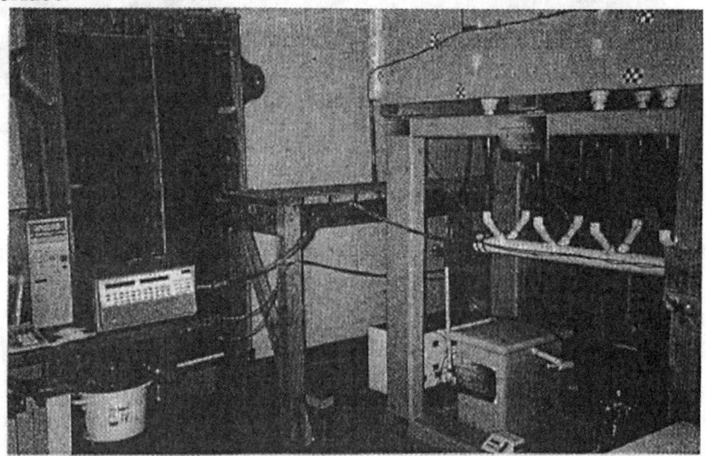

FIGURE 3. Landfill bioreactor bin view showing leachate collection manifold, leachate reservoir, and data acquisition system.

The instrumentation system allows temperature to be monitored on a continual basis. Temperatures within the solid matrix of the landfill material are monitored using Type T thermocouples and logged every hour using a Hewlett Packard 3852 data acquisition system. The thermocouples are housed inside stainless steel sleeves for protection against the corrosive environment. A total of twelve thermocouples are located within the matrix of each bioreactor bin at three different depths.

Load cells located under the four corners of each bin measure weight of the bin and its contents. A scale under the leachate storage reservoir measures weight of leachate awaiting recirculation. Together with gas collection and weighing, these devices allow mass balances to be performed on the bioreactor system.

RESULTS AND DISCUSSION

The moisture content of the laboratory scale bioreactor columns and bins is currently at field capacity, and has been for almost two months. After consumption of oxygen introduced during construction and loading was completed, the bins have achieved anaerobic conditions and appear to be in the initial lag phase or acclamation period of microbial development. This deduction is based on the observation of no appreciable gas production (methanogenesis) occurring yet. The preliminary weights for the two bioreactor systems are 3003 lbs (1362 kg) for the bin containing MSW, and 3104 lbs (1408 kg) for the bioreactor bin containing MSW and sludge. The preliminary temperature results indicate that temperatures range from 25.06°C to 28.87°C in the bioreactor bin containing MSW, and range from 23.79°C to 27.08°C in the bioreactor bin containing MSW and sludge. Although the sludge amended MSW is slightly cooler, the small temperature difference between the two bins is not considered significant at this early stage. Warith and Sharma (1998) reported that temperatures of 30°C to 45°C could be expected in full-size landfills. It is expected that temperatures will rise as the microbial activity increases and moves into the exponential growth phase. We plan to begin recirculation of leachate once methanogenesis is occurring at sustainable rates.

CONCLUSIONS

Although the described research is still in its preliminary stages, it is expected that comparison of results between the column studies and the larger bioreactor bins will allow relationships between parameters at the different scales to be determined. Planning is underway to design and construct pilot scale field test cells at the local municipal landfill. Future work will involve monitoring those cells and comparing the data with results from the laboratory column and bin studies. By comparing across the scales, design scale-up factors can be developed. This will allow bench scale evaluations to be made to aid the design and construction of full-size bioreactor landfills, in a manner similar to existing biotreatability bench studies in the bioremediation field. The benefits of operating landfills as bioreactors include increased capacity and working life of landfills, reduced post closure cost, and increased revenue by accepting sewage sludge. Additionally, landfill bioreactors treat the wastes *in situ*, thereby lessening the risk of contaminant transport off-site, reducing environmental risk, and enhancing public perception of landfills.

ACKNOWLEDGEMENTS

Support was provided by the City of Columbus; the Ohio Agricultural Research and Development Center; and the Department of Food, Agricultural, and Biological Engineering; the Department of Microbiology; and the Soil Science program at the Ohio State University. Thanks are due to Solid Waste Authority of Central Ohio, Laidlaw, Beinhower Brothers Drilling, and American Aggregates for their material contributions.

REFERENCES

Barlaz, M.A., D.M. Schaeffer, and R.K. Ham. 1989. Bacterial population development and chemical characteristics of refuse decomposition in a simulated sanitary landfill. *Applied and Environmental Microbiology* 55: 55-65.

Daniel, D.E., and R.M. Koerner. 1995. *Waste Containment Facilities: Guidance for Construction, Quality Assurance and Quality Control of Liner and Cover Systems.* American Society of Civil Engineers, New York, NY.

Martin, D.J., L.G.A. Potts, and A. Reeves. 1997. Small-scale simulation of waste degradation in landfills. *Biotechnology Letters* 19: 683-685.

Pohland, F.G. 1979. Leachate recycle as a management option. *Journal of Environmental Engineering ASCE* 106: 1057-1069.

Reinhart, D.R., and T.G. Townsend. 1997. *Landfill Bioreactor Design and Operation.* Lewis Publishers, Boca Raton, FL.

Suflita, J., C.P. Gerba, R.K. Ham, A.C. Palmisano, W.L. Rathje, and J.A. Robinson. 1992. The world's largest landfill. *Environmental Science and Technology* 26: 1486-1495.

Tchobanoglous, G., H. Theisen, and S. Vigil. 1993. *Integrated Solid Waste Management: Engineering Principles and Management Issues.* McGraw-Hill, New York, NY.

Townsend, T.G., W.L. Miller, H.J. Lee, and J.F.K. Earle. 1994. Acceleration of landfill stabilization using leachate recycle. *Journal of Environmental Engineering ASCE* 122: 263-268.

Wall, D.K. and C. Zeiss. 1995. Municipal landfill biodegradation and settlement. *Journal of Environmental Engineering ASCE* 121: 214-224.

Warith, M.A., and R. Sharma. 1998. Technical review of methods to enhance biological degradation in sanitary landfills. *Water Quality Research Journal of Canada* 33: 417-437.

BIOREMEDIATION POTENTIAL OF PHTHALATE ESTERS IN SEDIMENTS FROM SUB-ARCTIC AND ARCTIC SITES

Anne G. Rike and Marion Borresen
(Norwegian Geotechnical Institute, Oslo, Norway)
Ove Bergersen and Tormod Briseid (SINTEF, Oslo, Norway)
Carina Lysebo and Elsa Lundanes (University of Oslo, Norway)

ABSTRACT: The effect of cold climate upon degradation of PAEs and the distribution of PAE degrading bacteria in such environments, is not well understood. Cold-climate sediments with different PAEs and anthropogenic loads were examined for the presence of phthalate degrading populations. Degradation with inoculum from the different sites, was conducted under aerobic and denitrifying conditions at low temperature. The results show that PAE degrading populations were present in all the sediments. Aerobic degradation was strongly retarded at low temperature, while under denitrifying conditions the degradation was less reduced.

INTRODUCTION

Phthalate esters (PAEs) comprise a group of compounds that are widely used as industrial chemicals. Approximately 95% of the PAEs are utilised as softeners in polyvinyl chloride (PVC). Altogether 18 PAEs are listed as commercial important compounds (Staples et al., 1997) and the most commonly occurring are classified as priority pollutants in several countries. The annual world production from 1980-1990 is estimated to 2 700 000 tonnes (Furtmann, 1996). Di-(2-ethylhexyl) phthalate (DEHP) is the predominant softener used in PVC production. The aqueous solubility is low (0.003 mg/L) and the compound can be characterised as a light (sp. gravity 0.986), non-aqueous phase liquid (LNAPL). Butylbenzyl phthalate (BBP) is utilised in plastic materials, paintings and jointing products and is a dense (sp. gravity 1.111), non-aqueous phase liquid (DNAPL). Because of the rather high aqueous solubility (1100 mg/L) of di-ethyl phthalate (DEP), this phthalate is not considered as a NAPL.

PAEs are not bound to the polymeric matrix in PVC products and may enter the environment directly through spills during production and after disposal of plastic materials. Phthalates are frequently detected in the environment, in water, soils and sediment samples, as well as in drinking water and food (Staples et al. 1997; Vitaly et al.; 1997; Yin and Su, 1996).

PAEs are generally considered to be biodegradable, but it is reported that degradation is strongly retarded under anaerobic conditions (Shelton et al., 1984; Painter and Jones, 1990), in nutrient poor environments (Rubin et al., 1982) and at low temperature (Ritsema et al. 1989). Neither the effect of cold climate upon degradation nor the distribution of phthalate degrading bacteria in such environment is well understood. A prerequisite for bioremediation of phthalate contaminated sites is that phthalate degrading bacteria are present and that the activity is of practical interest.

In this pilot project, sediments from three sites in Norway (Aalesund, Svalbard and Oslo) with different phthalate and anthropogenic loads, were examined for the presence of phthalate degrading bacterial populations. Primary degradation of DEP, BBP and DEHP by acclimated inoculum from the sites was conducted under aerobic and denitrifying conditions, at low temperature.

Aalesund and Oslo represent sub-arctic sites with seasonal frost. Svalbard represents the arctic site with continuous permafrost. The permafrost is covered with a shallow active layer that thaws each summer (June-August).

At the Aalesund site, 4000 L DEHP were spilled from an above ground storage tank in April 1995. The liquid spread rapidly to the lake Lillevann, where the water and the shore were contaminated. Sediments from the site were used in the experiments. Sediments from the river Alna in Oslo were also examined. The river runs through the industrial part of the city and is influenced by effluents and general anthropogenic activities. On Svalbard sediments were taken from the lake Tvillingvann, the freshwater supply to the research station of Ny-Aalesund.

MATERIAL AND METHODS

The sediment samples were collected the summer 1998 from 2-10 cm below the surface. Temperature and dissolved oxygen were measured in the porewater in the field. Total organic carbon (TOC), pH, specific conductivity and the nutrient concentration of $N\text{-}NH_4$, $N\text{-}NO_3$ and $N\text{-}NO_2$ were determined by laboratory analysis. The concentration of the most common PAEs was determined in the sediments by GC-MS. The samples were screened by GC-MS for the presence of 100 organic priority pollutants.

The most probable number (MPN) technique was used to estimate the aerobic populations of DEP, BBP and DEHP degrading bacteria. Colony forming units (CFU) of total heterotrophs were determined on nutrient agar (NA) after serial dilution of the samples in mineral medium. The incubation temperature was 20°C.

Degradation of DEP, BBP and DEHP under aerobic and denitrifying conditions were performed at low temperature (7 °C and 8-10 °C,) and at 20°C in mineral medium (MM) pH 7.3. The MM with final concentration of 0.5% KNO_3 was used for denitrifying cultures. The sediments were acclimated in the presence of 0.25 g/L of DEP, BBP or DEHP for 7 days in precultures. The time course of degradation was followed in cultures with 0.25 g/L of DEP, BBP or DEHP after inoculation of 1% of the respective preculture. Aerobic cultures were incubated under shaking. All bottles used in the anaerobic experiments were completely filled up with medium and sealed with airtight screw caps with Teflon lining. Duplicate flasks were withdrawn and frozen until analysed.

PAEs in the culture media were determined by reversed-phase high performance liquid chromatography (HPLC). The samples were added ethanol (sample/ethanol, 50/50, v/v). The well-mixed solutions were filtered through 0.45 µm PVDF filters before subjected to HPLC-analysis. Samples containing high concentrations of the PAE were diluted before analysis.

RESULTS AND DISCUSSION

Characterisation of the samples. The freshwater sediments differed with respect to physical and chemical properties (Table 1), and to the exposure of phthalates and organic pollutants (Table 2). Common to the sediments was pore water temperatures about 10°C. The Svalbard temperature was, however, supposed to be higher than normal, because of the unusual warm summer in 1998. The mean air temperature in the summer month normally varies between 4 and 6°C.

In the marshy peat sediments from Aalesund, oxygen was depleted (0.4 mg/L). Three years after the phthalate spill, the DEHP concentration was 12 000 mg/kg and DEP, BBP and di-octyl phthalate (DOP) were detected. The Svalbard sediment consisted of sandy silt with low organic content. The oxygen concentration was 5.2 mg/L. Neither PAEs nor other priority pollutants were detected in the samples from the remote lake. The sediment from Oslo was characterised as alluvial silt. No PAEs, but low concentrations of PAHs were present. The TOC concentration was low in the aerobic sediments (2.3 mg/L oxygen).

TABLE 1. Physical and chemical characteristics of the sediments.

Site	Temp. (°C)	pH	O_2 (mg/L)	Sp. cond. (µS/cm)	TOC (%)	N-NH4 (mg/kg)	N-NO3+NO2 (mg/kg)
Aalesund	11.4	6.0	0.4	353	26.9	1.06	<0.039
Svalbard	8.7	6.0	5.2	70	0.39	0.94	<0.17
Oslo	7.9	7.2	2.3	128	1.7	1.47	0.48

TABLE 2. PAEs and organic priority pollutants in the sediments.

	Phthalates (mg/kg)					PAH (mg/kg)
Site	DEP	BBP	DEHP	DOP	Total	Total
Aalesund	53	29	12 000	32	12 000	<1.6
Svalbard	<0.02	<0.02	<0.02	<0.02	<0.1	<1.6
Oslo	<0.02	<0.02	<0.02	<0.02	<0.1	4.2

The aerobic heterotrophic and phthalate degrading populations in the sediments were measured (Table 3). Phthalate degrading populations were present in all the sediments, but the size of the different populations varied. All the populations were enhanced in the phthalate contaminated sediments from Aalesund, compared to the Svalbard and Oslo samples. The greatest difference was seen in the DEP populations where 3.5 10^5 cells were estimated in the Aalesund sediment, but only 2 and 43 respectively, in the Svalbard and Oslo samples. In Aalesund, PAEs have been present in the sediments for at least 3 years and enrichment of phthalate degrading bacteria have probably taken place.

TABLE 3. Microbial populations in the sediments.

Site	Heterotrophs (CFU/g) NA	DEP	PAE Degraders (MPN/g) BBP	DEHP
Aalesund	6×10^8	3.5×10^5	$>2.4 \times 10^6$	$>2.4 \times 10^6$
Svalbard	1×10^6	2	1.1×10^5	6.1×10^3
Oslo	2×10^6	43	1.5×10^4	1.3×10^5

Degradation of phthalates at low temperature. Primary degradation of DEP, BBP and DEHP under aerobic and denitrifying conditions by bacteria from the acclimated sediment samples were studied (Figure 1).

Under aerobic conditions at 20°C (Figure 1A), 4% or less of the initial phthalate concentration (0.25 g/L) was left after 2 weeks in cultures with inoculum from Aalesund and Svalbard. Despite the acclimation, a typical lag was seen in the BBP cultures with the Svalbard inoculum.

When the temperature was reduced to 7°C the degradation was retarded in all the cultures. After 14 days, 95, 68 and 92 % of the initial DEP, BBP and DEHP concentrations, respectively, were recovered in the cultures with the Aalesund inoculum. In the cultures with the Svalbard inoculum, 80, 106 and 85 % of DEP, BBP and DEHP, respectively, were present at the end of the experiment. The phthalates were not degraded in the cultures with the Oslo inoculum.

Many aerobic bacteria can use nitrate in place of oxygen as a final hydrogen acceptor if the conditions are anaerobic. When organic matter is decomposed in e.g. sediments and oxygen is exhausted as a result of aerobic microbial respiration, certain of these aerobes will continue to respire the organic matter if nitrate is present.

Degradation of DEP, BBP and DEHP in nitrate amended medium, was followed in cultures inoculated with the acclimated sediments. The cultures were incubated at 8-10°C or 20°C and degradation followed for 90 days (Figure 1B). DEP, BBP and DEHP were degraded in decreasing order by denitrifying bacteria at 20°C with the Aalesund inoculum. The degradation was retarded at 20°C compared to aerobic conditions and the longest delay was seen for DEHP. At 8-10°C the phthalate concentrations decreased in all the cultures, except for DEHP inoculated with sediments from Oslo. Under denitrifying conditions less well-defined temperature dependent degradation trends were seen. It is possible that the nature of the inoculum has a stronger influence on the phthalate degradation course under denitrifying conditions as compared to aerobic conditions.

The three sediments examined, all contained bacterial populations able to degrade phthalates. Higher populations were found in sediments previous exposed to PAEs. Degradation under both aerobic and denitrifying conditions was detected. The results from the aerobic experiments showed that the degradation activity was considerably reduced at low temperature in microbial populations adapted to cold climate environments. Under denitrifying conditions low temperature do not seems to reduce the degradation rate to the same extent.

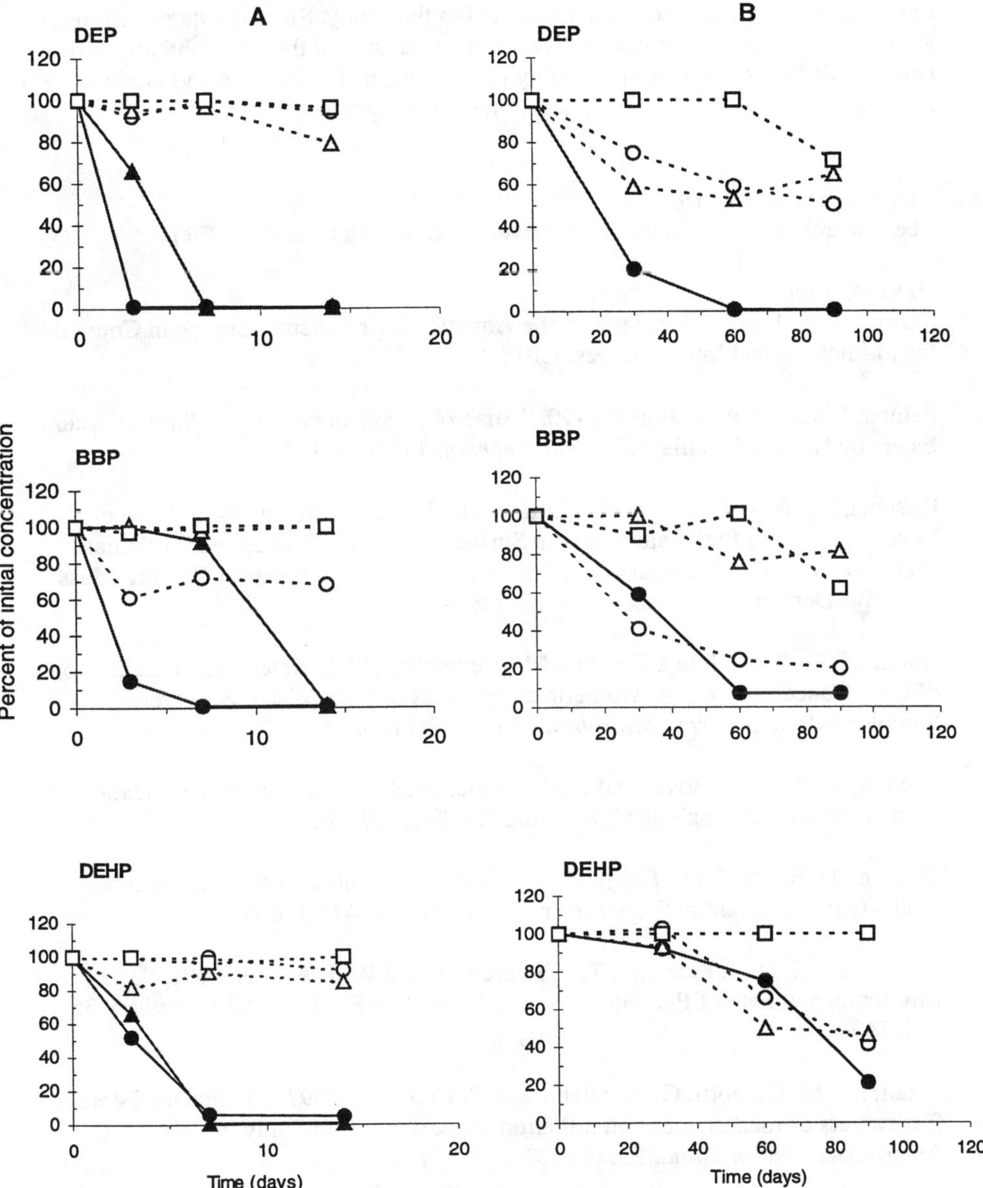

Sites: Aalesund (●), Svalbard (▲) and Oslo (■).
Temperature: 7°C (A) and 8-10°C (B), (---) open symbols and 20°C (A and B), (—)closed symbols.

FIGURE 1. Phthalate degradation under aerobic (A) and denitrifying (B) conditions by sediment inocula from cold climate sites.

Based on the results, we conclude that the potential for bioremediation is present in all the cold climate sediments examined in this study. Since the degradation activity is low, also at summer temperatures measured at the sites, phthalates from spill and diffuse sources will probably persist longer in sub-arctic and arctic regions, compared to temperate, before they are degraded.

ACKNOWLEDGEMENTS
The research was funded by The Norwegian Research Council (NFR).

REFERENCES

Furtmann, K. 1996. "Phthalates in the Aquatic Environment." European Council for Plasticisers and Intermediates, CEFIC.

Painter, S. E. and W. J. Jones. 1990. "Anaerobic Bioconversion of Phthalic Acid Esters by Natural Inocula." *Environ. Technol.* 11: 1015-1026.

Ritsema, R., W. P. Cofino, P. C. Frintrop and U. A. Brinkman. 1989. "Trace Level Analysis of Phthalate Esters in Surface Water and Suspended Particulate Matter by Means of Capillary Gas Chromatography with Electron Capture Mass Selective Detection" *Chemosphere* 18: 2161-2175.

Rubin, H. E., R. V. Subba-Roa and M. Alexander. 1982."Rates of Mineralization of trace Concentrations of Aromatic compounds in Lake Water, Sewage Samples." *Appl. Environ. Microbiol.* 43: 1139-1150.

Shelton, D. R., S. A. Boyd, and J. M. Tiedje. 1984. "Anaerobic biodegradation of phthalic acid esters in sludge." *Environ. Sci. Technol.* 18: 93-97.

Shelton., D. R. and J. M. Tiedje.1984. "General method for determining anaerobic biodegradation potential." *Appl. Environ. Microbiol.* 47(4): 850-857.

Staples, C. A., D. R. Peterson, T. F. Parkerton, and W. J. Adams. 1997. "The Environmental Fate of Phtalate Esters: A Litterature Review." *Chemosphere* 35: 667-749.

Vitali, M., M. Guidotti, G. Macilenti and C. Cremisin. 1997. " Phthalate Esters in freshwaters as markers of contamination sources – A Site Study in Italy." *Environment International* 23: (3) 337-347.

Yin, M. C. and K. H. Su. 1996. "Investigation on Risk of phthalate Esters in Drinking Water and Market Foods" *J. Food and Drug Anal.* 4: (4) 313-318.

INNOVATIVE IN SITU TREATMENT OF ACID MINE DRAINAGE USING SULFATE-REDUCING BACTERIA

Marietta Canty (MSE Technology Applications, Inc., Butte, Montana)

ABSTRACT: Results are presented that were gathered during field-scale testing of an innovative technology, sulfate-reducing bacteria (SRB), designed to treat and control acid mine drainage (AMD). The project was performed under the Mine Waste Technology Program (MWTP), which is funded by the U.S. Environmental Protection Agency (EPA) and jointly administered by the EPA and the U.S. Department of Energy through an Interagency Agreement. The MWTP is implemented by MSE Technology Applications, Inc., located in Butte, Montana. Sulfate-reducing bacteria, a common group of anaerobic bacteria, produce hydrogen sulfide and bicarbonate when supplied with sources of carbon and sulfate. Hydrogen sulfide reacts with metal ions in AMD, precipitating them as metal sulfides; the bicarbonate serves to help neutralize the drainage. A field-scale test is described in which the subsurface workings of an abandoned mine were used as an "in situ biological reactor."

INTRODUCTION

Within this report, results are presented that were gathered during the first 4 years of monitoring the field demonstration of the sulfate-reducing bacteria (SRB) project under the Mine Waste Technology Program (MWTP). The purpose of this field demonstration is to evaluate the ability of SRB to treat and control metal-contaminated wastewaters in situ at the Lilly/Orphan Boy Mine near Elliston, Montana. Data gathered from the onset of the field demonstration through July 1998 is presented within this report.

Acid Generation. Because natural acid generation typically accompanies sulfide-related mining activities, it is a widespread problem. In the United States, 10,000 miles of streams and 29,000 surface acres of impoundments are estimated to be seriously affected by AMD (Hunter, 1989).

AMD results when metal sulfide minerals, particularly pyrite (FeS_2), come in contact with oxygen and water. Acid generation occurs when metal sulfide minerals are oxidized according to the following general overall reaction equation:

$$FeS_2 + 15/4\ O_2 + 7/2\ H_2O <\text{---}> Fe(OH)_3 + 2SO_4^{-2} + 4H^+ \qquad (1)$$

This reaction is one of many that results in increased metal mobility and increased acidity (lowered pH) of the water.

The oxidation of sulfide minerals is accelerated by bacterial action. *Thiobacillus ferrooxidans* is a naturally occurring bacteria that at pH 3.5 or less can rapidly accelerate the conversion of dissolved iron(II) [Fe(II) (ferrous iron)] to iron(III) [Fe(III) (ferric iron)], which can act as an oxidant for the oxidation of FeS_2 (Cohen and Staub, 1992). This bacterial activity may cause up to 80% of the acid production in AMD (Welch, 1980). Ferric ions, as well as other metal ions, and sulfuric acid have a deleterious influence on the biota of streams receiving AMD (Dugan et al., 1968).

Site Description. As an initial step in this project, several mine sites were screened and prioritized according to five selection criteria: physical accessibility, legal accessibility, physical shaft characterization, hydrogeologic characterization, and shaft water characterization (MSE, 1994).

The Lilly/Orphan Boy Mine, located in the Elliston Mining District of Powell County, Montana, was selected as the demonstration site. The mine is situated on a patented claim on Helena National Forest Land approximately 11 miles south of Elliston, Montana. This abandoned mining operation consists of a 250-foot shaft, four horizontal workings, and some stoping. Prior to the field implementation of the SRB technology, the shaft was flooded with AMD to the 74-foot level and discharged approximately 3 gallons per minute (gpm) of pH 3 water from the portal associated with this level. Total aluminum, arsenic, cadmium, copper, manganese, iron, and zinc concentrations; sulfate concentrations; and pH measured in the Lilly/Orphan Boy Mine water are all shown in Table 1.

TABLE 1. Lilly/Orphan Boy shaft water chemistry [milligrams per liter (mg/L)].

As	Al	Cd	Cu	Fe	Mn	Zn	Sulfate (SO_4^{2-})	pH
1.07	9.69	0.33	0.32	27.7	6.21	26.1	277	3

TECHNOLOGY DESCRIPTION

The following section provides a description of the SRB technology available in the literature that was used as the technical basis from which the field-scale testing was designed.

Although the purpose of the field testing was to evaluate the use of SRB to mitigate metal-contaminated wastewaters in situ, other removal mechanisms will also be associated with an organic-based system. Wildeman et al. (1993) lists the removal processes in the following sequence of decreasing priority: 1) exchange of metals by an organic-rich substrate, 2) biological sulfate reduction with precipitation of metal sulfides, 3) precipitation of metal hydroxides, 4) adsorption of metals by ferric hydroxides, and 5) metal uptake by living plants. This last mechanism can be disregarded because plants were not associated with the design.

Biological Sulfate Reduction. Biological sulfate reduction is defined as the chemical reduction of dissolved sulfate by the action of biological processes (Dvorak et al., 1991). Biological sulfate reduction requires SRB (members of the *Desulfovibrio* genus are likely the principal microorganisms involved), dissolved sulfate as the electron acceptor, and a carbon source as the electron donor. Certain environmental conditions, such as an anaerobic environment, a pH of 5 to 8, and a redox potential below -100 mV, are also required for optimal growth (Cohen and Staub, 1992). Sulfate reduction generates hydrogen sulfide, which is then available for reaction with metal ions to form metal sulfides. The formation of metal sulfides, most of which are quite insoluble at a low E_H and a neutral pH, is very rapid.

Biological sulfate reduction improves the quality of acid mine water in four ways. First, the hydrogen sulfide that is produced will react with dissolved metals to form insoluble metal sulfides that will precipitate from solution (Equations 2 and 3). Second, the reaction has a neutralizing effect on the pH of the AMD because hydronium ions are consumed by the reduction of sulfate. Third, this reaction also produces alkalinity in the form of bicarbonate from the oxidation of the organic nutrients. Finally, sulfate is removed from aqueous wastestreams to produce hydrogen sulfide.

$$SO_4^{2-} + 2CH_2O \xrightarrow{SRB} H_2S + 2HCO_3^- \tag{2}$$

$$S^{2-} + M^{+2} \longrightarrow MS, \text{ where } M = \text{metal} \tag{3}$$

Several studies have been performed in recent years to research the process by which SRB can remediate metal-contaminated wastewater. These studies range from bench-scale experiments, such as SRB growth in chemostats, to field applications, such as constructed wetlands. The use of wetlands, or passive mine drainage treatment systems (PMDTS), to treat AMD evolved from the observation that the water quality of AMD flowing through natural sphagnum moss bogs improved. The Tennessee Valley Authority has the most experience in constructing wetlands for the treatment of AMD from coal mines, which are typically aerobic systems designed for iron removal (Brodie et al., 1988). Historically, PMDTS were constructed as shallow ponds resembling natural wetlands focusing on oxidation of metals as an important role in metals removal of these systems (Brodie et al., 1989).

For AMD from precious metal mines, however, an important method of metals removal has become recognized as biological sulfate reduction in the anaerobic zone of a treatment system. Consequently, the focus of PMDTS for precious metal mines is metals removal and generation of alkalinity by biological sulfate reduction through optimization of an anaerobic, reducing environment.

FIELD-SCALE TESTING

The design of the SRB field demonstration involved using the flooded subsurface mine workings of the Lilly/Orphan Boy Mine as an "in situ biological reactor" (see Figure 1). Two platforms were suspended by cables in both sides of the two-compartment shaft 30 feet below the static water level and were secured at the surface. An organic substrate used to nourish the SRB was placed in the shaft and supported by the platforms. In addition, two injection wells were drilled into

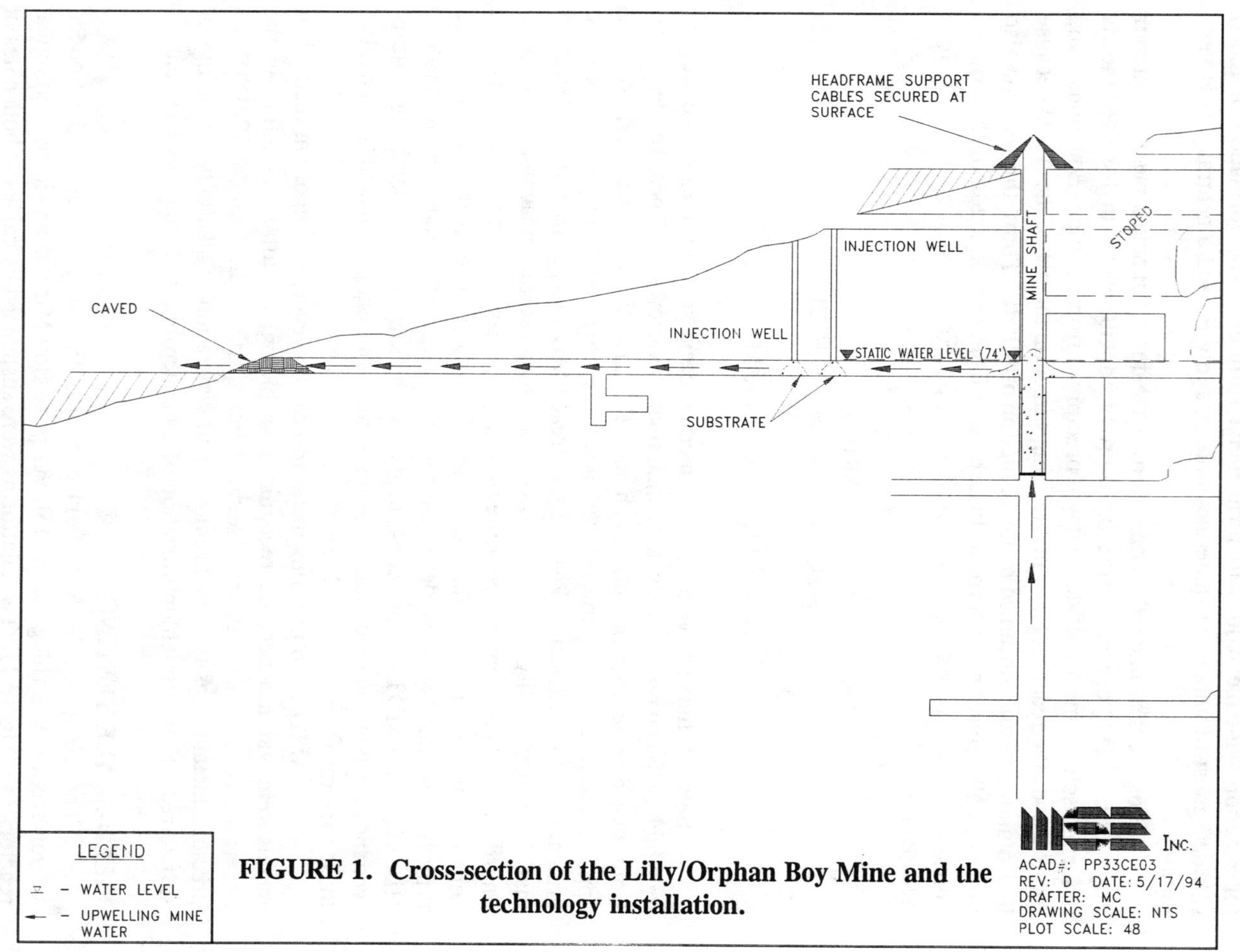

FIGURE 1. Cross-section of the Lilly/Orphan Boy Mine and the technology installation.

the main tunnel of the mine (Lilly Tunnel) that is associated with the portal so the substrate could be placed into this underground space. Consequently, the AMD flows upward through the substrate in the shaft (artesian effects have been observed at this site) and horizontally through the substrate in the tunnel. The biological reaction takes place in the substrate regions, and the treated water subsequently flows out of the mine through the portal. Because the technology causes the shaft water pH to rise and the E_H to fall, the amount of acid generation within the mine is decreased. Monitoring of the field demonstration began once the substrate was placed in late August 1994 and will continue for at least a 4-1/2-year period. This report presents data collected during the first 4 years of monitoring.

The main objective of conducting this field demonstration was to determine the effectiveness of the in situ use of bacterial sulfate reduction in the mitigation of metal contamination. This objective of the field testing is being facilitated by monitoring the SRB system within the mine over the testing period through the collection and analysis of multiple samples, principally metals concentrations in the mine water.

Dissolved metal removal efficiencies and pH data for samples collected from 1) within the horizontal mine tunnel (via an injection well) and 2) at the portal of the mine during the first 4 years of monitoring are presented within the following sections. Data collection included the following parameters: dissolved metals, total metals, alkalinity, physical parameters (flow rate, temperature, dissolved oxygen, pH, E_H), sulfate, sulfide, biochemical oxygen demand (BOD), chemical oxygen demand, and volatile fatty acids. Sampling is generally performed on a monthly basis. A trend can be observed for nearly all analytical parameters in which more desirable results are observed within the tunnel of the mine in comparison to the portal.

Flow Rate. The flow of water discharging from the portal of the mine remained fairly constant during the first 4 years of monitoring with most measurements recorded under 2 gpm. However, spring runoff caused the flow rate to climb to highs of 7 to 8 gpm each spring.

pH. The pH of the mine water prior to the installation of the field demonstration was near 3. After the implementation of the technology, the pH in the tunnel rose to approximately 7 and remained there during the first 4 years of monitoring. The pH at the mine portal rose to approximately 6 after the technology implementation but fell to approximately 3.5 during each spring runoff. After flow rates returned to normal levels (after each spring runoff), the pH at the portal rose to nearly 6. Within the tunnel of the mine, however, the pH remained near 7 during each spring runoff.

Dissolved Metals. Dissolved metals samples were collected on a monthly basis from both the tunnel and the portal and were analyzed for the following parameters: aluminum, arsenic, cadmium, copper, iron, manganese, and zinc. These metals were chosen for analysis based on their concentration in the mine water of the Lilly/Orphan Boy Mine. The dissolved portion of the sample is defined as that which passed through a 0.45-micron filter.

Dissolved metal removal efficiencies measured at the mine portal are presented in Figure 2. Dissolved metals concentrations were compared to background concentrations in the mine water prior to the technology implementation. Aluminum, cadmium, and copper had high removal efficiencies (85% to near 100%) during the first 8 months of the field demonstration. All three of these metals underwent a drastic reduction in metal removal efficiency during the high spring runoff observed in May and June of 1995.

As flow rates returned to normal levels, these metals were again removed, reaching removal efficiencies representative of the months prior to spring runoff. This cycle was repeated during each subsequent spring runoff.

Zinc and manganese were not as effectively removed as aluminum, cadmium, and copper. Prior to the first spring runoff, zinc was consistently removed at approximately 70% while the manganese removal efficiency was less than 20%. Zinc and manganese also underwent a reduction in removal efficiency during each spring runoff. Zinc and manganese were again removed as flow rates returned to normal levels. However, zinc removal has not achieved the efficiency observed prior to the first spring runoff.

After the addition of the organic substrate, the data indicated that dissolved iron concentrations (not shown) actually increased within samples taken from the mine portal in comparison to background data. This trend appeared to be reversed during the high flow rate that occurred during the first spring runoff. An increase in dissolved arsenic (not shown) was observed shortly after the implementation of the technology at the mine as well. Although only a slight increase in arsenic concentration was observed in the tunnel, a large increase was observed at the portal (from near the instrument detection limit of 0.0336 mg/L to a high of 14.7 mg/L). An analysis of iron and arsenic chemistry helps explain these phenomena. Under oxidizing conditions, Fe (III) precipitates as $Fe(OH)_3$ (ferrihydrite), and ferrihydrite precipitation very effectively adsorbs arsenic (Robins and Huang, 1988). However, under reducing conditions, Fe(II) becomes the predominant iron species, which is much more soluble than Fe(III). Fe(II) is then released into solution along with the previously adsorbed arsenic. Fe(III) was reduced to Fe(II), and arsenic was released when the underground workings of the mine converted from an oxidizing environment to a reducing one. The transformation of this underground environment, which was achieved through the addition of organic substrate, was necessary to provide suitable conditions to promote SRB growth.

Increases in iron and arsenic concentrations during spring runoff events indicate a reversal of the conditions imposed by the SRB technology and similar to the historical conditions of the mine. As the flow rate increased in the mine system and the retention time decreased, the SRB were likely not able to produce enough soluble sulfide to precipitate the metals, maintain a low reduction potential, and neutralize the water through bicarbonate production. Consequently, the pH fell, and the reduction potential increased while the system became more oxygenated. Fe(II) oxidized to Fe(III) under these conditions, adsorbing the arsenic and

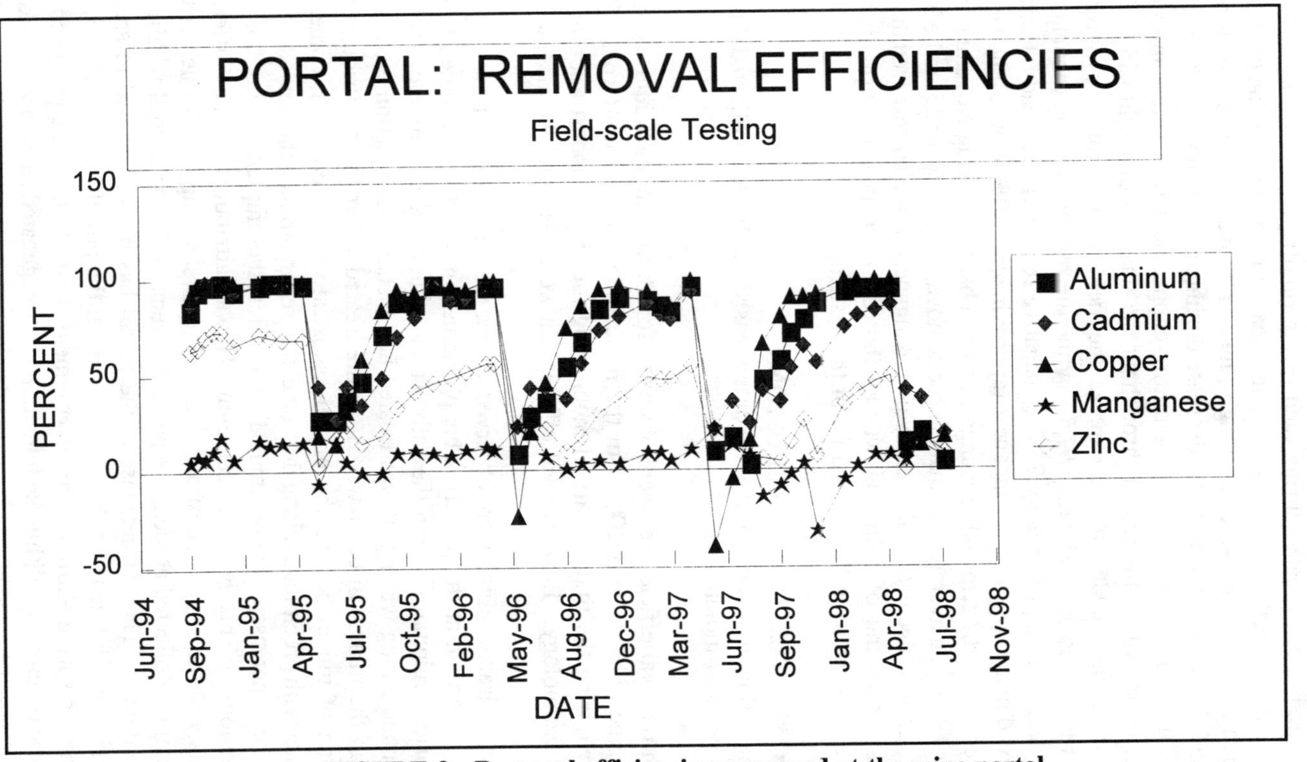

FIGURE 2. Removal efficiencies measured at the mine portal.

precipitating. Conversely, as the system recovered from high spring runoff, the reduction potential decreased, and the pH and iron concentration increased. This cycle appeared to be repeated during each spring runoff.

Dissolved metals removal efficiencies measured within the mine tunnel are presented in Figures 3 and 4. Interestingly, metal removal efficiencies within the mine tunnel did not display a drastic decrease during spring runoff as observed at the portal of the mine. Aluminum, cadmium, copper, and zinc were removed at high efficiencies within the tunnel throughout the duration of monitoring. The removal of arsenic tended to increase as the field demonstration progressed (negative removal to approximately 97%), while the removal of iron (not shown) tended to decrease (approximately 75% to negative removal). A steady decrease was observed in the removal of manganese (approximately 80% to 30%); however, a sharp increase in the removal of manganese was observed during November and December 1995. On average, the manganese removal efficiency tended to be 30% to 40%. A gradual decrease in the removal of manganese was also observed during the pilot-scale testing of SRB and was attributed to the lack of formation of manganese compounds at the pH and E_H of the system.

FIELD-SCALE TESTING CONCLUSIONS

The following conclusions were drawn based on the data collected during the first 4 years of monitoring:

1. High removal efficiencies were observed for aluminum, cadmium, copper, and zinc (70% to near 100%) in the mine water of the Lilly/Orphan Boy Mine during the first 4 years of monitoring the field demonstration of the SRB technology. Low removal efficiencies were observed for arsenic and iron.
2. More desirable results were observed within the tunnel of the Lilly/Orphan Boy mine than at the portal. This phenomenon has been attributed to the historic precipitates within the Lilly Tunnel that act as a source of metals, recontaminating water that was treated in the shaft and injection well areas within the mine. In other words, as the treated water travels through the section of tunnel beyond the organic substrate to the portal, it comes in contact with precipitates along the tunnel that recontaminate the treated water.
3. The pH increased almost immediately after the implementation of the technology. This initial increase in pH was attributed to the buffering capacity of the organic substrate. During the first spring runoff, the pH fell in samples collected at the mine portal but remained near neutral within the mine tunnel. The pH decrease at the portal has been attributed to spring runoff events influencing the water quality at the portal. Higher flow rates during the spring allowed a larger volume of water to be recontaminated by historic precipitates within the tunnel. Also, oxygenated surface water likely penetrated the ground near the portal where there is a short depth to the tunnel. This surface water then solubilized metal contaminants as it passed through the ground and tunnel.

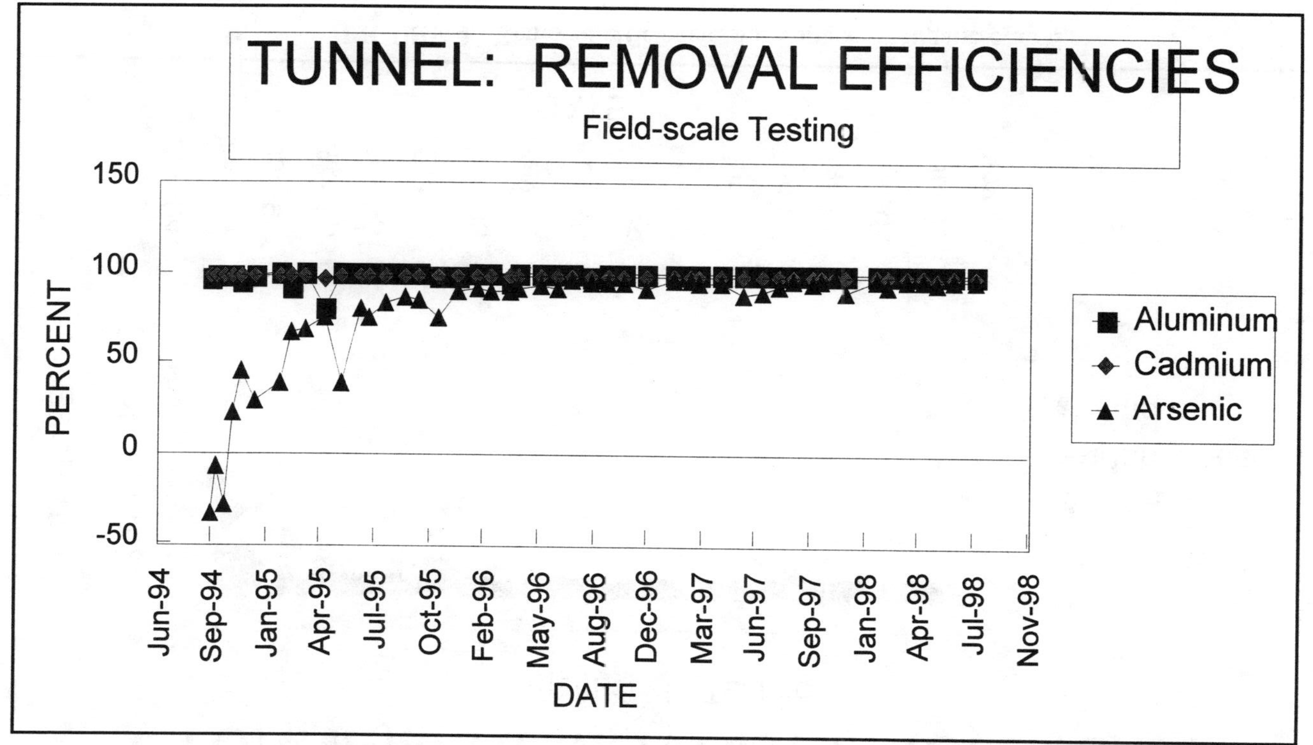

FIGURE 3. Removal efficiencies for aluminum, cadmium, and arsenic within the tunnel.

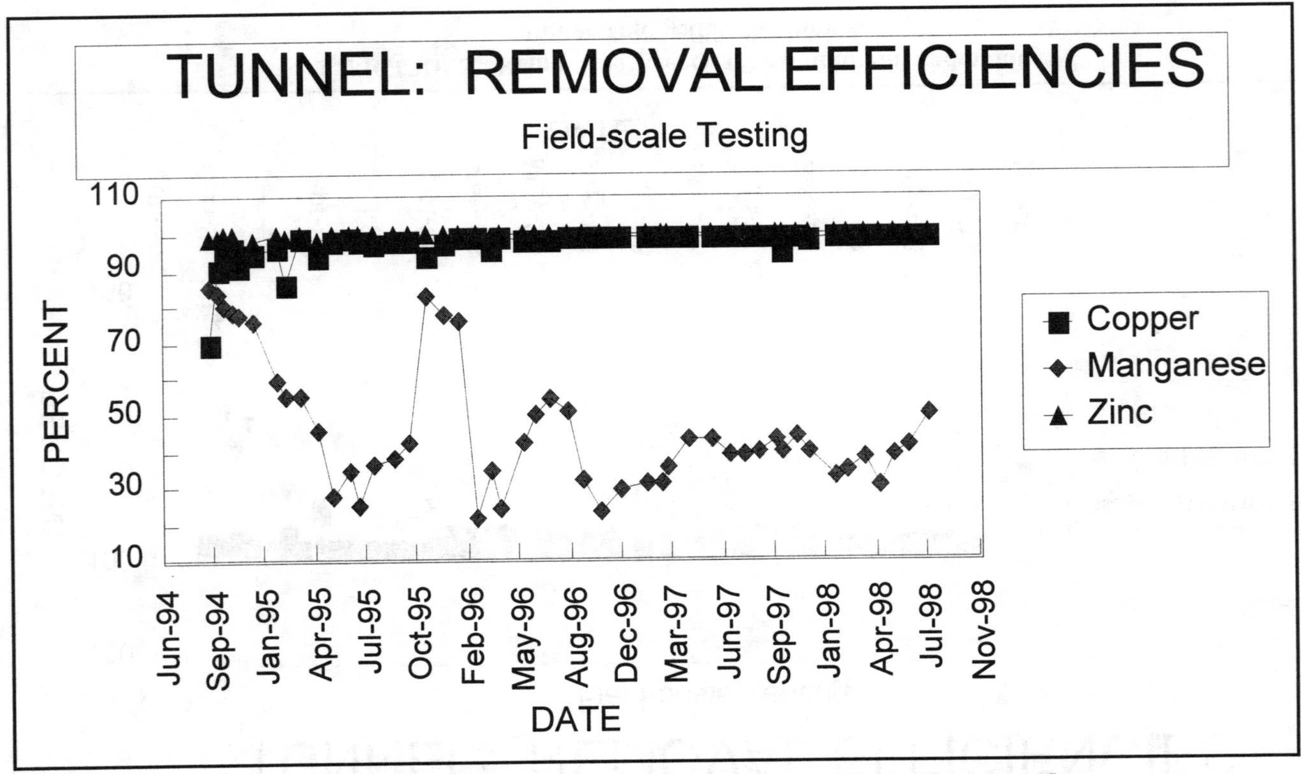

FIGURE 4. Removal efficiencies for copper, manganese, and zinc from the tunnel.

4. During spring runoff, metal concentrations generally rose in samples collected at the portal. Samples collected from within the tunnel did not show an increase in metal concentrations during this time. A large flow of oxygenated surface water running into the portal area likely caused the dissolution of historic precipitates.
5. Sulfate reduction was observed within the underground mine system. Sulfate reduction was evident by measured decreases of sulfate and the detection of soluble sulfide within the mine water.
6. An increase in iron and arsenic concentrations was observed in the portal discharge water shortly after implementation of the technology at the Lilly/Orphan Boy Mine. This phenomenon was explained by the reduction of ferric iron. Historically, ferrihydrite within the system was reduced to ferrous iron when the technology induced a reducing environment within the water in the mine. Arsenic adsorbed to ferrihydrite was released when ferric iron reduced to ferrous. Ferrous iron was also released because of its soluble nature compared to ferric iron.

ACKNOWLEDGEMENTS

Work for this project was conducted through the DOE Federal Energy Technology Center at the Western Environmental Technology Office under DOE Contract Number DE-AC22-96EW96405.

REFERENCES

Brodie, G. A., D. A. Hammer, and D. A. Tomljanovich. 1988. *Constructed Wetlands for Acid Mine Drainage Control in the Tennessee Valley*. Mine Drainage and Surface Mine Reclamation, U.S. Bureau of Mines Information Circular 9183, pp. 235-331.

Brodie, G. A., D. A. Hammer, and D. A. Tomljanovich, 1989. "Treatment of Acid Drainage with a Constructed Wetland at the Tennessee Valley Authority Coal Mine." In *Constructed Wetlands for Wastewater Treatment*, pp. 201-209. Lewis Publishers, Chelsea, MI.

Cohen, R. R. H. and M. W. Staub. 1992. *Technical Manual for the Design and Operation of a Passive Mine Drainage Treatment System*. Prepared for the U.S. Bureau of Reclamation by the Colorado School of Mines, Golden, CO.

Dugan, P. R., C. I. Randles, J. H. Tuttle, B. McCoy, and C. MacMillan. 1968. *The Microbial Flora of Acid Mine Water and Its Relationship to Formation and Removal of Acid*. State of Ohio Water Resources Center, Ohio State University.

Dvorak, D. H., R. S. Hedin, H. M. Edenborn, and S. L. Gustafson. 1991. "Treatment of Metal-Contaminated Water Using Bacterial Sulfate Reduction: Results from Pilot-Scale Reactors." U.S. Bureau of Mines Pittsburgh Research Center, International Conference on Abatement of Acidic Drainage. Montreal, Canada.

Hunter, R. M. 1989. "Biocatalyzed Partial Demineralization of Acidic Metal Sulfate Solutions." Doctoral Thesis, Montana State University, Bozeman, MT.

MSE, Inc. 1994. *Activity I, Volume 1, Appendix C—Issues Identification and Technology Prioritization for Sulfate-Reducing Bacteria*. MSE Technology Applications, Inc., Technical Report, MWTP-15, Butte, MT.

Robins, R. G. and J. C. Y. Huang. 1988. "The Adsorption of Arsenate Ion by Ferric Hydroxide." In *Proceedings of the Arsenic Metallurgy Symposium*, TMS/AIME Annual Conference, Phoenix, AZ.

Welch, E. B. 1980. *Ecological Effects of Waste Water*. Cambridge University Press, New York, NY.

Wildeman, T.; J. Dietz; J. Gusek; and S. Morea, *Handbook for Constructed Wetlands Receiving Acid Mine Drainage*, Risk Reduction Engineering Laboratory, Office of Research and Development, pp. 3-1 to 3-20, 1993.

DESIGN AND CONSTRUCTION OF BIOREACTORS WITH SULFATE-REDUCING BACTERIA FOR ACID MINE DRAINAGE CONTROL

Marek Zaluski, Martin Foote, Kenneth Manchester, Marietta Canty,
Michael Willis, James Consort, John Trudnowski, Michael Johnson,
Mary Ann Harrington-Baker
(MSE Technology Applications, Inc., Butte, Montana, USA)

ABSTRACT: At many abandoned mine sites in the Western U.S., conventional treatment of AMD is not feasible due to the of lack of power and limited site accessibility. Therefore, three bioreactors were built at an abandoned mine site in Montana to demonstrate feasibility of treating AMD using sulphate reducing bacteria (SRB) in a passive water treatment train. The SRB are capable of increasing the pH and reducing the load of dissolved metals in the effluent. The reactors, constructed in the Fall of 1998, were designed to evaluate the SRB technology applied under different environmental conditions. Each bioreactor was designed with mechanisms to enable simulation of seasonal dry and wet climatic conditions. Two bioreactors were placed in trenches and one was constructed above the ground to investigate impact of seasonal freezing and thawing on SRB activity. Two bioreactors contain a passive pretreatment section to increase pH of water before the AMD enters the bioreactor chamber.

INTRODUCTION

An abandoned mine site, called the Calliope Mine, was chosen to demonstrate the remediation of acid mine drainage (AMD) using sulfate-reducing bacteria (SRB) on-site bioreactors. The demonstration is funded by the U.S. Environmental Protection Agency (EPA) and jointly administered by the EPA and the U.S. Department of Energy (DOE) through an Interagency Agreement (IAG). The project was implemented by MSE Technology Applications, Inc., Butte, Montana.

At many abandoned mine sites in the Western U.S., conventional treatment strategies for AMD (e.g., lime neutralization) are not feasible due to the of lack of power and limited site accessibility in winter. Therefore, Calliope site bioreactors were constructed to demonstrate feasibility of treating AMD using SRB in a passive water treatment train that can be applied at remote mine sites.

Principles of the SRB Technology. AMD is a typical result of mining sulfide-rich ore bodies. Acid mine water is formed when sulfide bearing minerals, particularly pyrite (iron bisulfide), are exposed to oxygen and water as described by the following overall reaction (Equation 1).

$$FeS_2 + 15/4\ O_2 + 7/2\ H_2O \dashrightarrow Fe(OH)_3 + 2SO_4^{2-} + 4H^+ \qquad (1)$$

This reaction results in increased acidity of the water (lowered pH), increased metal mobility, and the formulation of sulfate.

SRB are capable of reducing the sulfate to sulfide by using sulfate as a nutrient along with an organic carbon source to produce soluble sulfide, acetate, and bicarbonate ions. The soluble sulfide reacts with the metals in the AMD to form insoluble metal sulfides. The bicarbonate ions increase pH and alkalinity of the water (Equations 2 and 3).

$$SO_4^{2-} + 2CH_2O \text{ -------> } H_2S + 2HCO_3^- \quad (2)$$

$$S^{2-} + M^{+2} \text{ ---> } MS\text{, where } M = \text{metal} \quad (3)$$

Site Features. The abandoned Calliope mine site, located in Silver Bow County, Montana, includes a collapsed adit discharging water into a large waste rock pile. This relatively good quality water percolates through the mine waste and reappears on the surface at the toe of the pile enriched in metals and with the pH of 2.6 on average. This AMD mixes with good quality surface water and accumulates in a pond (Lower Pond). As a result, the pH of water in the Lower Pond is in the range of 3 to 5.5 depending on the surface water and AMD mixing ratio which varies seasonally. Discharge from the Lower Pond affects the water quality of local surface waters. Table 1 includes analytical information on the water quality of the Lower Pond which is also the AMD supply source for the SRB bioreactors.

TABLE 1. AMD Analytical Data and Target Concentrations for Bioreactors' Effluent

Analyte	Concentration (µg/l)		Comments
	AMD	Target Effluent	
Aluminum	5,670	1,000	SMCL*=50 to 200 µg/l
Cadmium	38	5	MCL**=5 µg/l
Copper	1,660	100	MCL=1,300 µg/l
Iron	8,670	1,000	
Manganese	3,030	2,000	
Zinc	7,920	4,400	
Sulfate	165		
pH	3.5 to 5.5	6 to 8	
Zinc	7,890	4,000	

*SMCL = Suggested maximum contaminant level
**MCL = Maximum contaminant level

Bioreactors Layout. All three SRB reactors (denoted Bioreactors II, III, IV) were designed and constructed in parallel (Figure 1) downstream from the Lower Pond, allowing the AMD to be piped to and treated by the respective reactors using gravity flow. The reactors, constructed in the Fall of 1998, were designed to evaluate the SRB technology applied in slightly different environmental conditions. Two bioreactors were placed below grade (Bioreactors II and III) and one above grade (Bioreactor IV). The below ground reactors were built to minimize temperature changes and prevent freezing. The above ground reactor was constructed to evaluate the effect of the cold weather and freezing on the system. In addition, bioreactors II and IV were built with a pretreatment section to evaluate the effect of inducing an optimal pH and E_H into AMD on the efficiency of the SRB.

Each reactor was filled with a combination of organic carbon, crushed limestone, and cobbles (Figure 2) placed in discrete chambers. Each of these media is expected to play a certain role in the treatment train. (a) Organic carbon is bacterial food supply and, because it was provided in form of cow manure, also the SRB source. (b) For the pretreatment section a chamber with cow manure was included to lower the E_H of AMD. (c) Crushed limestone provides buffering capacity to increase the pH of AMD in the pretreatment section. (d) Cobbles placed in the "reactive," primary treatment section of the bioreactor constitute stable substrate for bacterial growth.

Bioreactors II and III, 71.5 feet and 61 feet in length respectively, were constructed in 14 feet wide trenches below grade; bioreactor IV, 72.5 feet in length, was constructed in a 12 feet wide metal half-culvert elevated above grade. Chambers filled with organic carbon or limestone are each 5 feet in length, whereas chambers filled with cobbles are 50 feet in length.

The first two chambers of bioreactors II and IV constitute the pre-treatment section and includes a chamber filled with a mix of cow manure and cut straw, and a chamber filled with crushed limestone. Following the pretreatment section is a primary treatment section that includes another chamber with the mix of manure and straw and the "reactive" chamber filled with cobbles. Bioreactor III does not include the pretreatment chambers, instead, this reactor has only the primary treatment section of organic carbon and cobbles. Such a configuration was chosen to evaluate its efficiency and compare it to the efficiency of bioreactors II and IV.

DESIGN REQUIREMENTS

Several functional and operational constraints and challenges were identified before the design of reactors commenced. The most important are listed below together with their design solutions:

A. The entire system must be passive. This condition was satisfied by incorporating site topography and flow control instrumentation into the design allowing gravity flow.

B. Construction of Bioreactors must allow for the simulation of seasonal droughts. It was achieved by constructing intake sumps where hydraulic head could be controlled through the system of valves and overflow piping (Figure 3).

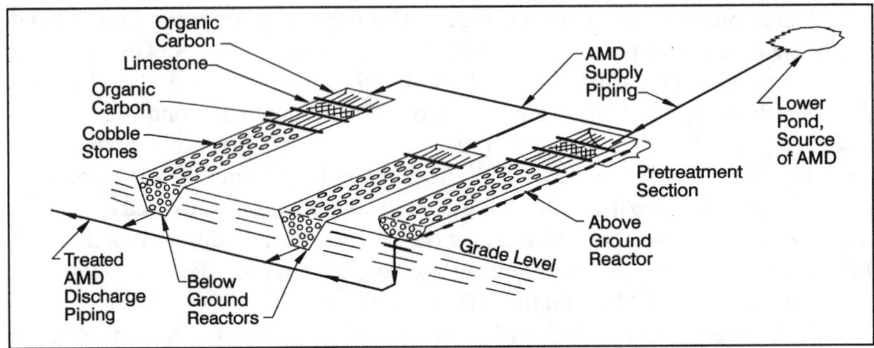

Figure 1 Layout of Bioreactors

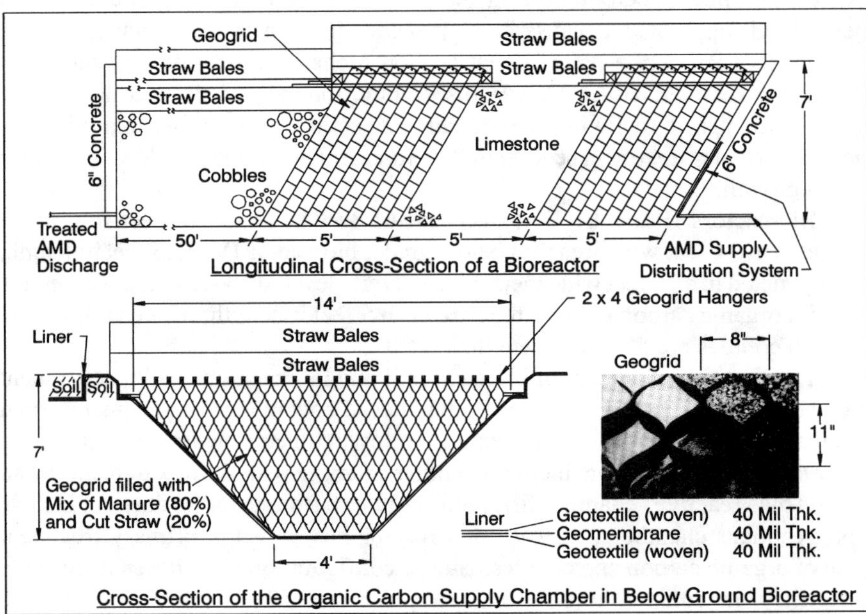

Figure 2 Bioreactors' Design

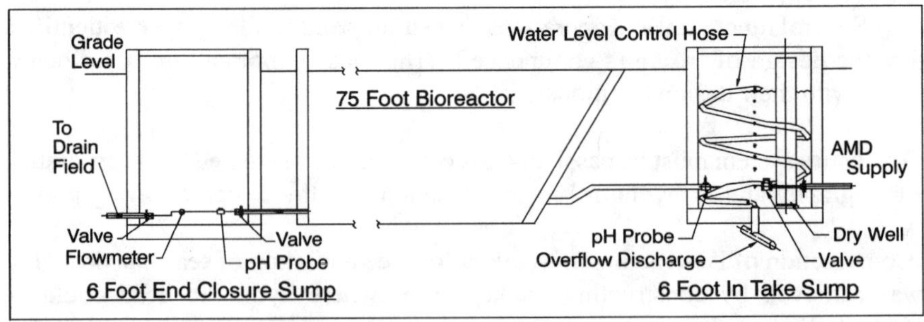

Figure 3 Flow Control System

C. The primary bioreactor chamber must be constructed in such a manner that will minimize fouling of the permeable medium. This was accomplished by filling the primary chamber with cobbles (3" to 6" in diameter).
D. The chambers with organic carbon must be designed so it fosters permeation of the AMD through the entire cross-sectional area (without "channeling"), and prevent settling of the substrate. The solution to this challenging requirement was threefold. (a) Cow manure was mixed (80% to 20% by volume) with cut straw to provide "secondary" porosity to the mix and make the substrate less compacted. (b) The mix of cow manure and straw was installed in the cellular containment system assembled of 10 lifts of Geogrid™ (Figure 2) that would limit settling (if occurred) of the organic mater to each individual cell. (c) The Geogrid™ lifts were positioned at 60^0 off the horizontal plane to facilitate packing it with the manure mix and to promote migration of AMD along a wavy-shape flowline[1].

CONSTRUCTION

Construction activities at the Calliope commenced in August 1998 and were completed in October 1998. Construction costs including labor, equipment, materials, and overhead averaged $70,000 each for the below ground bioreactors and $79,000 for the above ground bioreactor.

Excavation and grading for construction of all reactors and associated piping was completed while maintaining the existing slope grade of 2.5 percent. This grade was maintained to provide for natural run-off at the Calliope site in the SRB construction area and more importantly to avoid the influx of surface run-off into the below ground bioreactors.

Materials. All bioreactors were constructed with similar materials. These materials consisted of 40-mil geotextile, 40-mil PVC liner, PVC piping, and non-treated finished lumber. Geogrid™ material, commonly used in landscaping for slope stabilization, made of HDPE, was used to support the mix of manure and straw. Each layer (lift) of Geogrid™ was six inches tall and contained 11" x 8.5" rhombohedral shaped cells. Straw bales placed on the top of each bioreactor were used to provide thermal insulation for the below ground bioreactors and general protection for the above ground bioreactor.

Pre-cast reinforced concrete was used for the inlet and outlet walls, the intake sumps, and the end enclosures required for construction of the below ground bioreactors. A heavy gauge, multiple section, galvanized steel half culvert with fabricated steel endwalls, and a pre-cast reinforced concrete intake sump was used for the construction of the above ground bioreactor. Whenever possible, off-the-shelf, but acid resistant, building materials were used in the construction of the SRB's at the Calliope Mine Site.

---

[1]Patent in progress

Below Ground Bioreactors. Appropriate excavations for the below ground reactors were made and the liner system installed. The liner system utilized in all bioreactors consisted of a 40-mil PVC geomembrane sandwiched between two layers of a 40-mil woven Geotextile. The latter was used to provide additional abrasion resistance to the top and bottom of the PVC liner for all subsequent construction activities.

End walls, with the appropriate piping penetrations, were precast at the site and installed on top of the liner system at the AMD inlet and behind the liner system at the SRB outlet. Unique to the construction of the below ground reactors was the embedding of a 2 inch inlet manifold into the 60 degree precast inlet endwall. Additionally, the endwall slope angle and manifold configuration allows AMD water to enter the bioreactors and be diffused throughout the Geogrid™ utilizing the maximum cross-sectional area of each individual chamber.

Precast reinforced concrete intake sumps to control water levels within the bioreactors were installed directly upgradient of each reactor. Water levels within the bioreactors were controlled via the intake sumps. A siphon hose, connected to an overflow drain line was fabricated in all intake sumps to raise or lower the water level in each bioreactor. Water flows from these intake sumps through the manifold distribution system to the bioreactors.

Precast reinforced concrete end enclosures were installed at the end of each bioreactor to control flow rate and to house equipment to monitor performance of bioreactors. Effluent from bioreactors is piped from the end enclosures to the drainage field.

Above ground bioreactor. This reactor was constructed using multiple sections of a heavy gauge, galvanized steel culvert with fabricated steel end walls. Individual sections were joined together using carriage bolts. The liner system, as described above, was installed in this reactor.

Embedding the manifold in this reactor was not possible because the end wall was fabricated from steel; therefore, the inlet manifold was placed on top of the liner system and supported with wooden blocking material placed on the 60 degree endwall under the liner system. An end enclosure was not required for this reactor as it was constructed above ground. Other features including flow controlling and monitoring equipment and and the intake sump were alike to those used for the below ground bioreactors.

CONCLUSIONS

- Construction of an SRB reactor and treatment of AMD is feasible at remote mine sites.
- Construction materials are readily available, common, and generally inexpensive.
- Two SRB designs are available depending upon climate; above ground system or below ground system.
- SRB system may be passive and gravity controlled.
- No utilities are required.
- Systems may be virtually maintenance free unless monitoring is required.

USE OF BIOACTIVATION AND BIOAUGMENTATION TECHNIQUES FOR TREATING ACIDIC METAL-RICH DRAINAGE

Sophie Beaulieu, Gérald J. Zagury, Louise Deschênes and Réjean Samson
NSERC Industrial Chair in site Bioremediation,
École Polytechnique de Montréal, Montréal (Québec) Canada

ABSTRACT: In the last decade, interest in passive remediation technologies used for acid mine drainage (AMD) treatment have been increasing considerably. Although passives systems involving sulfate reducing activity offer many advantages, treatment performances still need to be improved. The application of bioactivation and bioaugmentation techniques on these systems can greatly increase their performances. In this study, bioactivation of a granular peat moss (GPM) was performed in 0.5 l laboratory batch reactors with two different sulfate-reducing bacteria (SRB) consortia. During these experiments, pH, lactate, sulfate and iron concentrations were measured periodically. After 1 to 3 spikes of sulfate, lactate and nutrients, the sulfate reduction rates increased from 80 mg/l*d to 217 and 230 mg/l*d. The pH in the reactors increased from 6 to 6.9 after 25 days. There was no SRB present in the GPM initially. A third consortium is under study and the most performant GPM will be used in bioaugmentation experiments.

INTRODUCTION

Acid mine drainage (AMD) is a persistent environmental problem at many active and abandoned mining sites. Sulfide minerals in contact with oxygen and water, combined with the presence of oxidizing bacteria, will be oxidised to produce AMD. AMD is characterized by high acidity and high concentration of metals and sulfate. Because of the negatives effects of AMD on streams and waterways, it usually requires treatment before being released in the environment.

With their minimal operation and maintenance requirements, passives systems technologies involving sulfate-reducing activity are very promising remediation techniques (Wildeman and Updegraff 1997). The main mechanism taking place is performed by SRB consortia, that can be found in natural environments where anoxic conditions prevail. SRB consume organic matter, produce HCO_3^- that raise the pH and reduce sulfate present in AMD to sulfide under anaerobic conditions. The sulfide then combines with metals cations to form insoluble metal sulfides. However, a certain amount of time (in most cases 3 to 6 weeks) is needed for the establishment of bacterial sulfate reduction, and subsequent drop in dissolved metals concentration

The use of bioactivation and bioaugmentation techniques can minimise the lag phase and improve the performances of passive systems involving SRB activity. Bioactivation consists in stimulating the growth of a SRB consortium on an appropriate support and increasing its density in a short period of time. Once the microbial consortium is stimulated, bioaugmentation is carried out by inoculating the media with the activated consortium.

Laboratory batch experiments have been performed on different SRB consortia to bioactivate granular peat. Granular peat moss was chosen for bioactivation due to its high porosity and expected high heavy metals adsorption capacity (Malterer et al., 1996). Results of the bioactivation experiments are presented.

Objective. The objective of this work is to minimise the lag phase and improve the performances of passive systems involving SRB activity using bioactivation and bioaugmentation techniques.

MATERIALS AND METHODS

SRB collection and bacterial count. The SRB consortia used in these experiments originate from bottom sediments taken from the anaerobic portion of small streams flowing through different mining sites. The first sediment (consortium D) was collected on an abandoned mining site and the second (consortium BH) on an active mining site. Evaluation of SRB concentration in the sediment was performed by the most probable number technique (MPN) (ASTM D4412-84, 1990). Concentration of total anaerobic bacteria was also estimated by the MPN method using a fluid thioglycollate medium U.S.P. (BBL, USA).

Experimental setup. Bioactivation was performed in 500 ml glass reactors sealed with teflon liner screw caps. Cubes of granular peat moss (1-2 cm^3) were saturated in Postgate medium B for 24 h. NaOH 5N was added (3 ml) to adjust the pH to 5.5. The solid portion of a 200 ml centrifugated suspension (11 000g) of sediment was then added to the reactor as the SRB source. Nitrogen was injected (1 l/min) into the reactor until the oxido-reduction potential (ORP) reached –100 mV. Mininert® Valves were installed on the reactors caps for sample collection. The reactors were incubated at room temperature (22 ± 1°C). All batch experiments were conducted using triplicates with abiotic controls. $HgCl_2$ and NaN_3 were added to the abiotic control and additional amount of $HgCl_2$ was supplemented as and when necessary.

Bioactivation monitoring. During the bioactivation, the reactors were spiked periodically with fresh sodium lactate, sulfate and nutrients. When sulfate concentration was near 500 mg/l or less, sulfate was spiked. Calcium, iron and magnesium sulfates were added with or without lactate when necessary. Essential nutrients, ammonium chloride and potassium phosphate, were added with every spike. Reactor preparation and sample collection was performed in an anaerobic glove box supplied with nitrogen.

Analytical methods. Samples were analysed periodically (1-5 days) for sulfate and lactate concentrations The filtered samples (Millipore 0.45 μm type HA) were then analysed by HPLC (ionic Dionex HPLC). The pH of every sample was measured with a Ross combination pH semi-micro electrode (model 8103BN). Total iron concentration in the liquid phase was determined by inductively coupled plasma (ICP) on filtered (0.45 μm) and acidified (HNO_3 16 N) samples.

RESULTS AND DISCUSSION

Figure 1(a) shows typical sulfate reduction rate observed in one of the triplicate of consortium D (reactor 2) which was spiked with sulfate 4 times. Results show that in reactor 2, initial sulfate reduction rate is –69 mg/l*d. Then,

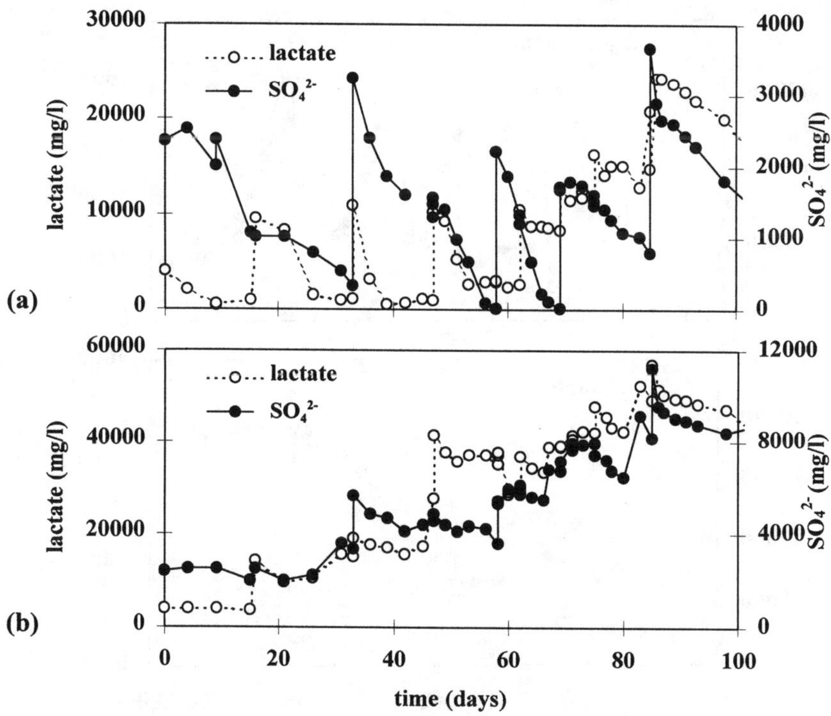

FIGURE 1. Consortium D : concentration of lactate and sulfate versus time (a) reactor 2. (b) reactor 4 (abiotic control).

it increases with time and reaches a maximum of -220 mg/l*d after two spikes of sulfate. Once a maximal sulfate reduction rate is reached, the following sulfate reduction rate are much lower (-67 mg/l*d for the third spike). However, the rate seems to increase again at the fourth spike (-113 mg/l*d), after the obtention of this maximum. Similar trends were observed in all triplicates of this consortium.

This phenomenom could be caused by the H_2S produced by the SRB consortium. High sulfate reduction rate increases the H_2S concentration in the reactor. Iron ($FeSO_4 \cdot 7H_2O$) was supplied to ensure precipitation of sulfide as FeS_2. Before each new spike, iron concentration was near zero, leaving a possible accumulation of H_2S in the reactor. After each spike, excess H_2S was removed by N_2 injection. Studies have shown that a certain level of H_2S and dissolved sulfide in SRB environment could have an inhibitory effect on the bacterial population, by slowing their metabolism (Postgate 1984, Maillacheruvu and Parkin, 1996).

Figure 1 (b) presents the evolution of lactate and sulfate concentrations versus time in the abiotic reactor. Unlike reactor 2 (figure 1a), concentrations of SO_4^{2-} and lactate increase with time, showing that there is a negligible bacterial activity.

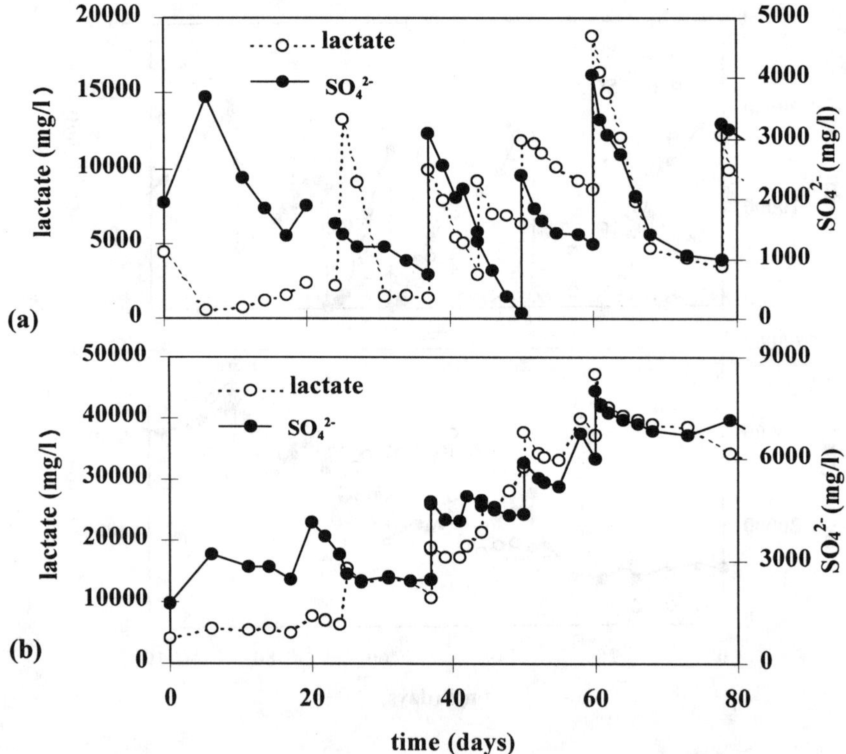

FIGURE 2. Consortium BH : concentration of lactate and sulfate versus time. (a) reactor 5. (b) reactor 8 (abiotic control).

As seen in figure 2 (a), the initial sulfate reduction rate for consortium BH (-72 mg/l*d) is similar to the one obtained for consortium D. Sulfate reduction rate observed in reactor 5 reaches maximum (-240 mg/l*d) after only one spike of sulfate. The sulfate reduction rate seems to follow the same pattern as the one noted for consortium D. The sulfate reduction rate decreases (-98 mg/l*d) after it has reached a maximum value. This was also observed in the other triplicates. In reactor 5 however the recovery seems to be more rapid, the rate of sulfate reduction observed after the fourth spike is near the maximum (-226 mg/l*d).

Figure 2 (b) presents the evolution of lactate and sulfate concentrations in the abiotic control (reactor 8). Sulfate and lactate accumulate in the reactor after each spike as in reactor 4 (figure 1b). It should be noted that $HgCl_2$ was added periodically in order to maintain abiotic conditions in these reactors.

Table 1 shows SRB concentrations in the consortia, maximum rates of sulfate reduction reached, number of sulfate spikes and time needed in each reactor during the bioactivation experiments. Sulfate reduction rates were estimated with linear regression performed on the experimental curves. Correlation coefficients (R^2) ranged from 0.83 to 0.98. Higher values of sulfate reduction rates were attained in consortium BH although the average sulfate reduction rate was greater for consortium D. SRB concentrations measured in sediment D and BH showed that consortium D contained more SRB before inoculation of the reactors. Average maximum sulfate reduction rates obtained with the two consortia are good (-217 mg/l*d and - 230 mg/l*d) in comparison with maximum rate obtained by Waybrant et al. 1995 (-154 mg/l*d) and by Dvorak et al. 1992 (less than –96 mg/l*d). The high sulfate reduction rates obtained in this study were expected, lactate being a preferred carbon source for many SBR (Posgate 1984, Dvorak et al. 1992).

The pH in the consortium D reactors (1, 2 and 3) was stable at 6.9 after 26 days. The pH of the consortium BH reactors (5, 6 and 7) was stable after 20 days, to a value of 6.8. This typical increase in pH due to HCO_3^- was expected (Posgate 1984, Gibson 1990). Average pH measured in the abiotics controls, 4.8 for consortium D and 5.4 for consortium BH, show that there is no increase of pH because there is no bacterial activity. Bioactivation is still going on in our laboratory.

TABLE 1. Maximal sulfate reduction rates reached in the reactors

Consortium **SRB conc.**	Reactor number	Maximum sulfate reduction rate	Time (days)	Sulfate spikes	Average sulfate reduction rate
D	1	-227 mg/l*d	47	1	
2.0*10³	2	-219 mg/l*d	69	2	-230 mg/l*d
cell./g sed.	3	-244 mg/l*d	85	3	
BH	5	-240 mg/l*d	50	1	
3.3*10²	6	-263 mg/l*d	60	2	-217 mg/l*d
cell./g sed.	7	-148 mg/l*d	73	3	

CONCLUSION

Up to this point, the following conclusions can be drawn from the first part of this study :

- Bioactivation seems promising
- Time needed for comsumption of 3000 mg/l of sulfate ranging from 30 to 40 days intially, was reduced to less then 10 days after bioactivation
- Sulfate reduction rates increased from –70 mg/l*d to –230 mg/l*d
- The most efficient GPM will be used in the bioaugmentation experiments

ACKOWLEDGMENTS

The authors acknowledge the financial support from the industrial chair partners: Alcan, Bodycote/Analex, Bell Canada, Browning-Ferris Industries, Cambior, Centre québécois de valorisation de la biomasse et des biotechnologies (CQVB), Hydro-Québec, Natural Science and Engineering Research Council (NSERC), Petro-Canada and SNC-Lavalin.

REFERENCES

Dvorak Darryl H., Hedin Robert S., Edenborn Harry M. and McIntire Pamela E. 1992. « Treatment of metal-contaminated water using bacterial sulfate reduction : Results from pilot-scale reactors. » *Biotech.and Bioeng.* 40(5) :609-616.

Gibson G.R. 1990. « A review : Physiology and ecology of the sulphate-reducing bacteria. » *Journal of applied bact.* 69 :769-797.

Maillacheruvu K. Y., Parkin G. F. 1996. « Kinetics of growth, substrate utilization and sulfide toxicity for proponiate, acetate, and hydrogen utilizers in anaerobic systems. » *Water Env. Research.* 68(7): 1099-1106.

Malterer T., McCarthy B. and Adams R. 1996. « Use of peat in waste treatment. » *Mining eng.* January: 53-55.

Postgate J.R. 1984. *The sulphate-reducing bacteria.* 2nd ed. Cambridge University Press, Cambridge, GB.

Standard test methods for sulfate-reducing bacteria in water and in water-formed deposits. ASTM 1990. Part D4412-84 : 533-535.

Waybrant K.R., Blowes D.W. and Ptacek C.J. 1995. « Selection of reactive mixture for the prevention of acid mine drainage using porous reactive walls » *Mining and the Environment.* Sudbury. Ontario. May 28th-June: 945-953.

Wildeman T. and Updegraff D. 1997. "Passive bioremediation of metals and inorganic contaminants." In D.L. Macalady (Ed*), Perspectives in environmental chemistry*, pp.473-495. Oxford university press, NY.

DESIGN, CONSTRUCTION AND OPERATION OF A 1,200 GPM PASSIVE BIOREACTOR FOR METAL MINE DRAINAGE

James Gusek (Knight Piesold LLC, Denver, CO)
Thomas Wildeman (Colorado School of Mines, Golden, CO)
Aaron Miller (Doe Run Co., Annapolis, MO)

ABSTRACT: An active underground lead mine produces water having a pH of 8.0 with 0.4 to 0.6 mg/L of Pb and 0.18 mg/L of Zn. A full-scale 1,200 gpm capacity bioreactor system was designed and permitted based on a phased program of laboratory, bench and pilot scale bioreactor testing; it was constructed in mid-1996. The gravity flow system, covering a total surface area of about five acres (2 ha), is composed of a settling basin followed by two anaerobic bioreactors arranged in parallel which discharge into a rock filter polishing cell that is followed by a final aeration polishing pond. The primary lead removal mechanism is sulfate reduction/sulfide precipitation. The discharge has met stringent in-stream water quality requirements since its commissioning. The system was designed to last about 12 years, but estimates suggest a much longer life based on anticipated carbon consumption in the anaerobic cells.

INTRODUCTION

Doe Run's West Fork Unit is an underground lead-zinc mine that discharges water from mine drainage to the West Fork of the Black River (West Fork) under an existing NPDES permit. The West Fork Unit is located in Reynolds County in central Missouri, in the New Missouri Lead Belt, about three hours from St. Louis. Flow rates in West Fork vary from about 20 cubic feet per second (cfs) to more than 40 cfs; water quality is relatively good, despite being located in an area with naturally high background levels of lead due to the bedrock geology. The mine discharges about 1,200 gpm on the average (2.7 cfs) or about 10 percent of the total flow in West Fork.

The adoption of water quality based discharge limits in its NPDES permit issued in October, 1991, prompted Doe Run to evaluate treatment methods for metal removal. Evaluations of alternative treatment processes determined that biotreatment methods were feasible and cost less than half as much as sulfide precipitation. The goal of the water treatment project was to ensure that the stringent water quality based limits in the permit would be consistently met.

Since 1987, a group from the Colorado School of Mines and Knight Piesold LLC has been active in developing passive treatment methods for metal-mine drainages. The primary treatment method is through the generation of hydroxides and sulfides through microbial metabolism. The biogeochemical principles are summarized in Wildeman et al.. 1995 and Wildeman and Updegraff, 1998. The design principles are explained in Wildeman, Brodie and Gusek, 1993. In the case of the West Fork Unit, biotreatment consists of two stages. First, an anaerobic unit generates sulfide through sulfate reduction and is responsible for the lead removal. Second, an aerobic unit is a rock filter/wetland. This unit is responsible for removing dissolved organic matter, and excess sulfide from the effluent from the anaerobic cell. The aerobic unit also reoxygenates and polishes the water before it enters the river.

Extensive laboratory, bench-scale, and pilot scale tests were made on the anaerobic unit. These are described in Wildeman et al. 1997, and Gusek et al., 1998. This paper concentrates on the design, construction and operation of the final full-scale system.

FULL-SCALE DESIGN

The large-scale system was designed based on the performance of the pilot scale system and the interim bench scale studies. The large-scale system was estimated to cost approximately $500,000 and require about two to three months of construction time, depending on the vagaries of weather and construction surprises. System operational costs include water quality monitoring as mandated by law. No additional costs for reagents are incurred; since the system uses gravity flow, moving parts are few and include valves, minor flow controls and monitoring devices. Based on carbon depletion rates observed in the pilot system, the anaerobic cell substrate life was projected to be greater than 30 years; the full-scale biotreatment system should be virtually maintenance-free. Should mine water quality deteriorate, the full-scale design included a 50 percent safety factor. The pilot-scale system was tested by operating for about 90 days at double the design capacity. Compliance effluent with respect to total lead concentration and other key performance parameters resulted from this test.

System Description. The biotreatment system is composed of five major parts a settling pond, two anaerobic cells, a rock filter, and an aeration pond (Knight Piesold, 1997). The system is fully lined. The design was also integrated into the mine's pre-existing fluid management system.

A rectangular-shaped, 40 mil HDPE-lined **settling pond** has a top surface area of 32,626 ft^2 (3030 m^2) and a bottom surface area of 20,762 (1930 m^2) ft^2. The sides have slopes of 2 horizontal to 1 vertical (2H:1V). The settling pond is nominally 3 m deep. It discharges through valves and parshall flumes into the two anaerobic cells.

Two anaerobic cells are used, each with a total bottom area of about 14,935 ft^2 (1930 m^2) and a top area of about 20,600 ft^2 (1930 m^2). Each cell is lined with 40 mil HDPE and was fitted with four sets of fluid distribution pipes and three sets of fluid collection pipes, which were subsequently modified (see Start Up discussion). The distribution/collection pipes were connected to commonly-shared layers of perforated HDPE pipe and geonet materials sandwiched between layers of geofabric. This feature of the design was intended to allow control of sulfide production in hot weather by decreasing the retention time in the cell through intentional short circuiting.

The spaces between the fluid distribution layers were filled with a mixture of composted cow manure, sawdust, inert limestone, and alfalfa, referred to hereafter as "substrate." The total thickness of substrate, piping, geonet and geofabric was about 2 m. The surface of the anaerobic cells was covered with a layer of crushed limestone. Water treated in the anaerobic cells flows by gravity to a compartmentalized concrete mixing vault and thereafter to a rock filter cell. The gravity-driven flows can be directed upward or downward .

The **rock filter** is an internally bermed, clay-lined shallow cell with a bottom area of about 63,000 ft^2 (5,900 m^2) and a nominal depth of one foot. It is constructed on compacted fill that was systematically placed on the west side of a pre-existing mine water settling pond. Limestone cobbles line the bottom of the cell and the cell is compartmentalized by limestone cobble berms. The discharge from the rock filter flows through a drop pipe spillway and buried pipe into a 40 mil HDPE lined **aeration pond**. The aeration pond surface covers approximately 85,920 ft^2 (8,000 m^2). The aeration pond discharges through twin 12-in (30 cm) HDPE pipes into a short channel that leads to monitoring outfall and thence into West Fork.

After the water pumped from the underground mine enters the settling pond, all flows are by gravity.

THE PERMITTING PROCESS

The permitting aspects of the project were very complex. Regulators needed to be convinced that an organic-based wetland-type substrate could remove dissolved lead from mine effluent. However, regulators were willing to listen to facts and the flow of communications was good. Nevertheless, cow manure as an ingredient in the anaerobic cell substrates was a special regulatory hurdle because its use raised issues of BOD, fecal coliform bacteria and other organic-related water quality criteria problems from a non-degradation of West Fork perspective.

Education of permit document reviewers was a key aspect of the permitting effort, supported by the results of the two years of pilot scale test results. The original permitting application was made after gathering one year's worth of pilot data; data acquisition continued throughout the permitting process. Missouri Department of Natural Resources (DNR) raised useful and valid concerns which were addressed with additional testing, including monitoring for fecal coliform, color, BOD, and other minor constituents. This additional testing raised the level of knowledge of passive treatment performance in general and improved the database utilized in the final design.

The closure and reclamation of the biotreatment system after its scheduled decommissioning at the end of the West Fork facility life was also a DNR concern. The system was constructed within the boundaries of the waste management areas as defined by the Metallic Minerals Waste Management Act and was, by definition, a waste management structure. Therefore, closure and reclamation activities would adhere to Section 5 of the Metallic Minerals Waste Management Permit issued to Doe Run's West Fork Unit in January, 1991.

The substrate material, made up primarily of sawdust, alfalfa hay, limestone and cow manure, was projected to accumulate metals over time through the operation of the water treatment system. Based on average flow and metal content of the mine water, it was estimated that the final metal loading in the substrate will be 1,866 mg/kg Pb as PbS. At the end of the active life of the biotreatment system, core samples of the substrate will be subjected to TCLP. If the substrate material fails TCLP, disposal will be in accordance with all applicable laws and regulations pertaining to characteristic hazardous waste. If the substrate passes TCLP, it will be used as an organic fertilizer to stimulate vegetation growth on the slope of a nearby tailings dam. Data from other sites have suggested that organic substrate containing metals will pass TCLP tests if it is allowed to oxidize first.

Odor control from the proposed facility was not expected to be a problem. Doe Run personnel conducted a reconnaissance air quality screening study at the site with chemically activated sniffer sampling of air immediately adjacent to the operating pilot scale biotreatment plant. Hydrogen sulfide concentrations were the focus of the survey. Air quality modeling suggested that the facility would be in compliance with applicable standards.

Other points ,which helped in the permitting process, were that the biotreatment method had been used at other Asarco facilities (in Colorado, Montana [which was issued an interim NPDES permit] and Canada) and it was accepted as a viable treatment method by agencies in other states and the USEPA. Some of the original research work into biotreatment was sponsored under the EPA's Emerging Technology Program. The following mine/mill sites are known to have included biotreatment in their record of decision: Clear Creek, Colorado, Buckeye Landfill, Ohio, Palmerton Site, Pennsylvania, Bunker Hill, Idaho. In these cases, biotreatment was the preferred alternative or a key component of the preferred alternative.

SYSTEM CONSTRUCTION

Following permitting, the biotreatment system was constructed in accordance with plans and specifications as submitted to and approved by the Missouri DNR Water Pollution Control Program. The construction was authorized under the Construction Permit issued on March 12, 1996. Work commenced on March 13, 1996. By July 10, 1996, the work was declared to be substantially complete in accordance with the Plans and Specifications. There were no change orders. Minor field changes in the design typically improved the facility. Some of these are discussed below.

The original recipe for the substrate included aged sawdust, low-manganese limestone, aged cow manure, and alfalfa hay in decreasing proportions. As specified, the alfalfa hay was assumed to be baled. A readily-available source of slightly moldy alfalfa hay cubes was substituted as a field change. The volumetric proportions of the substrate components changed slightly (the substrate became denser) and additional sawdust was used to make up the total volumetric deficit. The addition of more organic carbon could increase projected cell life, already in excess of the required operational time.

The construction was sequenced so that the settling pond was built and commissioned first so that the mine and mill could continue to operate during construction. Subsequently, the old settling pond was backfilled in part to become the foundation of the rock filter. The portion of the remaining settling pond was lined with HDPE geomembrane and became the aeration pond.

START-UP EXPERIENCE

Bench-scale test results suggested that the anaerobic cells be incubated with settled mine water for about 36 hours or less before fresh mine water was introduced at full flow to minimize initial levels of BOD, fecal coliform, color and manganese. For about two weeks, pumps recycled the water within the two anaerobic cells. Based on data collected in the field, and subsequent laboratory confirmation, the water from the anaerobic cells was routed to the tailings pond for temporary storage. At that point, the rock filter and aeration ponds were brought on-line. In the meantime, the mine discharged according to plan through an overflow pipe from the settling pond as it had during construction of the other components. Plumbing was available to temporarily discharge to an adjacent tailings pond, if necessary, where it would be stored for later treatment and release.

After about six weeks of full scale operation, the apparent permeability of the substrate was found to be lower than expected and the system was operating nearly at capacity. Study found that H_2S gas, generated by the sulfate reducing bacteria, was being retained in the substrate in the anaerobic cells; this created a gas-lock situation that prevented full design flow. A temporary solution was obtained by periodic "burping" of the cells using the control valves. However, the "burping" had to be performed at 24-hour intervals and it was determined that this solution was too labor intensive.

The sulfide gas lock problem was investigated in December, 1996 by installing vent wells in the substrate and measuring the gas pressures. Observations indicated that the gas was a factor in apparent short circuiting of the water passing through the cell. The layered geotextiles, (geonet and geofabric) originally intended to promote horizontal flow, appeared to be trapping the sulfide gas beneath them and vertical flow was being restricted. The permeability of the substrate itself was for the most part unaffected. However, construction practices in the south anaerobic cell could have contributed to the situation. Here, a low ground bearing bulldozer was used to place substrate in nominal six-inch lifts. This could have created a layering effect that may have trapped gas as well. Substrate layers in the north anaerobic cell were placed in a single lift and no layering effect was observed during subsequent excavation. It is noteworthy that the mid-cell geotextiles had not been a feature of the pilot test cell design.

The first phase of a permanent solution was implemented with a trenching machine that ripped through the geonet/geofabric layers in the south anaerobic cell. This disrupted the gas-trapping situation. Subsequently, the substrate from the entire south anaerobic cell was excavated and the cell refilled without the geotextiles in June, 1997. Identical action was taken on the north anaerobic cell in September, 1997. These actions have apparently solved the gas lock problem.

Table 1. West Fork Water Quality Data; Sources: Doe Run, Inc., 1997, and Knight Piesold LLC, 1995.

Parameter	Typical Average Influent Water Quality in mg/L	Range of Water Quality Discharge in mg/L (June - November 1997)
Pb	0.4	0.027 - 0.050
Zn	0.36	0.055 - 0.088
Cd	0.003	<0.002
Cu	0.037	<0.008
Oil and Grease	--	<5.0
H_2S	--	0.011 - 0.025
Total Phosphorus	--	<0.05 - 0.058
Ammonia as N	0.52	<0.050 - 0.37
Nitrate and Nitrite	2	<0.050 - 1.7
True Color	--	10 - 15
BOD	1.7	<1 - 3
Fecal Coliform	—	<1 - 2
pH	7.94	6.63 - 7.77
TSS	—	<1 - 4.2

RESULTS AND CONCLUSIONS

The average influent water quality can be compared with discharge water quality (Table 1) during the June through November, 1997 period. Discharge levels of Pb and other metals were reduced substantially from average influent levels. For Pb, the level was reduced from a typical average of 0.40 mg/L to between 0.027 and 0.050 mg/L. Zn, Cd and Cu effluent concentrations were also reduced.

The following conclusions can be made about the design, construction and operation of the West Fork treatment system.

1) A practical design has been developed to bring Pb values down to stringent water quality standards.
2) Bacterial sulfate reduction is the major Pb removal process.
3) An aeration step is needed to polish for Mn, BOD, fecal coliforms removal and re-oxygenation.
4) Pilot testing should include as many features of the final design as possible to minimize start up difficulties.
5) Education of regulators on innovative water treatment techniques can facilitate permit approvals.

REFERENCES

Knight Piesold LLC. 1997. "Engineer's Construction Report, Mine Water Biotreatment Facility, Asarco West Fork Unit, Bunker, Missouri."

Wildeman, T.R., et. al. 1993. "Passive Treatment Methods for Manganese: Preliminary Results from Two Pilot Sites." In: Proceedings of 13th National Meeting of American Society for Surface Mining and Reclamation, pp. 665-677. Princeton, WV.

Wildeman, T. R., Gusek, J. J., Cevaal, J., Whiting, K., and Scheuring, J., 1995, Bio Treatment of acid rock drainage at a gold-mining operation. In R. E. Hinchee, J. L. Means, and D. R. Burris, Eds., *Bioremediation of Inorganics*, pp. 141-148. Battelle Press, Columbus Ohio.

Wildeman, T.R. J.J. Gusek, A. Miller, J. Fricke, 1997. Metals, sulfur, and carbon balance in a pilot reactor treating lead in water. *In Situ and On-Site Bioremediation, Volume 3. pp. 401-406. Battelle Press, Columbus, OH.*

Wildeman, T., D. Updegraff, 1998, Passive bioremediation of metals and inorganic contaminants. In: D.L. Macalady, Ed, *Perspectives in Environmental Chemistry*, pp. 473-495. Oxford University Press, New York.

Gusek, J. J., T. R. Wildeman, A. Miller, and J. Fricke, 1998, The challenges of designing, permitting, and building a 1,200 gpm passive bioreactor for metal mine drainage, West Fork, Missouri. In: Proceedings of 15th Annual Meeting of American Society for Surface Mining and Reclamation, pp.203-212, Princeton, WV.

REMEDIATION OF ACID MINE DRAINAGE BY SULFATE REDUCING BACTERIA IN BIOFILM REACTORS

Åsa K Kolmert, Kevin B Hallberg and D Barrie Johnson
(University of Wales, Bangor, Gwynedd, UK)

ABSTRACT: Acid mine drainage (AMD) is an environmental hazard in many parts of the world. Processes using sulfate reducing bacteria (SRB) have proven to be successful in the treatment of metal contaminated water. Known SRB strains, however, are sensitive to the low pH of AMD. Therefore, we have isolated acidophilic/acidotolerant SRB strains from acidic sites and begun to study their potential for use in AMD remediation. In order to develop a simple and robust low-cost system for treatment of AMD, a biofilm reactor system was constructed. Three different SRB consortia were tested (various combinations of acidophilic and neutrophilic SRB) and three carbon and energy sources (lactic acid, glycerol and ethanol) were evaluated for their potential use. All three SRB-consortia reduced sulfate when challenged with acidic influent, at rates of about 0.3 g SO_4^{2-}/l*day.

INTRODUCTION

Acid mine drainage (AMD) is characterised by high metal concentrations and, in many cases, low pH. It is an environmental hazard in mining areas worldwide (Banks et al., 1997). The genesis of AMD is due to the oxidation of sulfide minerals, which is greatly accelerated by the activities of certain (mainly acidophilic) microorganisms (Johnson, 1995). This process generates acidity in the form of sulfuric acid. The solubility of many cationic metals is increased at low pH, and extremely high concentrations of metals (including some toxic metals) can occur in AMD. Improvement of water quality may be achieved by passage of AMD through natural or constructed wetlands. Wetlands, however, usually require large land areas. In countries such as Wales, where many abandoned mines are found in mountainous regions, suitable land areas are not always available for construction of wetlands. Therefore, we are studying more intensive biological treatment processes for the remediation of AMD. To treat metal-contaminated water, sulfate reducing bacteria (SRB) have been shown to be useful. SRB generate hydrogen sulfide, which results in the precipitation of chalcophilic metals present in AMD and also has the benefit of producing net alkalinity. A number of large-scale plants based on THIOPAQ process developed by Paques Biosystems are already in operation (De Vegt et al., 1997).

This study aims to develop a cost-effective process for efficient remediation of AMD using SRB biofilm reactors. Biofilms are robust to environmental changes and hence may be suitable for the harsh AMD environment. One problem with the use of SRB to treat AMD is the fact that characterised strains are sensitive to low pH and therefore elaborate bioreactor systems are generally required to treat acidic solutions. To circumvent this problem, we have isolated acidophilic/acidotolerant SRB (aSRB) and have begun to study their potential for use in AMD remediation (Johnson 1995). Porous glass beads were used as biofilm carrier in tubular-flow bioreactors.

MATERIALS AND METHODS

Three laboratory-scale (2 L) column reactors, filled with porous glass beads, were inoculated with different populations of SRB to compare their abilities to remediate acidic, metal-rich waters. The first was inoculated with a mixed population of aSRB (three different populations, two obtained from acidic sites in the U.K. and the other from Montserrat, W.I.), the second with a mixture of nSRB and aSRB and the third with neutrophilic SRB (nSRB). The nSRB population used was dominated by a *Desulfovibrio* strain isolated in the Netherlands. A biofilm of the respective culture was established on the glass beads over a two month period by recirculating a growth medium containing basal salts (0.45 g $(NH_4)_2SO_4$, 0.05 g KCl, 0.5 g $MgSO_4 \cdot 7H_20$, 0.014 g $Ca(NO_3)_2 \cdot 4H_2O$, 0.05 g KH_2PO_4, 1.5 g K_2SO_4, 0.28 g $Fe_2SO_4 \cdot 7H_20$ per l distilled water) and trace elements. Three different carbon and energy sources were included - glycerol, ethanol and lactic acid, all at 5 mM. The medium was sparged with nitrogen and autoclaved before use. The pH of the medium was adjusted to 4 for the aSRB and the aSRB/nSRB columns and to 5 for the nSRB column. Initially, the medium was recirculated through the columns in order to enhance mass transfer whilst the biofilm was established. During the experimental runs, however, the medium was percolated in a tubular flow mode to imitate the proposed full-scale treatment system. The target flow rate was 40 ml/h, equivalent to a retention time of approximately 50 h.

The carbon and energy sources were evaluated to determine the most efficient, simple and cost-effective material or mix of materials that could be used with minimal loss of sulfate reduction capacity. Glycerol was first removed, followed by lactic acid, leaving ethanol as the sole remaining carbon and energy source at 5 mM. The ethanol concentration was thereafter increased to 15 mM. Later, glycerol was re-introduced to the medium. The medium was used for 17-23 days after each variation, with the exception of 5 mM ethanol, which was maintained for only 14 days.

The sulfate reduction rate was followed daily by measuring sulfate concentrations and pH in reactor samples. Later, the sulfide concentrations in the reactor samples were also assayed, allowing the determination of sulfur mass balance for each reactor. Sulfate was assayed according to a standard turbidometric method (Rand et al., 1975) and sulfide was assayed according to the method described by Cord-Ruwisch (1995). Growth of the biofilm on the glass beads was visualised using scanning electron microscopy (SEM).

RESULTS AND DISCUSSION

SEM of the beads showed that the surface of the beads was covered in biofilm whereas the pores were sporadically colonised after three months (Figure 1). All three reactors were effective at reducing sulfate when percolated with mixed substrate medium; the pH of the effluent from the reactors was also

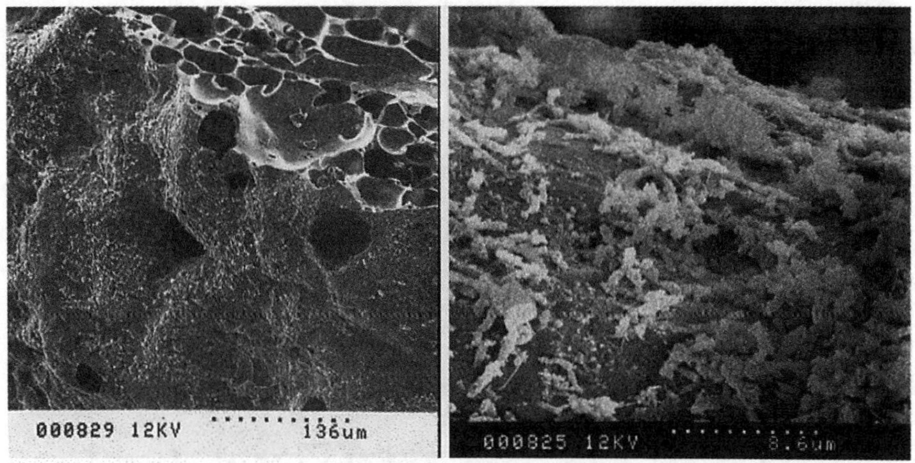

FIGURE 1. SEM pictures of SRB biofilm on Poraver beads at 220x and 3500x magnification.

significantly greater (pH 5.5- 6.5) than of the influent liquor (Figures 2-4). Figures 2-4 show the sulfate reducing capacity and the sulfur mass balance for each reactor as well as the increase in pH. Ideally, the H_2S+SO_4-line should be 1 mM below the "SO_4-medium"-line since the medium contained 1 mM $FeSO_4$ and all iron was precipitated as FeS in the reactor. Figure 5 shows the sulfate conversion rates in each system, i.e. how much sulfate was reduced per liter reactor volume per day. The average conversion rate was 0.25-0.30 g sulfate reduced per litre and day reactor (g SO_4^{2-}/l*day).

The substrate requirement of the SRB reactors has been evaluated by sequential removal of each of the three carbon sources in the medium. The diagrams for all three reactors (Figures 2-5) show that there was no major difference in sulfate reducing capacity between the reactors. When lactic acid and glycerol were removed from the medium, and the sole energy and carbon source was ethanol, the sulfate reducing capacity decreased to zero in both reactors. When the ethanol concentration in the medium was increased to 15 mM at day 48, the sulfate reducing capacity slowly returned to the level before lactic acid removal. The initial decrease in sulfate reducing capacity when the aSRB and the aSRB/nSRB bioreactors were changed from a "mixed substrate" input to one where only ethanol was present could have been due to the bacteria utilising only one of the available substrates (lactic acid) in the original medium. However, the ability of the aSRB and aSRB/nSRB populations to use ethanol as sole carbon source was clearly demonstrated when the systems were run with a medium containing 15 mM ethanol. There was evidence of incomplete oxidation of carbon substrates in the bioreactors, in that acetic acid was detected on both occasions on which it was assayed. The stoichiometry of ethanol oxidation to acetate is such that two ethanol molecules are required to reduce one sulfate to sulfide ($2\ C_2H_5OH + SO_4^{2-} + 2H^+ \rightarrow 2C_2H_4O + H_2S + 2H_2O$), which would imply that a maximum sulfate reduction of 2.5 mM would occur in the presence of 5 mM ethanol.

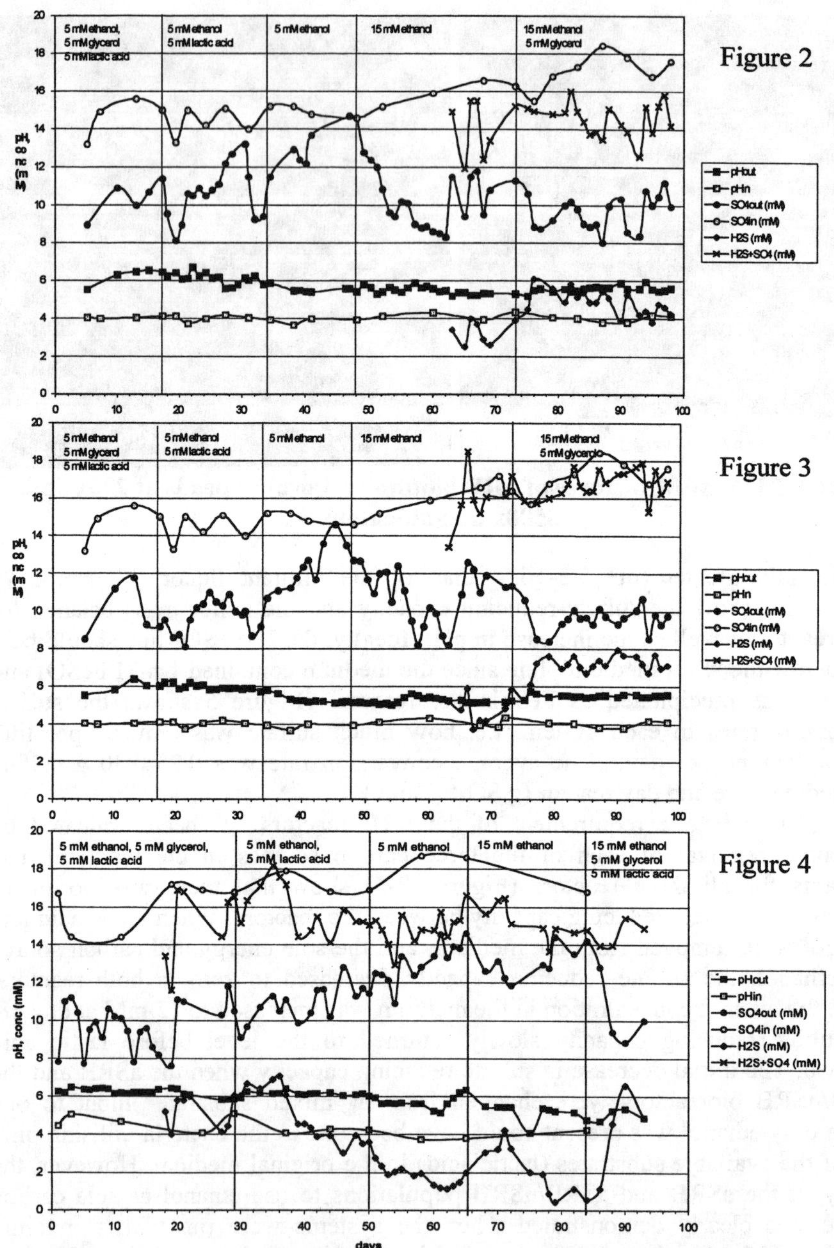

FIGURES 2-4. The sulfate reducing capacity of (2) the aSRB reactor, (3) the mixed aSRB/nSRB reactor and (4) the nSRB reactor.

Interestingly, the nSRB reactor (which is lagging behind the other two) did not display the same dramatic decline in sulfate reduction when the system was run on 5 mM ethanol. The latter system has been somewhat more effective in terms of sulfate conversion rates than the other two systems, though it was initially not subjected to such acidic influents as the other two reactors (Figures 4-5).

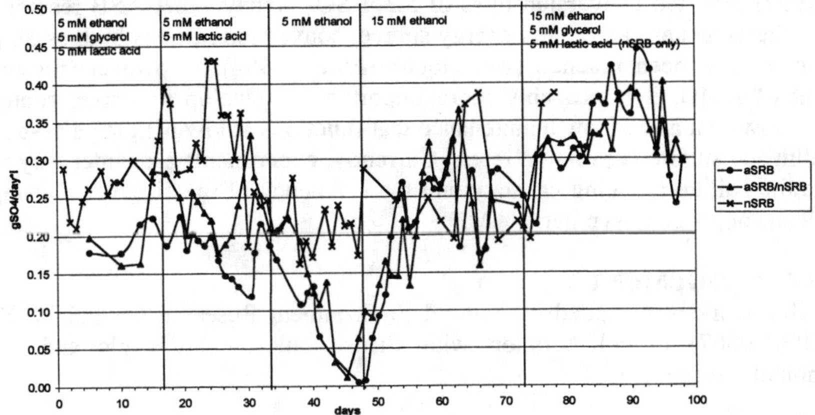

FIGURE 5. Sulfate conversion rates for all three reactors. Day 1 for the nSRB corresponds to day 13 in figure 4.

CONCLUSIONS

The aim of this study is to develop a simple, robust, low-cost system that demands little maintenance for treatment of AMD. A biofilm reactor was chosen for the treatment of AMD because it satisfies these requirements. The use of different consortia of SRB in biofilm reactors was compared to determine their robustness. No significant difference in sulfate reducing capacity was seen between the system with only aSRB and the system with aSRB/nSRB with an influent at pH 4. The nSRB system has been marginally more efficient. We have chosen to use a consortium of aSRB and nSRB in one bioreactor in the event that the aSRB will become established at the front end of the column where the AMD will enter the bioreactor and the nSRB will become established in the rear end of the bioreactor, by which time some amelioration of the percolating liquor (in terms of pH increase) would have occurred.

The use of glass beads as a biofilm carrier is suitable in many aspects. They have a high surface/volume ratio, they are chemically inert and they have been proven by SEM to be an effective support for biofilm growth (Kolmert et al., 1997). Furthermore, they are cost-effective and they are made of recycled glass, which is an attractive concept in an environmental remediation application.

The carbon and energy sources utilised were initially chosen since they had proved effective for the strains used as inoculum. Lactic acid is often utilised by neutrophilic SRB but could cause problems at lower pH because of the toxicity of organic acids to acidophilic microorganisms. Glycerol has previously been used as substrate for aSRB (Johnson, 1995), but would be too costly in a large-scale process. Ethanol is more cost-effective and should not be toxic even at low pH.

In the tubular flow biofilm reactors utilised here, average sulphate conversion rates of 0.3 g SO_4^{2-}/l*day were achieved. This is considerably lower than has previously been reported. With a similar reactor system operated at neutral pH using neutrophilic SRB and with recirculation of the medium, Kolmert *et al.* (1997) reported conversion rates of 5.3 g SO_4^{2-}/l*day. In UASB reactors, using synthesis gas as carbon and energy source, conversion rates as high as 30 g SO_4^{2-}/l*day have been reached (van Houten *et al.*, 1994). However, for the treatment of AMD, it is probably more important to develop a simple, robust system of low cost and of low maintenance that functions at low pH than a system with optimised sulfate conversion rates. Currently, experiments are under way to examine the sulfate reducing capacity and the efficiency of metal removal using authentic, as opposed to synthetic, AMD.

ACKNOWLEDGEMENTS

This work was funded by Natural Environment Research Council, U.K. (ref. GR3/C0007) in collaboration with British Nuclear Fuels plc and the Environmental Agency, U.K.

REFERENCES

Banks, D., P. L. Younger, R.-T. Arnesen, E. R. Iversen, and S. B. Banks. 1997. "Mine Water Chemistry: the Good, the Bad and the Ugly." *Environmental Geology*. 32(3): 157-174.

Cord-Ruwisch, R. 1985. "A Quick Method for the Determination of Dissolved and Precipitated Sulfides in Cultures of Sulfate-Reducing Bacteria." *Journal of Microbiological Methods*. 4: 33-36.

van Houten, R.T., L. W. Hulshoff Pol and G. Lettinga. 1994. "Biological Sulphate Reduction Using Gas-Lift Reactors Fed with Hydrogen and Carbon Dioxide as Energy and Carbon Source." *Biotechnology & Bioengineering*. 44: 586-594.

Johnson, D.B. 1995. "Acidophilic Microbial Communities: Candidates for Bioremediation of Acidic Mine Effluents." *International Biodeterioration & Biodegradation*. 41-58.

Kolmert, Å., T. Henrysson, R. Hallberg and B. Mattiasson. 1997. "Optimization of Sulphide Production in an Anaerobic Continuous Biofilm Process with Sulphate Reducing Bacteria." *Biotechnology Letters*. 19: 971-975.

Rand, M. C., A. E. Greenburg and M. J. Taros. 1975. In: *Standard Methods for the Examination of Waste Water*, 14th ed., pp. 496-498. U.S.S. American Public Health Association.

de Vegt, A. L., J. P. Krol and C. J. Buisman. 1997. "Biological Sulfate Removal and Metal Recovery from Mine Waters." In: *Proceedings of the International Biohydrometallurgy Symposium, Sydney, Australia*.

NEMATODES AS SOIL QUALITY MIRROR

P. Doelman (IWACO B.V., Rotterdam, The Netherlands)
A.G. Veltkamp (NAM, Assen, The Netherlands)
A.M.T. Bongers (Wageningen Agricultural University, Wageningen, The Netherlands)

ABSTRACT: The impact of persistent residues in soil, such as crude mineral hydrocarbons, polycyclic aromatic hydrocarbons (PAHs), heavy metals, and organochlorinated pesticides, has been discussed relative to the environmental health of the topsoil layer. The fundamental background of risk analysis has been related to practical use. Based on literature and our own field data, it has been proposed that the structure of the nematode community is the best ecologically and the most economically feasible method to evaluate soil quality.

INTRODUCTION

This paper will address the question of when remediation of a contaminated soil profile can be considered successful. The soil layer to consider is the arable, upper layer, containing residues of persistent contaminants such as crude mineral hydrocarbons, polycyclic aromatic hydrocarbons, chlorinated pesticides, and heavy metals. The answer to this question is provided by the structure of the nematode community in the soil. Figure 1 shows the relationship between the mineral hydrocarbon content in soil, and the Maturity Index (MI 2-5) of those soils. Here, the MI is a convincing indicator of the degree of hydrocarbon contamination.

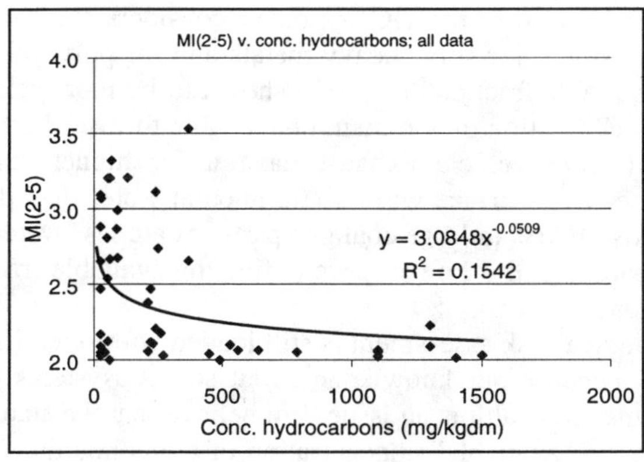

Figure 1. The relation between the structure of the nematode community, here expressed as a Maturity Index (MI), and the mineral hydrocarbon content in soils.

Before focusing on nematodes as a soil ecosystem biomonitor, some background information on the official risk approach is presented. Then practical field data are presented and discussed.

Risk assessment. In The Netherlands the multifunctional soil remediation approach from the early 1980s, initially used as a guideline by many countries, was replaced in 1996 by the functional approach. This new approach focuses on the risk of contamination to human and environmental health This agrees with the EPA approach of "natural attenuation," which is defined as "the biodegradation, diffusion, dilution, sorption, volatilization and/or chemical or biochemical stabilization of contaminants to effectively reduce toxicity, mobility or volume to levels that are protective of human health and the environment."

By "environmental health" we mean the health of the ecosystem soil. In short, we indicate how the process of "dealing with ecological risk" has been developed and refer to key publications.

Ecotoxicological effect assessment or risk assessment for the biotic part of the environment has been based on dose-effect (DE) data, or effective concentration (EC) data (Van Leeuwen, 1990). For example, an EC-5 value indicates that 5% of the earthworms have been killed due to a certain dose of a contaminant (Van Gestel & Ma 1988). For one contaminant, many data on different representative soil microflora and fauna have been evaluated and applied to set limits (Van Straalen & Denneman, 1989). In particular, the NOEC (no ecological effect concentration), the EC-5, and the EC-50 values have been extrapolated to standard values (Alderberg & Slob, 1993). Many data have been derived from well-defined rather homogeneous laboratory experiments, whereas the field is extremely heterogeneous, with spatial and temporal scales, and probably 2 to 5 times less vulnerable. Moreover, safety factors were suggested to consider the differences in sensitivity among species (Kooijman, 1987).

Currently, the HC-5 (hazardous concentration) value and the MAR (Maximal Allowed Risk) play a role in evaluating the expected ecological risk. Statistically but also ecologically, there are uncertainties and limitations with this approach (Smith & Cairns, 1993). For example, within the HC approach only one compound is causing stress. Furthermore, the HC approach considers the total content of the contaminants. However, for some heavy metals and for polycyclic aromatic hydrocarbons, the "available fraction" has been shown to be more relevant. The present "bound-residue" fraction may remain unavailable forever. Particles as small as diffuse lead from vehicle exhaust have a far higher surface/volume ratio and a larger contact surface with environmental water than do particles such as lead bullets. Also PAHs in large coal grit particles are less water-soluble than PAHs from dust particles. But how can we define the available fraction or the bound residue fraction?

The framework for ecological risk assessment is still loaded with scientific considerations (Renner, 1996), because our knowledge about soil ecosystems is limited. However, soil environmental health is an issue. We believe that the status of soil risk should be based on the "available" concentration of the contaminants and the biological in-situ status of the soil itself. Supported by well-respected soil ecologists such as Odum and Coleman (Coleman et al., 1998), we believe that the structure of the nematode community of a soil is the key factor, the best mirror of soil quality.

Nematodes. On earth 80% of all multicellular animals are nematodes. Nematodes are small worms (circa 0.4 to 5 mm in length) that live in every habitat where oxygen, water, and organic matter are available. They occupy key positions in soil food webs (De Ruiter et al., 1995). Soil nematodes live in the waterfilms between soil particles in various types of niches and come into contact with available contaminant fractions. Moreover their isolation is very efficient. Nematodes are ubiquitous. Throughout the world, 15,000 species have been described. In the Netherlands, 1 kg of soil contains on average 30,000 nematodes, consisting of 30 to 50 species.

Nematodes are diverse in their survival mechanisms and feeding methods (Bongers & Bongers, 1998). They consist of plant parasites (plant feeders), hyphal (fungi) feeders, bacterial feeders, substrate ingestors, carnivores (animal predators), and omnivores. Moreover, in their strategy for life, i.e., their ecological behavior, they can be differentiated in "colonizers" as enrichment opportunists (CP-1 group) and general opportunists (CP-2 group), and "persisters"(CP-3, CP-4, and CP-5 groups) which are sensitive to stress. In Table 1, CP values are given for terrestrial and freshwater nematodes (Bongers & Bongers, 1998).

Table 1 CP values for terrestrial and freshwater nematode communities.

Family	CP	Family	CP	Family	CP
Alloionematidae	1	Aulolaimidae	3	Teratocephalidae	3
Bunonematidae	1	Bastianiidae	3	Tobrilidae	3
Diplogasteridae	1	Criconematidae	3	Tripylidae	3
Diplogasteroididae	1	Cyatholaimidae	3	Alaimidae	4
Diploscapteridae	1	Desmodoridae	3	Anatonchidae	4
Neodiplogasteridae	1	Diphtherophoridae	3	Bathyodontidae	4
Odontopharyngidae	1	Diplopeltidae	3	Choanolaimidae	4
Panagrolaimidae	1	Dolichodoridae	3	Dorylaimidae	4
Rhabditidae	1	Ethmolaimidae	3	Ironidae	4
Tylopharyngidae	1	Halaphanolaimidae	3	Leptonchidae	4
Anguinidae	2	Hemicycliophoridae	3	Mononchidae	4
Aphelenchidae	2	Heteroderidae	3	Nordiidae	4
Aphelenchoididae	2	Hoplolaimidae	3	Qudsianematidae	4
Cephalobidae	2	Hypodontolaimidae	3	Trichodoridae	4
Monhysteridae	2	Leptolaimidae	3	Actinolaimidae	5
Myolaimidae	2	Linhomoiedae	3	Aporcelaimidae	5
Neotylenchidae	2	Meloidogynidae	3	Belondiridae	5
Ostellidae	2	Microlaimidae	3	Chrysonematidae	5
Plectidae	2	Odontolaimidae	3	Discolaimidae	5
Tylenchidae	2	Onchulidae	3	Longidoridae	5
Tylenchulidae	2	Pratylenchidae	3	Nygolaimidae	5
Xyalidae	2	Prismatolaimidae	3	Thornenematidae	5
Achromadoridae	3	Rhabdolaimidae	3		

The abundance of the CP groups leads to the Maturity Index.

$$MI = \sum_{i=1}^{n} v(i) \cdot f(i)$$

Where v(i) the CP value of family i and f(i) the frequency of this family in a sample.

Moreover, their isolation is very efficient. Quantification of nematodes is done routinely and costs ca. $200 US per soil sample. Bongers and Ferris (1999) provide in an overview of the reasons why nematodes can be used as bioindicators in an in situ environmental assessment system. For heavy metals the use of nematodes as bioindicators has been extensively described by Korthals et al. (1996b).

MATERIALS AND METHODS

Mineral crude hydrocarbon soils were collected two or three times a year from June 1994 to July 1998 from an experimental field (MS-1) where extensive landfarming was monitored as ecological recovery (NOVEM, 1998). The PAH-contaminated soils were collected from all over the Netherlands as part of nematode biomonitoring project (NOBIS, 1999). Soils containing chlorinated pesticides were collected from a location where pesticides had been produced many years ago (IWACO, 1998).

The soils were sampled in the same way. Every sample of a defined location consisted of 10 subsamples, taken with a soil core at a depth of 5 to 15 cm. The collection of mineral hydrocarbon soils took place in the upper 20-cm soil layer. The soil samples were packed in polyethylene bags and stored for a maximum of 2 days in a cooler (4 to 8°C). Between 100 and 500 grams of soil was used for nematode extraction by the "Oostenbrink-technique" (s'Jacob & van Bezooijen, 1984). The principle of this sampling technique is based on collecting the nematodes in water by using their specific gravity and a defined upflow waterstream in a funnel system. Extraction takes place when the nematodes pass through a cotton wool filter. The extraction period can last 1 or 2 days. Then the nematodes are counted and identified or processed (killed and stored) for counting and identification elsewhere. "Elsewhere" can be the BLGG in Oosterbeek (The Netherlands), where this characterization routinely takes place on samples from all over the world.

Other soil characterization concerned relevant physico-chemical data such as moisture content, pH, organic matter content, lutum fraction, nutrient status, and the contaminants involved. Often the vegetation of the site was described and the soil management (landfarming, soil along the railroad, wasteland, etc.) was characterized.

RESULTS

The following data regarding the nematodes were collected: (1) the total number per 100 grams of soil, (2) the number of bacteria-feeders, (3) the number of fungi-feeders, (4) percent omnivorous, (5) percent carnivorous, (6) the number in the

CP-1 group, (7) the number in theCP-2 group, (8) the number in the CP-3 to -5 groups, (9) the Maturity Index for the CP-1 to CP-5 groups (MI 1-5), and (10) the Maturity Index for the CP-2 to CP-5 groups (MI 2-5). When the number of bacteria-feeders (CP-1) was extremely high, as in nutrient-rich agricultural soils and in soils with easily degradable organic matter, this group was omitted and the MI 2-5 was used to omit the enrichment effect. In this presentation we focus on some selected items from the comprehensive data set.

Mineral hydrocarbon soils. Figure 2 presents the field data of the MI 2-5 in relation to the mineral hydrocarbon concentrations in the soils.

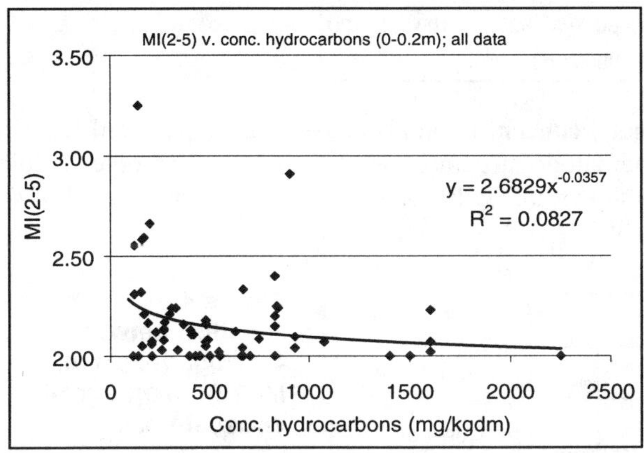

Figure 2. The MI (2-5) in relation to the mineral hydrocarbon concentration in the soils.

The number of bacteria-feeders was extremely high in the soils studied. This is explained by the activity of hydrocarbon degradation and the increase of oli-degraders and also because extra nutrients (nitrogen and phosphorous) were applied. There is a positive correlation between the hydrocarbon content and the percentage of bacteria-feeders. The rhabditus nematode species *Cuticularia oxycerca* has in those soils a special biomonitoring position because its numbers increase with increasing hydrocarbon content and decomposition. When the concentration decreases, *C. oxycerca* also disappears. The Maturity Index shows a nice correlation with the hydrocarbon content. The limiting factors for ecological recovery are more or less completely gone, in this loamy sandy soil with pH ca. 6.5, when the hydrocarbon concentration is lower than 1000 mg/kgdm.

Polycyclic aromatic hydrocarbons. Figure 3 presents the MI (2-5) relative to the PAH concentrations in the soil. Those data are not so clear. Here, the MI(2-5)-index is so high that it appears as if PAHs have no toxic effect in those soils. This is in accordance with the research of Hund & Traunspurger (1994), who concluded that PAHs in soils quite quickly become harmless residues. The Maturity Index generally is higher than in the mineral oil soils. The PAH soils studied vary much more in their intrinsic physico-chemical characteristics than the hydrocarbon soils studied. In general, clay soils have a more abundant nematode community structure than do sandy soils or peat soils.

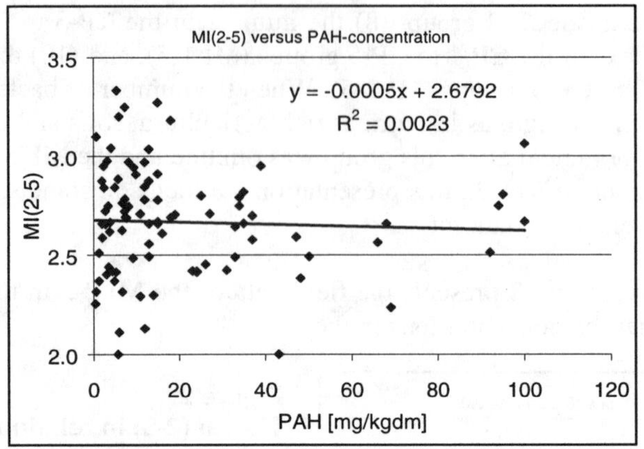

Figure 3. The MI (2-5) in relation to the PAH concentration in the soils.

The pH and heavy metal concentration also have to be considered for their negative influence on the nematode structure, as already found in other studies (Korthals et al., 1996a) and shown in Figure 4. This aspect will be studied in detail by multiple variance analysis.

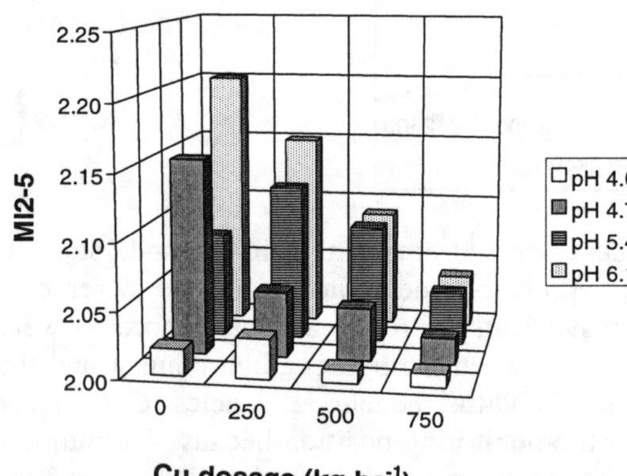

Figure 4. Relation between Cu dosage, pH and MI(2-5) in an agroecosystem.

Persistent chlorinated compounds. Figure 5 presents the MI (2-5) relative to concentrations of organo-chlorinated pesticides (OCP), (i.e., residues of DDT, DDE, and DDD).

According to Dutch standard values, concentrations of those persistent chlorinated compounds were considered too high for safe use of the soils. With the nematode "crop" as the mirror we tend to conclude that those soils are ecologically safe. Additional availability determinations by comparing the solubility in water and octanol and accumulation experiments with earthworms will demonstrate whether the nematode approach is appropriate for those compounds.

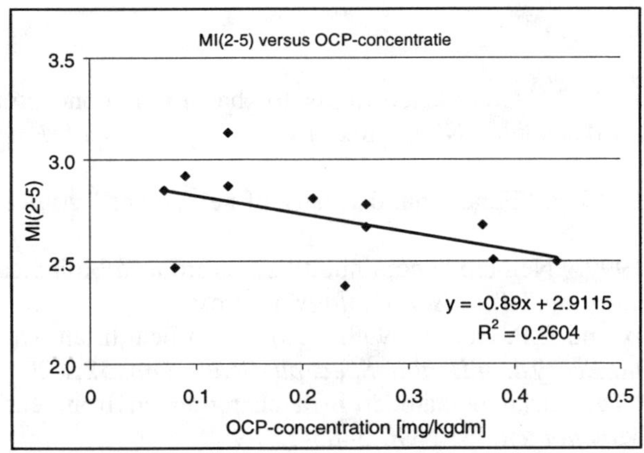

Figure 5. The MI (2-5) in relation to the OCP concentration in the soils.

Figure 6 presents the CP triangle of the OCP-contaminated soils.

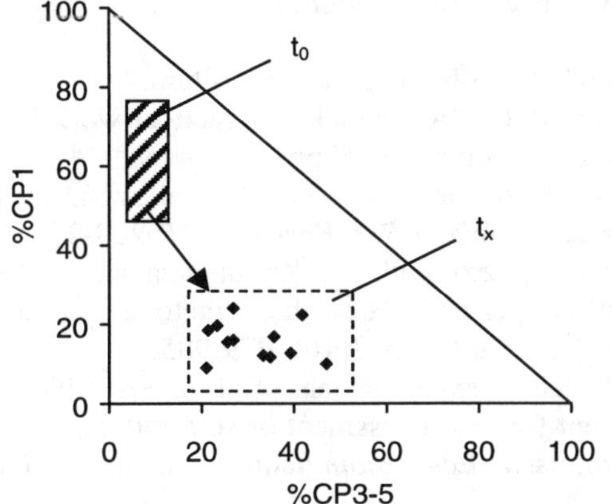

Figure 6. The CP triangle of the OCP soils.

For monitoring the ecological recovery during years, this triangle is very appropriate. We speculate that proportional shifts took place from the CP-1 area (the rectangular area) at time zero towards the CP 3-5 area at time x. The improved quality of the soils can be shown with those short-term or long-term successional changes (De Goede, 1993). Nematode data in relation to soil type, vegetation, and stress provide good insight into soil quality.

REFERENCES

Alderberg, T. and W. Slob. 1993. "Confidence limits for hazardous concentrations based on logistically distributed NOEC toxicity data." *Ecotoxicol. Environ. Safety 25*: 48-64.

Bongers, T. and M. Bongers. 1998. "Functional diversity of nematodes." *Applied Soil Ecology 318*: 1-13.

Bongers, T. and H. Ferris. 1999. "Nematode community structure as a bioindicator in environmental monitoring." *Trends in Ecology*. In press.

Coleman, D.C., P.F. Hendrix and E.P. Odum. 1998. "Ecosystem health: an overview." *Soil Chemistry and Ecosystem Health. Spec. publication no. 52*:1-20.

De Goede, R.G.M. 1993. "Terrestrial nematodes in a changing environment." *PhD thesis, University Wageningen*. ISBN 90-5485-130-9.

De Ruiter, P.C., A.M. Neutel, and J.C. Moore. 1995. "Modelling food webs and nutrient cycling in agro-ecosystems." *Trends in Ecological Evolution 9*: 377-383.

Hund, K. and W. Traunspurger. 1994. "ECOTOX — evaluation strategy for soil bioremediation exemplified for a PAH-contaminated site." *Chemosphere 29*(2): 371-390.

IWACO. 1998. "Alternative remediation option, location Kollum."

s'Jacob, J.J. and J. van Bezooijen, 1984. "A manual for practical work in nematology." Wageningen Agricultural University. 77pp.

Kooijman, S.A.L.M., 1987. "A safety factor for LC 50 values allowing for differences in sensitivity among species." *Water Res. Research 21*: 269-276.

Korthals, G.W., A.D. Alexiev, Th. M. Lexmond, J.E. Kammenga and T. Bongers. 1996a. "Long-term effects of copper and pH on the nematode community in an agroecosystem." *Environ. Tox. and Chem. 15*(6): 979-985.

Korthals, G., de Goede, R.G.M., Kammenga, J.E. and T. Bongers. 1996b. "The maturity index as an instrument for risk assessment of soil pollution." *In*: Van Straalen, N.M. & D.A. Krivolutsky (Eds.) *Bioindicator Systems for Soil Pollution*. Kluwer: 85-93.

NOBIS. 1999. "Risk assessment of PAH soils by nematodes." In press.

NOVEM. 1998. "Ecological recovery of landfarming."

Renner, R. 1996. "Ecological risk assessment struggles to define itself: practitioners are debating how to turn this young field into a rigorous discipline." *Environ. Sci. Technology* 30: 172-174.

Smith, E.P. and J. Cairns. 1993. "Extrapolation methods for setting ecological standards for water quality: statistical and ecological concerns." *Ecotoxicology* 2: 203-219.

Van Gestel, C.A.M. and W.C. Ma. 1988. "Toxicity and bioaccumulation of chlorophenols in earthworms, in relation to bio-availability in soil." *Ecotoxicol. Environ. Safety* 15: 289-297.

Van Leeuwen, K. 1990. "Ecotoxicological effects assessment in the Netherlands: recent developments." *Environ. Management 14*: 779-792.

Van Straalen, N.M. and G.A.J. Denneman. 1989. "Ecological evaluation of soil quality criteria." *Ecotoxicol. Environ. Safety 18*: 241-251.

PHYTOREMEDIATION OF PERCHLORATE CONTAMINATED WATER: LABORATORY STUDIES

Valentine A. Nzengung and C. Wang (University of Georgia, Athens, GA., USA)
Harvey, G. (AEM, Wright Patterson AFB, Dayton, OH, USA)
McCutcheon, S. and L. Wolfe (USEPA-NERL, Athens, GA., USA)

Abstract: The uptake and degradation of perchlorate by three woody plants (willow (*Salix ssp.*), Cottonwoods (*Populus ssp.*) and *Eucalyptus cineria*) species was investigated. Perchlorate was removed by willow from bioreactors dosed with 10, 20 and 100 mg/L of perchlorate to below the method detection limit of 2 µg/L by willows and Eucalyptus. Non-woody plants, such as spinach, French tarragon, and *Myriophyllum* (an aquatic plant) also removed perchlorate from the contaminated water. Two phytoprocesses identified as important in the remediation of perchlorate-contaminated water included: (1) uptake and phytodegradation in the plant organs, and (2) rhizodegradation. Experiments conducted with growth media taken from the rhizosphere of willow trees confirmed that rhizosphere-associated microorganisms mediated the rhizodegradation of perchlorate to chloride. High nitrate concentrations interfered with the rhizodegradation of perchlorate because nitrate acted as a competing terminal electron acceptor (TEA). Uptake of perchlorate by woody plants was minimized in medium with ammonium as the N-source. Since perchlorate does not volatilize from water readily, a perchlorate remediation scheme may involve an intensively cultivated plantation of trees with phraetophytic characteristics, and irrigation with the contaminated water.

Introduction

Perchloric acid and its salts have been used extensively in a number of commercial applications such as wet digestions, organic syntheses, electropolishing of metals, animal feed additives, explosives, pyrotechnics, missile propellants and herbicides. In addition, it is found as a contaminant in certain fertilizers and bulk water treatment chemicals (Schilt, 1979). Perchlorate explosives and propellants are less sensitive to shock and environmental changes than dynamite. Accordingly, the United States military and National Aeronautical Space Administration (NASA) has extensively used ammonium perchlorate as solid rocket propellant. Recently, perchlorate has been found in ground water at numerous hazardous waste sites in California, Nevada, Utah, Texas, and at low levels (5 to 9-µg/L) in the Colorado River. The highest perchlorate concentration detected in surface and groundwater are 1.7-mg/L (Lake Mead inlet) and 8-mg/L, respectively (Herman and Frankenberger, 1998). Being a salt, ammonium perchlorate is readily soluble in water and remarkably stable.

The anion perchlorate poses potential environmental concerns because its ionic radius and charge are similar to that of iodine, which allows perchlorate to competitively block thyroid iodine uptake. At relatively high doses, perchlorate is known to interfere with the thyroid's ability to produce hormones and regulate metabolism (Cao, 1994). A severe deficiency can cause cretinism. It is considered the world's most readily preventable cause of mental retardation. In iodine deficient populations a mild form of cognitive impairment occurs five times as frequent as cretinism and the IQ curve of the entire population can be shifted to the left (Ekdahl, 1948). Although there is no National Primary Drinking Water Regulation (NPDWR) or Health Advisory established for perchlorate at this time, the California Department of Health Services has established an action level for perchlorate in drinking water of 18 μg/L (Nzengung et al., 1999).

Perchlorate is not air strippable and so is not a candidate for conventional pump-and-treat remediation. Various approaches have been investigated for the remediation of perchlorate contamination including adsorption by activated carbon, reverse osmosis, anion exchange, and bioremediation. Phytoremediation is an emerging biotechnology that is rapidly gaining interest because it promises effective and inexpensive cleanup of certain hazardous waste sites. Plants may take up and assimilate contaminants (phytoaccumulation), volatilize the contaminants into the atmosphere (phytovolatilization), or degrade the contaminants within plant tissues using enzymes (phytodegradation). In the rhizosphere of some plants, released plant exudates and enzymes that stimulate biochemical activity may enhance the biodegradation of environmental contaminants (rhizodegradation). Also, plants may be used to absorb and precipitate large quantities of toxic metals in soils, thus reducing their bioavailibility and preventing their entry into groundwater and food chains (phytostabilization) (Flathman and Lanza, 1998).

The general goal of this study was to investigate the feasibility of using plants for decontaminating water polluted with perchlorate. The specific objectives were to: (1) determine the primary phytoprocesses, (2) determine the role of plant nitrogen source and the effect of NO_3^- (a likely competing TEA), used as N-source in the nutrient media, and (3) verify if bacteria isolated from the rhizosphere the study plants have perchlorate degrading ability.

MATERIALS AND METHODS

Woody plants used in this study included rooted cuttings of willow (*Salix* spp.) and Eastern Cottonwood (poplar) and one-year old *Eucalyptus cineria* plants. The harvested cottonwood and willow cuttings [40-cm (length) by 4-mm (stem diameter)] were rooted hydroponically in half-strength Hoagland's solution (Burken and Schnoor, 1998) for three months. Initial perchlorate degradation studies, including screen tests, were conducted in 2-L screw-cap sand bioreactors. Because perchlorate degradation was mass-transfer limited in the sand bioreactors, subsequent studies were performed only in hydroponic bioreactors. The growth solutions were diluted Hoagland's solution and Miraclegro™. Three initial perchlorate concentrations, as ClO_4^-, of approximately 10, 20, and 100 mg/L, were used in these experiments. Accompanying each set of experiments were dosed blanks (no plant) and undosed-planted controls. A daily record of the volume of water taken up by each tree was maintained over the duration of each study. A 1-mL aliquot of the solution was also withdrawn for perchlorate analysis once every 24 h. The distribution of perchlorate in different plant fractions was determined by sacrificing the study trees for extraction and analysis at the termination of the experiments. Each fraction was extracted several times by blending for 30 - 60 min with a solution of 1 mM NaOH (pH = 11). The extract was analyzed for perchlorate, chlorate, chlorite and chloride ions.

Degradation of perchlorate by herbs and microbial mats was investigated in 60-mL serum bottles. French tarragon and spinach were each minced before used in the perchlorate studies. One gram of the minced plant or microbial mats were added to the vials, filled with deionized water, dosed with perchlorate to obtain an initial solution concentration of 10-mg/L and mixed on a rotary shaker. The pellets were separated from solution by centrifugation, and the liquid-phase was analyzed by Ion Chromatography (IC).

Determination of Nitrate Effects. Experiments were performed in diluted Hoagland's solution supplemented with NO_3^- to achieve three desired concentration ranges, <1.6 mM, 1.6 – 6.4 mM, and >6.4 mM. The nitrate concentration in these experiments was maintained at the specified concentrations by adding sodium nitrate as needed to achieve the targeted NO_3^- levels. Each set of reactors was dosed multiple times with about 100-mg/L perchlorate. The chloride, nitrate, and acetate concentrations of the growth solution were measured during the course of each experiment.

Homogeneous Studies. 50-mL of the media withdrawn from the rhizosphere of the willow reactor after perchlorate was completely degraded to the IC method detection limit of 2-µg/L was placed in eight serum bottles. One set of vials was amended with either 3.2-mM of NO_3^-, or 6.8 mM acetate or solution of 3.2 and 6.8 mM of NO_3^- and acetate, respectively. One pair of vials contained the unamended media. The samples and controls were each dosed with perchlorate to obtain an initial solution concentration of 100 mg/L and incubated at ambient temperature.

The microbial contribution to perchlorate degradation in the rhizosphere was determined in two ways. Subsamples of the previously described rhizosphere solution were boiled for 2-h or filtered with a 0.45-µm membrane filter before dosed with perchlorate. Equal volumes (50-mL) of the autoclaved, filtered and untreated media were each placed in three sterilized serum bottles dosed immediately with perchlorate and sealed. The vials were continuously mixed on a shaker until sacrificed for analysis by IC (Nzengung et al., 1999). Bacteria were isolated from the tree rhizosphere and used to degrade perchlorate.

RESULTS AND DISCUSSION

Rooted cuttings of three woody plants (cottonwood, *Eucalyptus* and willow) removed perchlorate ions from aqueous solution. Tables 1 shows that *Eucalyptus* was relatively more effective in phytoremediation of perchlorate than the cottonwood and willow trees. *Eucalyptus* had the largest fraction root (6.6%), leaf (46%), and evapotranspiration rate than the cottonwood and willow. Comparing the perchlorate mass fraction extracted from the different plant organs, it is evident that the perchlorate taken up by all three woody plants was mostly accumulated in leafs and branches (Table 1). Sorption studies confirmed that perchlorate was not sorbed by the sand.

Table 1: Distribution of perchlorate in woody plants grown in sand bioreactors for 26 days.

TREE TYPE	LOCATION	MASS (mg)	PERCENT
Cottonwood	Roots & Stems	0.1	0.2
	Leaves & branches	4.2	4.9
	Bioreactor	78	92
	Phytoremediated	**2.6**	**3.1**
Eucalyptus	Roots & Stems	0.007	0.009
	Leaves & branches	22.8	29.6
	Bioreactor	23.3	30.2
	Phytoremediated	**31.1**	**40.2**
Willows	Roots & Stems	0.6	0.7
	Leaves & branches	0.5	0.6
	Bioreactor	76.2	87.7
	Phytoremediated	**9.5**	**11**

Figure 1 shows that following an initial exposure of the woody plants to perchlorate-dosed nutrient media the rate of perchlorate reduction in the medium increased by several orders of magnitude. The very fast kinetics observed after the initial spike suggested that the predominant reaction mechanism was degradation in the root zone (rhizodegradation) and not phytoextraction. This was directly confirmed by monitoring the chloride concentration in the rhizosphere. The estimated zero-order rate constants for the 2nd, 3rd, 4th, and 5th spikes were 15.3, 20.5, 32.6, and 25.9 μM/h. The average water uptake rate by the willow during the latter experiment was 64.5, 129, 93, 95, 80 mL/day for the 1st, 2nd, 3rd, 4th, and 5th spikes, respectively. The perchlorate concentration in the unplanted reactor did not change and control and sample plants continued to grow at the same rate over the course of this study.

Direct evidence of phytodegradation was obtained from experiments conducted with crude extracts of French Tarragon and Spinach and the minced plants. The degradation of perchlorate by the crude extract and minced herbs provided additional evidence that the reactions in the plant tissues were enzymatically catalyzed. Possible transformation products of perchlorate such as chlorate, chlorite, hypochlorite and dichlorooxide were analyzed for but none was identified in the medium or plant extract. However, the chloride concentration in the rhizosphere increased as each dose of perchlorate was completely degraded to below the IC detection limit (Fig. 1).

Effect of Nitrate Concentration on Perchlorate Degradation. The fastest perchlorate degradation kinetics were observed at NO_3^- concentrations of

<1.6 mM and decreased with increasing NO_3^- concentration Fig. 2. The utilization rates of NO_3^- and perchlorate were described by first and zero-order kinetics, respectively, at the lowest NO_3^- concentrations. Average perchlorate utilization rates of 23.6 ± 3.7 and 2.3 ± 0.2 μM/h were obtained for experiments with the NO_3^- concentration maintained at <1.6 and 1.6 – 6.5 mM, respectively. We attributed the slow reduction rates of perchlorate at high NO_3^- concentrations in Fig. 2 to competing reactions in which both anions were utilized as TEAs. This shortcoming was overcome by replacing the NO_3^- N-source (Hoagland's solution) with an ammonium N-source (MiracleGro) (Fig. 3).

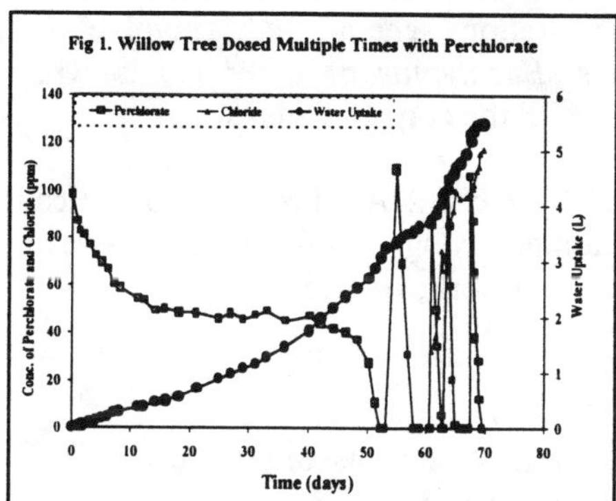

Fig 1. Willow Tree Dosed Multiple Times with Perchlorate

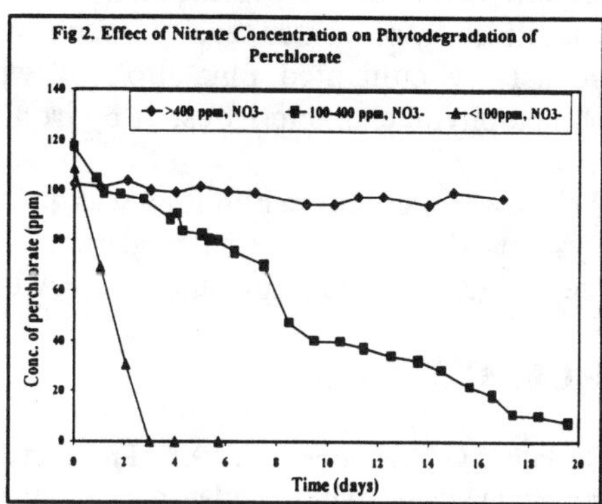

Fig 2. Effect of Nitrate Concentration on Phytodegradation of Perchlorate

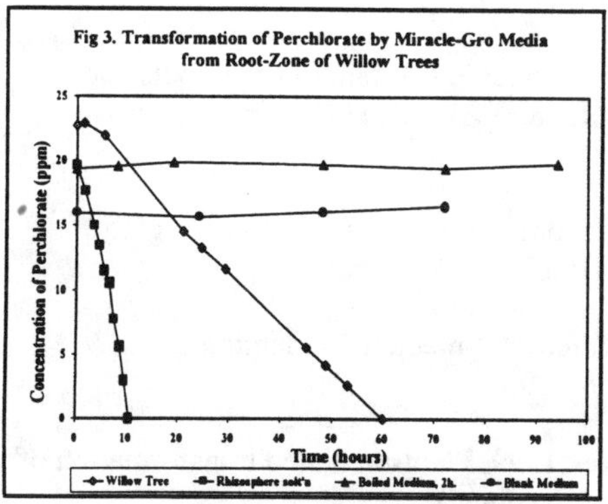

Fig 3. Transformation of Perchlorate by Miracle-Gro Media from Root-Zone of Willow Trees

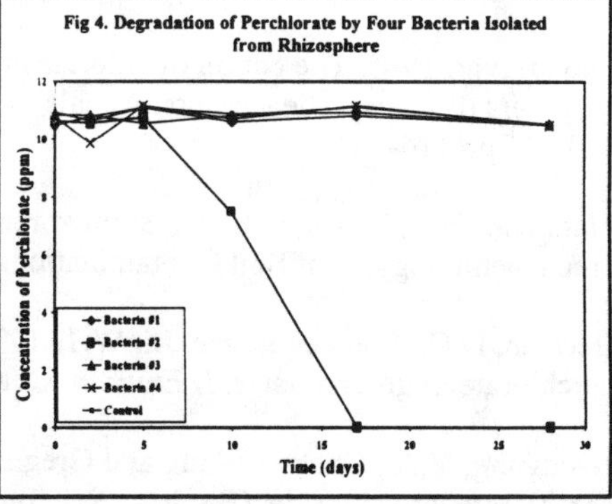

Fig 4. Degradation of Perchlorate by Four Bacteria Isolated from Rhizosphere

Growth media withdrawn from the rhizosphere of willow trees previously used in hydroponic experiments degraded perchlorate to chloride (Fig. 3). Nutrient media supplemented with $NaNO_3$ to achieve 3.2 mM NO_3^- showed no activity. However, the unamended media or media amended with acetate or both acetate and NO_3^- degraded perchlorate to chloride. This suggests that a high concentration of the carbon (electron) source is not inhibitory and may reverse the inhibitory effects observed at high NO_3^- activity in the root zone. Fig. 4 shows that one of four bacteria isolated from the rhizosphere of willow trees degraded perchlorate and most likely mediated the observed rhizodegradation reactions.

Our data shows that selected plants can be used to detoxify environments contaminated with perchlorate and the effectiveness of the approach depends on the N-source and the concentration of competing TEAs, such as NO_3^-. The initial or prolonged exposure of rooted willow trees to perchlorate-dosed media stimulated the growth of perchlorate-degrading microorganisms in the rhizosphere. Two phytoprocesses were identified as important in the remediation of perchlorate-contaminated water: (1) uptake and phytodegradation of perchlorate in the tree branches and leafs, and (2) rhizodegradation. Pilot testing of the approach should provide data needed to validate the feasibility of phytoremediation of perchlorate-contaminated sites. Since perchlorate does not volatilize from water easily, such a remediation scheme may involve an intensively cultivated plantation of willows, *Eucalyptus* or other plants with phraetophytic characteristics, or by irrigation with the contaminated water.

Acknowledgements: Funding was provided by the US Air Force Aeronautical Systems Environmental Management Directorate through a subcontract between the University of Georgia and USEPA-NERL in Athens, Georgia.

REFERENCES

Burken, J.G., Schnoor, J.L. 1998. Predictive relationships for uptake of organic contaminants by hybrid poplar trees. Environ. Sci. Technol. 32, 3379.

Cao Xi. 1994. Timing of Vulnerability of the Brain to Iodine Deficiency in Endemic Cretinism. New England Journal of Medicine. 331, 1739.

Ekdahl Ivar. 1948. The action of chlorate and some related substances upon roots and root hairs of young wheat plants. Annals of the Royal Agricultural College of Sweden Vol 15, pp 114-168.

Flathman, P.E., Lanza, G.R. 1998. Phytoremediation: current views on an emerging green technology. J. of Soil Contamination. 7, 415.

Herman, D.C., Frankenberger. Jr., W.T. 1998. Microbial-mediated reduction of perchlorate in groundwater. J. Environ. Qual. 27, 750.

Nzengung, V.A., Chuhua Wang and Greg Harvey.1999. Plant-mediated transformation of perchlorate into chloride. (Environ. Sci. Technol. accepted).

Rikken, G. B., Kroon, AGM., van Ginkel, C.G. 1996. Transformation of (per)chlorate into chloride by a newly isolated bacterium: reduction and dismutation. Appl. Microbiol. Biotechnol., 45, 420.

Schilt, Alfred A. 1979. Perchloric acid and perchlorates. The G. Frederick Smith Chemical Co., Columbus, Ohio.

Van Ginkel, C.G., Plugge, C.M., Stroo, C.A. 1995. Reduction of chlorate with various energy substrates and inocula under anaerobic conditions. Chemosphere. 31, 4057.

AEROBIC IN SITU STABILIZATION OF A COMPLETED LANDFILL

Daewon Pak and Soochul Kim (KIST, Seoul, Korea)
Dongheng Cho (Inha University, Incheon, Korea)

ABSTRACT: Active aeration was tested as a way to accelerate the conversion of a landfill body to the status of low biological activity. During the test, the development of the changes within the landfill body was documented through changes in the composition of landfill gas, the temperature, and the in situ respiration rate. The biological activity, determined by the mean in situ respiration rate, was high in the area where active aeration was applied. Due to high biological activity, temperature increased up to 70°C. A geophysical survey of the landfill showed that the electrical resistivity of the landfill body decreased as stabilization proceeded.

INTRODUCTION

In Korea, approximately 590 landfills are in operation and more than 900 completed landfills and old sites are known in total. Most of those completed landfills and old sites do not have a bottom liner and, thus, an environmental impact may occur via emissions of gas and leachate. A study on the long-term behavior of landfills has reported that it takes long periods of 100 to 300 years for emission concentrations of especially leachate to drop to completely harmless limiting values (Heyer et al., 1996).

To minimize emissions from landfills, a landfill capping system can be considered to secure the landfill body. However the cap may fail to reduce emissions while maintaining the emission potential in the landfill. To reduce the emission potential, the landfill body should be transferred as soon as possible to the status of low biological activity. This can be done by accelerating the microbial conversion processes in the landfill. To achieve this, active aeration is a suitable measure.

In this study, active aeration was applied to a completed landfill. The active aeration system was designed at pilot scale and operated to accelerate the microbial conversion processes in the landfill. The goal of the project was to demonstrate that active aeration is effective in converting a landfill body to the status of low biological activity. The development of the changes within the landfill body was documented through changes in the composition of landfill gas, the temperature, and the in situ respiration rate. Geophysical surveys were carried out on the landfill during its stabilization. Electrical resistivity tomography was used in an attempt to verify that the landfill was becoming stabilized.

MATERIALS AND METHODS

Site Description. The in situ test was conducted on the completed landfill with 63,180 m^2 of area located 10 km to the northeast of Seoul. The landfill had been built in 1991 and operated for 4 years. The waste volume deposited from 1991 to 1994 was estimated to be about 125,000 m^3. The site investigation indicated that the landfill was actively generating landfill gas and leachate. Methane was estimated to be 30 to 65% of the total landfill gas. The proportional composition of the landfill body is presented in Figure 1.

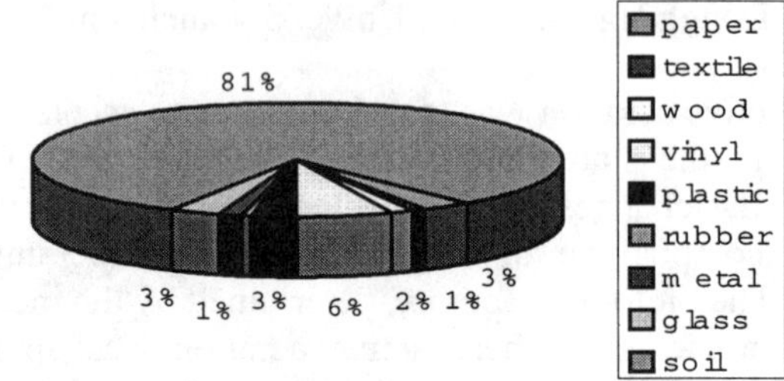

FIGURE 1. Proportional composition of the waste in the landfill

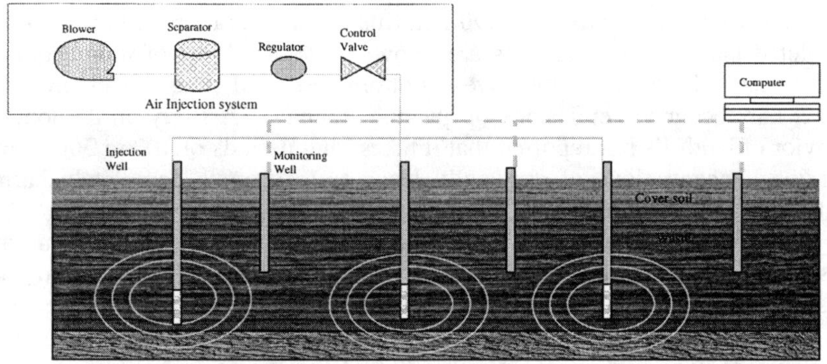

FIGURE 2. Active aeration system for stabilization of completed landfill

Active Aeration System. The active aeration system presented in Figure 2 was designed in pilot scale and operated to activate the biological degradation processes. It consists of an air injection system, air injection wells, and monitoring wells. In the air injection system, a blower was used to supply air. Air is distributed to the injection well by the air distributor designed to operate at both low pressure and high pressure. Three 4-in-diameter wells were constructed of stainless steel for air injection. Nine 2-in-diameter wells were constructed of polyvinyl chloride (PVC) for monitoring the landfill gas composition, temperature, and pressure. Active aeration was initiated in December 1997 and continued until the end of 1998.

RESULTS AND DISCUSSION

Changes in Landfill Gas Composition. Once air was introduced into the landfill, methane in the landfill gas collected from the monitoring wells decreased immediately to a low level and oxygen started to increase. The oxygen concentration of landfill gas in #1 monitoring well 5 m distant from #1 injection well reached 16%. The lower oxygen level in landfill gas collected from #3 monitoring well 15 m distant from #1 injection well indicated that oxygen was being consumed after air introduction. The oxygen level in the monitoring wells was maintained at greater than 2% of the landfill gas. The radius of air influence, defined as the distance from the air injection well to the monitoring well where oxygen can be detected, was estimated to be 15 m.

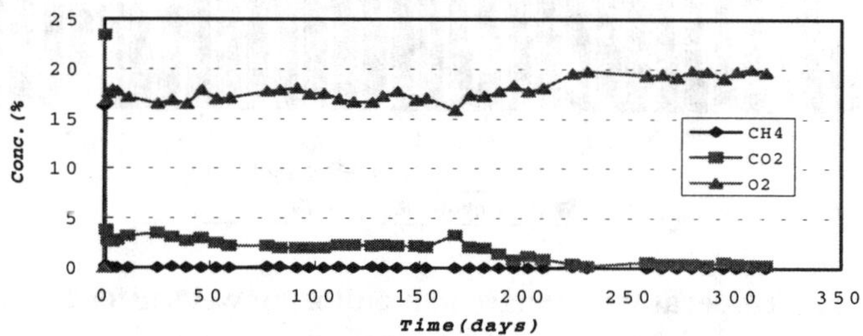

FIGURE 3. Changes in landfill gas collected from #1 monitoring well 5 m distant from #1 injection well after the initiation of active aeration

In situ Respiration Test. In situ respiration test was conducted periodically in the injection well as a mean of rapid field measurement of biological activity. The in situ respiration rate was estimated by monitoring the oxygen concentration over time after air injection was turned off. The results are plotted in Figure 4. The oxygen utilization rate was estimated to be 0.2% oxygen per hour before initiation of active aeration. After three months of active aeration, the oxygen utilization rate increased up to 0.8% oxygen per hour in #2 monitoring well. The increased rate indicated that the waste deposited around #2 monitoring well was highly degradable. The oxygen utilization rate decreased slowly as stabilization of the landfill proceeded further.

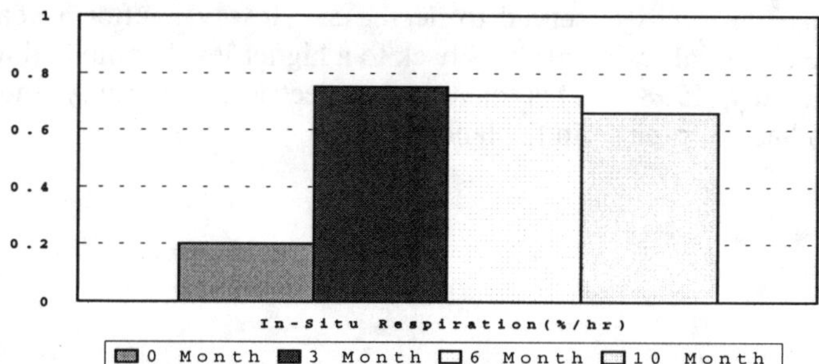

FIGURE 4. The results of in situ respiration test conducted at injection well

Changes in Temperature of Landfill Body. Due to high microbial activity verified by the high in situ respiration rate, the temperature started to increase as stabilization of the landfill proceeded. After three weeks of active aeration, temperatures in all monitoring wells increased from 20°C to between 40 and 60°C. After two months of active aeration, a temperature of 70°C was observed in #2 monitoring well. The high temperature coincided with the high respiration rate observed in #2 monitoring well.

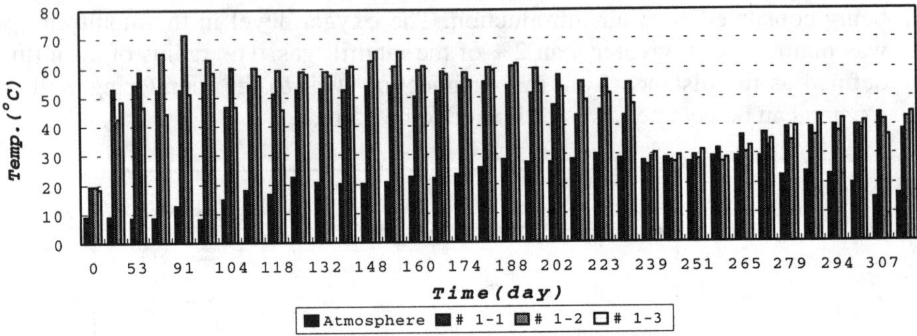

FIGURE 5. Temperature changes in monitoring wells after the initiation of active aeration

Change in Electrical Conductivity. The geophysical properties of typical refuse can be used to distinguish refuse from native soils (Jensen et al., 1993). For example, mapping theelectrical conductivity change is comparatively easy and can be done with high resolution and little labor input. Electrical conductivity is known to be a function of the following parameters: degree of saturation, gas generation, internal temperatures, leachate generation, mobility of leachate, and compaction density (McCann, 1994). These parameters are related with the changes that can be observed during stabilization of landfill. We attempted to relate the changes in the landfill body during its stabilization to the changes in electrical conductivity. Figure 6 shows pseudosections created from electrical resistivity data measured on the landfill. Red indicates low electrical resistivity and blue high electrical resistivity. As stabilization of the landfill proceeded, the electrical resistivity was observed to decrease. However, after a year of active aeration, the electrical resistivity was back to a higher level compared to the levels observed in July 1998. This change in electrical resistivity indicates that stabilization has taken place in the landfill.

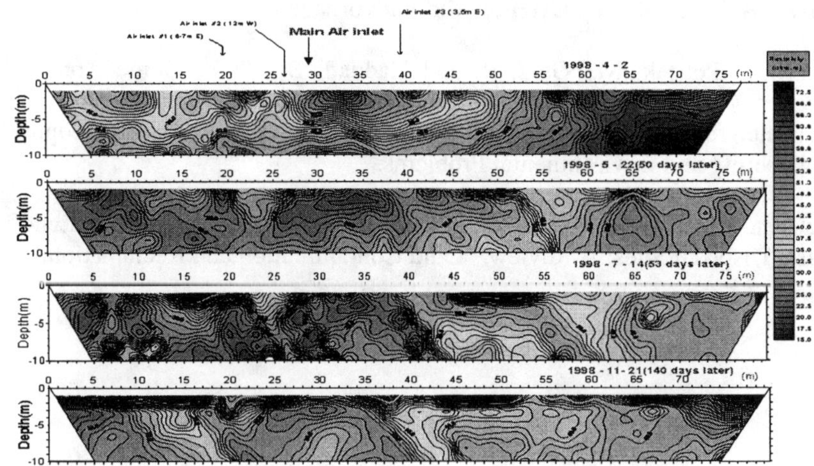

FIGURE 6. Electrical resistivity change observed in landfill during its stabilization

CONCLUSIONS

The long-term emission potential of landfill can be reduced by accelerating the microbial conversion processes. Active aeration was demonstrated to be effective as a mean to accelerate the conversion of a landfill body to the status of low biological activity. During the test, the development of the changes within the landfill body was documented through changes in the composition of landfill gas, the temperature, and the in situ respiration rate. Once air was introduced into the landfill, methane in the landfill gas collected from the monitoring wells decreased immediately to a low level and oxygen started to increase. According to the change in landfill gas composition, the radius of air influence, defined as the maximum distance from the air injection well to a monitoring well where oxygen can be detected, was 15 m in a given condition. The biological activity, determined by means of an in situ respiration test, was high in the area where degradable waste had been deposited. Due to high biological activity, the temperature increased up to 70°C. A geophysical survey carried out on the landfill shows that the change in the electrical resistivity of the landfill body can be an indication that stabilization has occurred.

REFERENCES

Bernstone, C. and Dahlin, T. 1997. "Characterization of Landfills with the Aid of DC-Resistivity." In Proc. Sardinia 97, Sixth International Landfill Symposium.

Ferguson, I.J., Taylor, W.J. and Schmigel, K. 1996. "Electromagnetic Mapping of a Saline Contamination at an Active Brine Pit." *Canadian Geotechnical Journal*, 33.

Heyer, K.U., and Stegmann, R.. 1996 Erweiterter Zischenbericht zum Teilvorhaben TV 4 der TUHH im Verbundvorhaben.

Jensen, J. Pencak, M., Gnat, R. and Haddad, B. 1993. "Some Application of Frequency Domain Electromagnetic Induction Survey for Landfill Characterization Studies." in Proc of Symp. on the Appl. of Geophysics to Engineering and Environmental Problems.

McCann, D.M. 1994. "Geophysical Methods for the Assessment of Landfill and Waste Disposal Sites: A Review." Land Contamination and Reclamation.

RAPID LANDFILL STABILIZATION AND IMPROVEMENTS IN LEACHATE QUALITY BY LEACHATE RECIRCULATION

Kenneth W. Kilmer, Gerv C. Griffin, and Ian D. MacFarlane EA Engineering, Science, and Tech., Inc., Hunt Valley, Maryland
John Tustin, Department of Public Works, Worcester County, Maryland

ABSTRACT: Leachate recirculation is an important new concept in municipal solid waste (MSW) leachate management, and the full-scale applications of the practice are only now being implemented. The Worcester County, Maryland, Central landfill Facility (CLF) has been recirculating leachate since opening in April 1990. EA Engineering, Science, and Technology, Inc. designed the landfill in 1988, and has been providing operational consulting on the process since1990.

Rapid stabilization of the waste mass is the primary objective of leachate recirculation. The Enhanced Natural Degradation process (ENDpro^{SM})is facilitated by wetting with the recirculated leachate. From a practical standpoint, this results in increased available disposal volume, reduced leachate disposal costs, and a dramatic reduction in the strength of the leachate.

The 8-year operational history at the CLF has shown leachate chemistry stabilization comparable to conventional MSW landfills typically with 30-year stabilization time frames. Analytical trends which demonstrate accelerated biological activity are rapid decreases in biological oxygen demand, metals and volatile organic compounds as well as increased pH to neutral. Other electron acceptors such as sulfate and nitrate are also assessed in relation to enhanced biodegradation processes.

OVERVIEW

This paper will demonstrate how the bioreactor landfill utilizes the recirculation of leachate to encourage and accelerate the natural decomposition of the solid waste.

The solid waste is decomposed by the natural processes of hydrolysis and methanogenesis. In this multiphase decomposition process, the leachate is treated to very low levels of contaminants and the waste mass is stabilized and consolidated.

Reductions in organic strength, chlorinated volatile organics and metals accompany the geochemical changes that landfill leachate goes through during decomposition of the solid waste.

The dramatic improvements in leachate quality to near drinking water levels for most constituents make a powerful argument for reduced post-closure ground water monitoring and financial assurance requirements.

THE LANDFILL BIOREACTOR PROCESS

The natural decomposition of waste in a landfill has been studied and characterized by several authors. The first to identify and publish this process were Pohland and Harper in 1986.

The rate of decomposition is dramatically affected by the moisture level of the waste as shown in the work of Rathjee in the well publicized "Garbage Project."

According to Pohland and Harper, a landfill cell, through most of its active life, exists as an anaerobic microbial process and undergoes five stabilization phases (See Figure 1). Principal events occurring in each phase are outlined below:

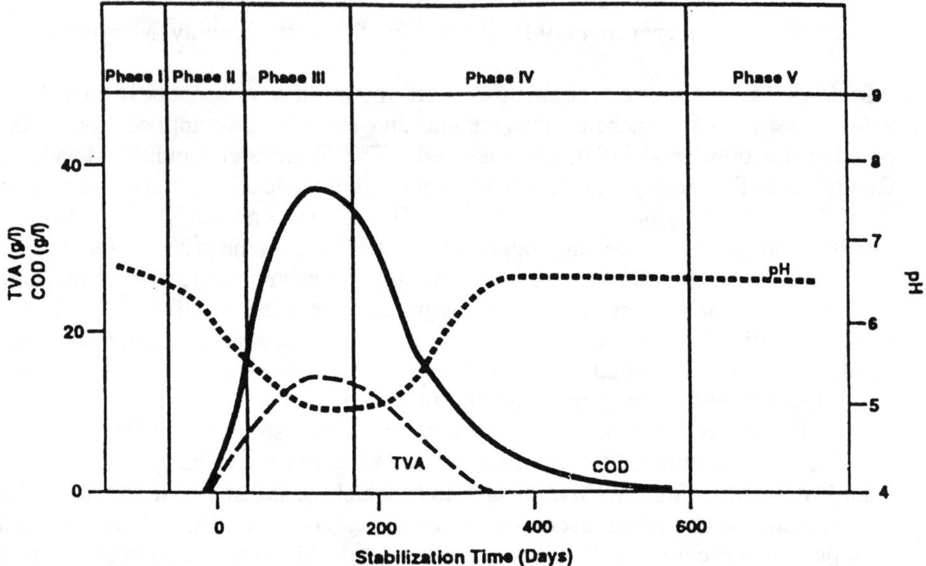

Figure 1: **Variations in Chemical Indicators - Leachate Recycle Study (Source: Pohland 1986).**

Phase I/II: Initial Adjustment/Transition

- Initial waste placement
- Field capacity is exceeded and leachate is formed.
- A transition from aerobic to anaerobic microbial stabilization occurs.

Phase III: Acid Formation

- Volatile fatty acids become predominant with hydrolysis and fermentation of waste
- A decrease in pH occurs with a concomitant mobilization of metals.
- Nitrogen and phosphorus are released and utilized by the biomass.

Phase IV: Methane Fermentation

- Fatty acids are converted to methane and carbon dioxide.
- The pH returns to near neutral.
- Nutrients are consumed.
- Complexation and precipitation of metals proceeds.
- Leachate organic strength is dramatically decreased.

Phase V: Final Maturation

- Nutrients become limiting.
- Measurable gas production all but ceases.
- Oxygen and oxidized species slowly reappear.

These phase changes can be identified by tracking changes in chemical parameters in the leachate as shown later.

The Worcester County Bioreactor Landfill. These processes have been implemented in a full-scale operating landfill at Worcester County in Maryland, since April of 1990. The Worcester County Central Landfill Facility is located in the Mid-Atlantic region on Maryland's Eastern Shore.

The facility serves the whole County and receives an average daily tonnage of 250 tpd with summer seasonal averages of 350-450 tpd.

The Central Landfill Facility (CLF) site plan in designed to accommodate the sequential filling of four landfill cells approximately 17 acres in footprint and permitted for about 80 feet of waste. Leachate storage/management facilities are located centrally to these landfill cells and consist of a 400,000 gallon above-ground storage tank with associated piping/pumping and transfer facilities.

The cell liner section is designed and constructed in full compliance with State and USEPA regulations as established by RCRA Subtitle D.

The leachate is collected, stored, and recycled back onto the landfill to facilitate Enhanced Natural Degradation processes (ENDpro^{SM})in the waste.

The recycled leachate provides the moisture and nutrients required by the microorganisms to break down the solid waste.

RESULTS/DISCUSSION

In the Worcester County facility, in April of 1990, the initial BOD values were less than 200 mg/l but by January of 1991, BOD had already risen to over 2000 mg/l (Figure 2) marking the Phase II transition period in which the fermentation and hydrolysis are breaking down the organic wastes. This transition period progresses rapidly to the acid formation (Phase III) dominance shown by a drop in leachate pH (Figure 3) around April of 1991 and a sharp rise in BOD beginning at the same time (Figure 2).

The entry into Phase IV, the methane fermentation phase, is marked by a culture being established within the landfill in which the fatty acids produced in Phase III are now the feedstock for the anaerobic methane-producing biomass. With the consumption of these acids, a shift in the trend of leachate pH back towards neutral becomes noticeable (Figure 3 between April and July of '91). This conversion process will, over the next several years, eventually drive the pH through neutral to the slightly basic range as the acid feedstock is exhausted and a bicarbonate buffering system becomes dominant.

Concurrent with the shift from an aerobic to an anaerobic decomposition there is a shift to the consumption of election acceptors such as sulfate. Figure 4 shows the rapid consumption of sulfate starting in July of '91 until its exhaustion in January of '93.

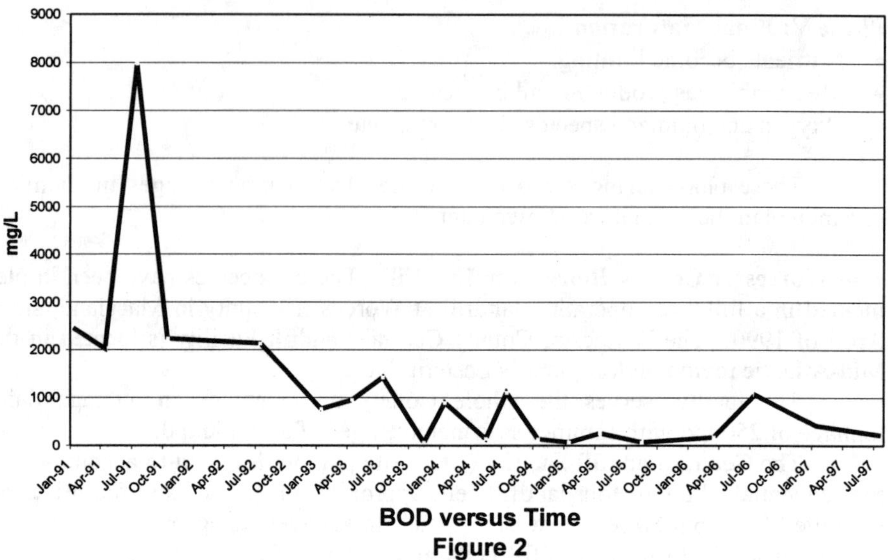

BOD versus Time
Figure 2

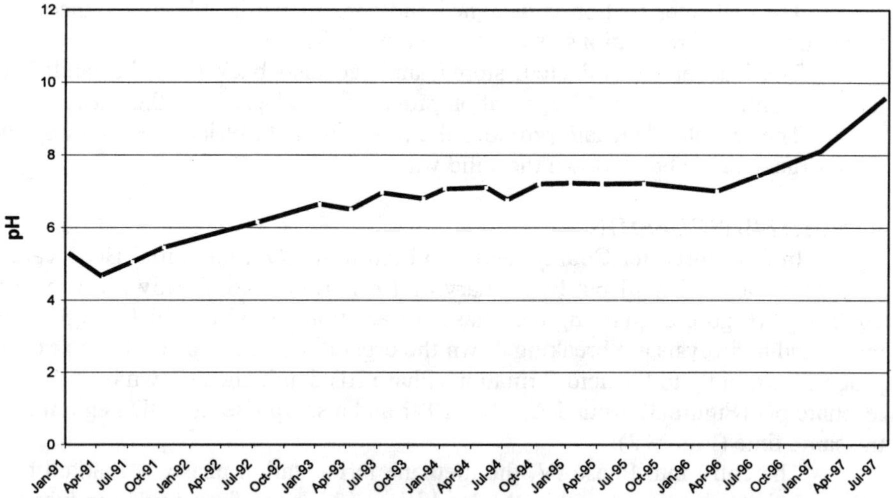

12
10
8
6
4
2
0
pH
Jan-91 Apr-91 Jul-91 Oct-91 Jan-92 Apr-92 Jul-92 Oct-92 Jan-93 Apr-93 Jul-93 Oct-93 Jan-94 Apr-94 Jul-94 Oct-94 Jan-95 Apr-95 Jul-95 Oct-95 Jan-96 Apr-96 Jul-96 Oct-96 Jan-97 Apr-97 Jul-97

pH versus Time
Figure 3

At the same time, conversion of nitrate through its intermediaries to ammonia starts an irregular increase in the ammonia concentration (Figure 5). It's not until mid '93 however, when the sulfate is exhausted, that the ammonia concentration starts a fairly regular climb that lasts through mid '97.

The changes in leachate chemistry that reflect the transition to a Phase V condition are clear.

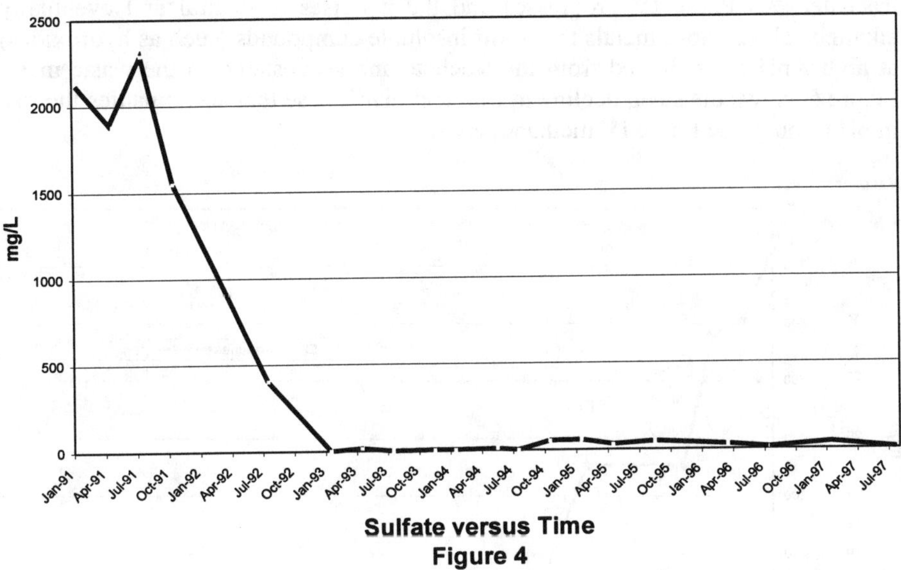

Sulfate versus Time
Figure 4

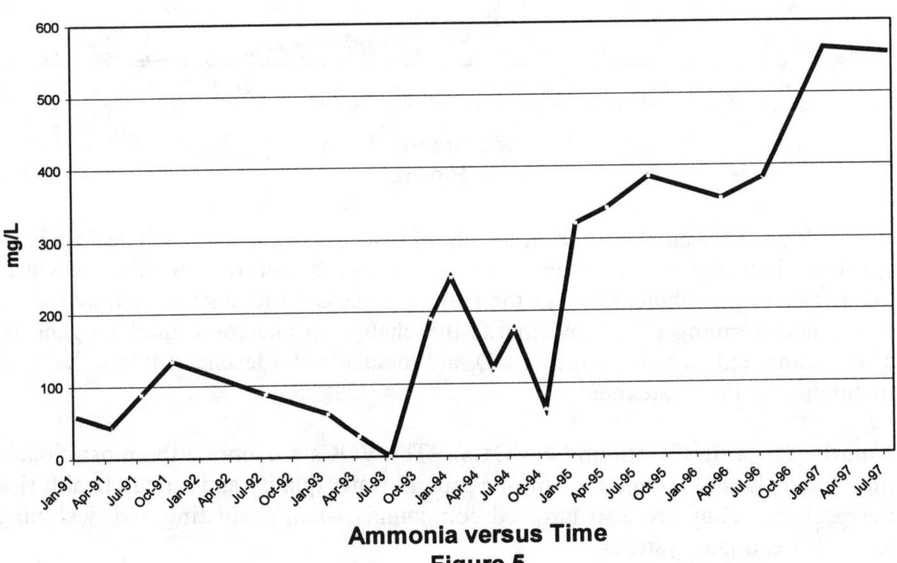

Ammonia versus Time
Figure 5

The BOD (Figure 2) declines to a low and relatively stable level. The BOD since early '94 has averaged about 380 mg/l with a standard deviation of ±380. This is an order of magnitude below its levels during the active phases of waste decomposition (avg. 3358 mg/l with a standard deviation of ±2315).

As a result of the acidic pH environment in the early phases of decomposition (Phase II, III and substantially into IV), metals are mobilized into the leachate. As Phase IV progresses and the pH rises to neutral and eventually alkaline values, those metals that form insoluble compounds (such as hydroxides) at higher pH are removed from the leachate and sequestered in the waste mass. Figure 6 shows the steep decline in iron and manganese that accompanies the rise in pH through the Phase IV methanogenesis.

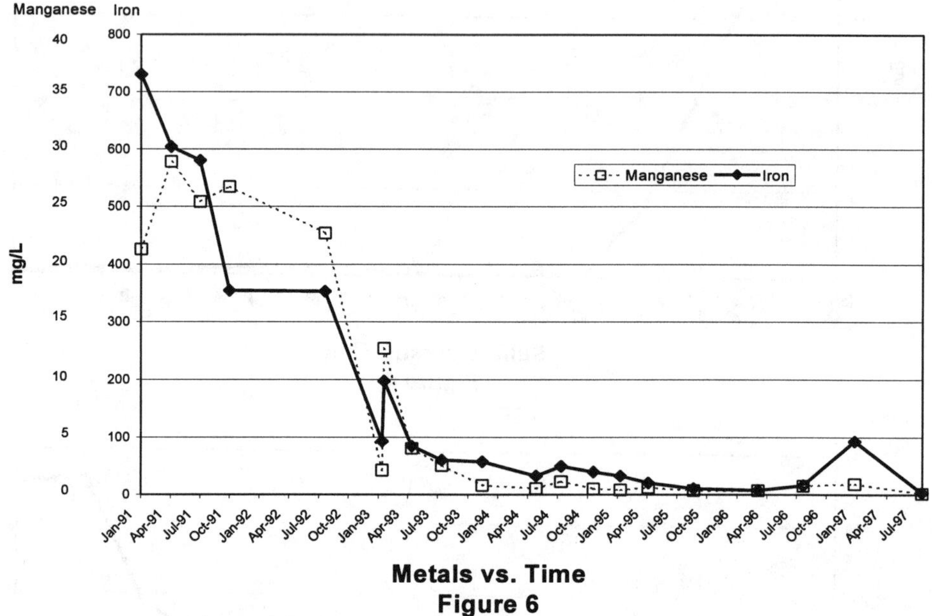

Metals vs. Time
Figure 6

It can be seen clearly from the above discussion that the leachate data is an excellent indicator of the decomposition processes occurring within the waste mass. Taken as a whole, it tracks the natural process through its transitional stage to the acid forming phase and marks the change to anaerobic methanogenesis. The decline and stabilization of BOD and metals in the leachate marks the final maturation of the waste mass.

Volatile Organic Compounds (VOCs). The VOCs are among the most closely monitored class of compounds from a ground water quality and human health risk perspective. They are also targeted compounds when evaluating and designing leachate treatment systems.

Data on the degradation of VOCs has been separated into the chlorinated compounds (including the chloroethanes and the chloroethenes) and the non-chlorinated compounds typified by the gasoline/diesel fuel constituents—benzene and ethylebenzene, and the more ubiquitous toluene.

Both classes of chlorinated compounds degrade to very low levels through the period of active methanogenesis stretching from July '91 through April of 1993, as shown in Figures 7 and 8. The chloroethenes remain low to non-detect

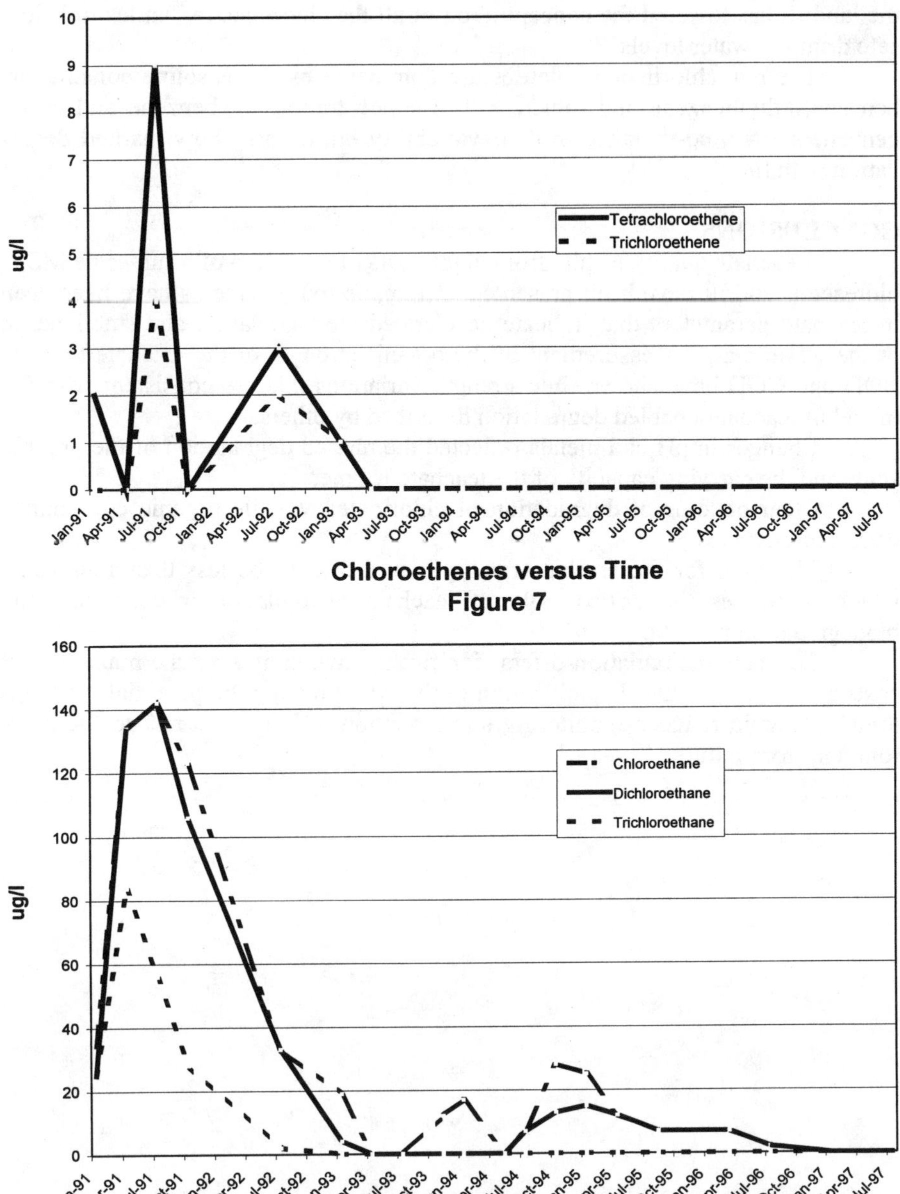

Chloroethenes versus Time
Figure 7

Chloroethanes versus Time
Figure 8

for the remainder of the period to date. The chlorethanes remain low but with some variability until reaching nondetect by July of '97.

From peak levels of up to 17 times the maximum contaminant level (MCL) established under the Safe Drinking Water Act, the degradation process in the landfill has lowered the concentration of all the chlorinated volatiles to below safe drinking water levels.

The non-chlorinated volatiles are dominated by the gasoline constituents benzene, ethylbenzene and toluene. The concentrations of benzene and ethylbenzene show a co-variance in their variability but do not show marked degradation with time.

CONCLUSIONS

Leachate quality results from eight years of operation of a full scale MSW bioreactor landfill have been presented. Dramatic improvements have been seen in leachate parameters that indicate accelerated biodegradation and stabilization of the waste mass. Measurement of the organic strength of the leachate through BOD and COD have shown time/strength variations that are consistent with the model of leachate-enabled degradation described by others.

Changes in pH and metals reflected the phased degradation of the organic acids and the carrying capacity of the leachate for metals.

Clear patterns of degradation of chlorinated volatile organic compounds were demonstrated.

The time for the leachate quality was shown to be less than four years which reinforces the critical role of leachate recirculation in managing the biodegradation process.

Leachate recirculation offers significant, savings in leachate management costs as well as advanced stabilization of the waste raising the potential for long-term savings in reduced monitoring and reduction of long-term risk to the environment from failure.

BIO-REMEDIATION OF LANDFILL LEACHATE WITH REMOTE MONITORING AND ON-SITE DISPOSAL

E. Craig Jowett, Waterloo Biofilter Systems, Guelph ON; J. David Redman, University of Waterloo, Waterloo ON; Frank C. Ford, Henderson Paddon Environmental, Owen Sound ON; Steen Klint, Waste Management, Simcoe County ON; George L. Prentice and Chris P. Hughes, City of Owen Sound ON.

Abstract: Treatment of landfill leachate with on-site disposal is being carried out at the Genoe and Nottawasaga-10 landfills in Ontario since 1994 and 1996 respectively. Raw leachate is sprayed onto the absorbent Waterloo Biofilter® in a rougher-polisher sequence, and is biochemically treated as it percolates slowly down through the medium. Spray irrigation or pressurized tile bed is used for disposal. Cost recovery is less than three years. Following an acclimatization-construction period, the Nottawasaga treatment results are as follows: BOD_5 from 453 to 17 mg/L (96%); TSS from 117 to 4 mg/L (97%); COD from 767 to 151 mg/L (80%); TKN from 99 to 10 mg/L (90%); TN from 101 to 51 mg/L (50%); and VOCs from 213 to 0 µg/L (>99%). Significant Fe, Mn, S, Al, Ba, and Sr, and lesser Pb, Zn, V, Sn, and Ni accumulate in the carbonate and hydroxide precipitate in the rougher filter. Treatment improves by insulating leachate manholes and by avoiding excessive hydraulic 'spikes', but is little affected by natural flow and concentration variations. A robust data collection and remote monitoring system is proving invaluable for troubleshooting and process optimization.

INTRODUCTION

The Waterloo Biofilter® is a single-pass, absorbent trickle filter developed since 1991 at the University of Waterloo to treat septic tank effluent, but has proven to be useful for wastewater treatment in general, including food processing, farm animal, immediate reuse, and landfill leachate wastewaters (Jowett 1997; Jowett et al. 1998).

The primary purpose of this paper is to present the results of a 6-year laboratory and field research program in landfill leachate renovation, focussing on the full-scale treatment and disposal of the Nottawasaga-10 leachate in Ontario.

ABSORBENT FILTER MEDIUM

Physical Properties. Filter media for sustainable biological degradation should combine high surface area for microbial attachment with large pores so that the microbial slime does not plug the medium and short-circuit the treatment process. High absorption provides microbes adequate contact time to degrade the contaminants. Typically there is a trade-off between the higher retention time of sand with a higher rate of plugging, and the lower retention time of conventional trickle filter media with lower plugging potential.

The Waterloo Biofilter® medium utilizes a foamed plastic with an interconnecting, three-dimensional reticulate solid framework with high surface area, high porosity, high absorption, and dual pathways for air and wastewater. (The medium performance is detailed in Jowett 1997 and Crites and Tchobanoglous 1998.)

LABORATORY STUDIES WITH LEACHATE

Clearview (Nottawasaga-10) Landfill, Simcoe County. Following successful column experiments in 1993 on three leachates, a rougher–polisher sequence of filters was tested in 1995. Following a month 'start-up', high removal rates of 99% BOD_5, 94% TSS, 97% TKN, 93% Fe, and 99% turbidity (NTU) are obtained (Table 1). Nitrite formed in the first 10 days, and thorough nitrification and 40% TN removal follows in 30 days. Typically, pH increases from 7.0 to 8.9.

Significant removals (>90%) of Fe, Mn, and Ba, and moderate removals (50–90%) of Al, Ti, Sr, V, and Ag were found (Table 2), with little removal of Pb, Cd, Co, Cr, Mo, Be, and Se, although many of these are close to non-detectable. Cu and B increased in the polisher.

FIELD STUDIES WITH LEACHATE

Genoe Landfill, Owen Sound. A 3.5 m^2 filter unit, 0.9 m deep, has been operating in Owen Sound since September 1994 at a loading rate of ~30 cm/day (0.3 $m^3/m^2/day$) (Jowett 1997). The first maintenance requirements came after 2 years of continuous operation and consisted of unplugging several nozzles and changing the fan due to encrustation by carbonate minerals. After 4.5 years, the upper layer of filter medium is cemented with iron-stained carbonate minerals but continues to operate well. The leachate is pumped on a timed basis from a toe drain manhole and disposed of downstream of the unit.

BOD removals average 78% (40–95% range) with effluent values of 1–30 mg/L, down from 40–290 mg/L, and became consistently better with time. Effluent COD values are 200–300 mg/L, down from 300–500 mg/L.

Suspended solids fluctuate between 9 and 80 mg/L in the leachate, but effluent values are typically 3–10 mg/L. Effluent phenols values are 30–80 μg/L, down from 100–200 μg/L. Why the phenol degradation is much less than the Clearview leachate (>90%) is not understood.

The effluent colour begins dark and 'coffee-like' but within days changes to clear 'tea-like', and within weeks changes to light yellow, clear (NTU of 1–10) and odourless. The coloration is due to dissolved organics which are photo-sensitive but not readily biodegradable (likely lignins and tannins, etc.), and is typical of leachate. The average pH increases from 7.4 in the leachate to 8.9 in the effluent.

Metals are not high in this leachate and most Fe, Al, Mn, and Ba is removed, likely by forming oxides, hydroxides, and sulphates or carbonates which are stable minerals in oxygenated environments. Typical influent/effluent analyses (January 26, 1998) are: 8.3 to 0.1 mg/L Fe, 1.2 to 0.7 mg/L Al, 0.30 to 0.02 mg/L Mn, and 0.12 to 0.05 mg/L Ba. Some Zn (0.02 to 0.001), Sr (0.99 to 0.72), and minor Cu, Ti, and As are removed, but no boron. Other metals such as Pb and Hg are not present in the leachate. Calcium decreases from 126 to 10 mg/L along with alkalinity from 1730 to 1280 mg/L CaCO, but magnesium is constant from 147 to 138 mg/L. Chloride is similar from 168 to 151 mg/L. Almost all removal of the metals and alkalinity is in the roughing filter.

A characteristic of the filter is very thorough removal of volatile organic compounds. Typical non-target hydrocarbon values at this field site are 0–2 μg/L, down from the main range of 60–200 μg/L. VOCs are non-detectable to trace in the effluent (rarely, one or two are detectable), down from 6 to 10 detectable in the leachate (>99% removal from 101 to <1 μg/L). Of the targeted VOCs, the most commonly occuring are 1,1 dichloroethane, cis 1,2 dichloroethylene, benzene, trichloroethylene, toluene, chlorobenzene, ethyl benzene, m+p xylene, o-xylene, and 1,4 dichlorobenzene. Non-target VOCs make up a significant portion of the total. Chloroform was the only VOC regularly detected in the effluent, and only in trace amounts at or below the blank level of the laboratory instrumentation.

Chloroform was not found in the leachate samples, however, even at trace amounts, suggesting it may be a product of biodegradation or a laboratory contaminant.

Experience with sewage treatment and laboratory studies indicates that nitrification improves with deeper beds and with a sequence of rougher–polisher. A 3.0 m^3 polishing filter was added in mid-1997 (~day 700) to treat the effluent from the rougher, primarily to complete nitrification. The polisher removes additional COD (10-20%) and increases pH, but does not contribute significantly to BOD or NH_4–N removal, likely because the rougher does a thorough job. Surface spray disposal has been approved on a test basis, and this began in 1997 along with studies on the effects of the effluent on soil and vegetation.

Clearview (Nottawasaga-10) Landfill. A major installation was constructed near Stayner to treat 16 m^3/day of leachate (Ford 1998). The unit consists of 3 concrete tanks serving as a roughing filter (36 m^2 area total, 2.4 m deep) and 2 tanks serving as a polisher to treat the rougher effluent (24 m^2 area total, 2.4 m deep).

A robust data collection and remote monitoring system (Telesafe), using Windows-based custom software, allows temperature and operational parameters such as pump on-off cycles, cumulative pump times, alarms, and flow rates to be reviewed remotely and problems interpreted from the record. Data is downloaded on demand from any of the administrative or maintenance computers, and the station file updated automatically. After logoff, the data can be viewed in table form or a flexible graphic format, using any short or long-term time range chosen. Anomalies such as excessive pumping of leachate from one cell to another, low temperatures in the tanks, or excessive pump run times give the operator an idea of problems to be anticipated.

The treatment plant was started up in mid-1996 and within a month the effluent was light clear yellow until the end of November, with >90% BOD removal, although nitrification had not started. Cold leachate below 5°C due to exposed manholes and ongoing construction deterioriated treatment in early 1997, but by early spring the effluent was clear yellow again, and by summer 1997, complete nitrification (>99%) and excellent removal of BOD (98%), TSS (98%), and COD (93%) were accomplished (Ford 1998). Rapid pumping of excessive volumes of leachate into the plant from other cells caused short-term treatment lapses, and this was resolved by regular pumping schedules.

Table 3 shows the summary of analyses taken from the raw leachate ('IN') and from the polisher effluent ('OUT'). Construction and operational difficulties were solved by September 1997, and the first of the four time divisions, Sept 96 - Aug 97, represents the 'starting up' period before the ongoing treatment phase. The second and third divisions represent 'summer' and 'winter' periods, and the fourth a 'four-season' period. Removal of BOD, TSS, and COD are fairly consistent through the seasons, but TKN removal noticably decreases during the winter, due to sensitivity of the nitrifying bacteria. Table 4 shows the averages and ranges of influent and effluent during the entire period after 'start-up'. This is what can be expected on an ongoing basis in future operations likely after 1–2 months initialization.

Removal of VOCs (>99%) and phenols does not suffer during winter operation. Of the targeted VOCs, the most common is toluene, with some m+p xylene, o-xylene, and little benzene and ethylbenzene. Others are typically non-detectable.

The three roughing filters have iron staining and carbonate precipitation on the upper surfaces and a heavy oily odour in them when the access hatches are

Table 1. Biochemical analyses of Nottawasaga-10 laboratory leachate tests from May to July 1995; all mg/L; n = number of samples.

Parameter	n	IN	OUT	%	Parameter	n	IN	OUT	%
BOD_5	5	201	3	99	TN	5	172	102	40
TSS	6	84	5	94	pH	7	7.04	8.85	-
COD	6	460	125	73	Fe	5	9.4	0.6	93
TKN	5	171	4	97	NTU	5	108	1	99

Table 2. Metals analyses of Nottawasaga-10 laboratory leachate tests from May to July 1995; all µg/L; n = number of samples.

Metals	n	IN	OUT	%	Metals	n	IN	OUT	%
Zn	5	23	13	43	Mo	5	6	4	38
Cu	5	9	32	++	V	5	31	15	53
Ni	5	26	29	0	Sr	5	1532	395	74
Pb	5	50	30	40	Be	5	0.5	0.3	42
Cd	5	3	2	41	Ag	5	20	10	50
Mn	5	1273	11	99	Ba	5	387	25	94
Fe	5	14,441	104	99	Ti	5	20	4	82
Al	5	177	34	81	B	5	1828	2186	+
Cr	5	8	7	13	As	5	2	2	11
Co	5	9	5	49	Se	5	2	1	30

Table 3. Nottawasaga-10 landfill leachate and effluent analyses 1996-1999; n = number of samples.

Parameter	Sept 96 – Aug 97				Sept 97 – Oct 97				Nov 97 – Apr 98				May 98 – Jan 99			
(mg/L)	n	IN	OUT	%	n	IN	OUT	%	n	IN	OUT	%	n	IN	OUT	%
BOD_5	5	877	187	79	7	1239	14	99	32	287	17	94	8	427	21	95
TSS	6	111	25	77	6	208	3	99	32	111	5	95	8	72	2	97
COD	5	935	450	52	6	1766	88	95	32	559	139	75	7	859	258	70
TKN	2	182	87	52	4	151	5	97	27	72	12	83	8	164	6	96
TN	2	182	170	7	4	151	69	54	30	75	48	36	8	171	54	68
VOCs (µg/L)	2	47	0	>99	3	332	0	>99	16	191	0	>99	-	-	-	-
Phenols (µg/L)	2	18	4	78	1	275	1	>99	2	-	0.2	-	-	-	-	-

Table 4. Overall Nottawasaga-10 averages from September 1997 to January 1999; n = number of samples.

Parameter (mg/L)	n	IN	IN Range	OUT	OUT Range	% Removal
BOD_5	47	453	35 – 2595	17	4 – 39	96
TSS	46	117	18 – 638	4	0 – 13	97
COD	45	767	92 – 3770	151	16 – 741	80
TKN	39	99	33 – 216	10	1 – 26	90
TN	42	101	41 – 218	51	31 – 117	50
VOCs (µg/L)	19	213	29 – 534	0	0 – 1	>99

Table 5. Analysis of calcareous crust from Nottawasaga-10 rougher filter sampled December 1998.

Metal	mg/L	Metal	mg/L	Metal	mg/L	Metal	mg/L
Ca	70.5%	Na	0.1%	Al	290	Zn	28
Fe	4.25%	S	0.1%	Ba	253	V, Sn	12
Mg	0.3%	Sr	526	P	206	Ni	4
Mn	0.2%	K	291	Pb	80	Zr	1

Ag, As, Be, Bi, Cd, Ce, Co, Cr, Cu, Hg, La, Mo, Nb, Sb, Se, Te, Th, Ti, U, W, Y are all <1 mg/L

opened. The distribution nozzles plug up occasionally with bituminous-like residue. The two polishing filters are relatively clean with no oily odour. The water analyses show that hydrocarbons and VOCs are removed in the roughing filter, as are most of the metals and other constituents. Analysis of the calcareous crust forming in the rougher filter is shown in Table 5, which indicates which metals are being removed from the wastewater stream. Iron, manganese, and lead are metals of concern which are readily removed in the filter. Other metals of concern (Hg, Cd, etc.) are typically present in very low amounts.

After hydrogeologic modeling studies of the effects of chloride and nitrogen on the groundwater, regulatory approval was granted for disposal of the treated effluent in a raised pressurized leaching field on a permanent basis (Ford 1998). Since August 1997, trucking of the leachate off-site to sewage treatment plants has not been necessary. The plume from the disposal bed shows elevated nitrate and chloride levels, but no addition of BOD, COD, or metals.

CONCLUSIONS

The Waterloo absorbent trickle filter is treating landfill leachate effectively enough to dispose of the effluent on-site in subsurface trenches and by surface spray irrigation.

Contaminants are efficiently removed from the leachate despite fluctuating concentrations, and with no addition of nutrients. The system works well in winter and summer, but performs better when the wastewater flow rates are varied gradually and when the leachate manholes are insulated from the winter cold.

The custom remote monitoring software is invaluable in understanding the process and explaining changes, especially at such an isolated landfill site. It is also useful in predicting the failure of pumps, level switches, and other mechanical components.

ACKNOWLEDGEMENTS

This research is supported financially or in-kind by the City of Owen Sound, County of Simcoe, and Ontario Ministry of Environment (MOE) London and Rexdale laboratories. Grants from the Natural Sciences and Engineering Research Council (Operating Civil Engineering OGP0153439 to Jowett), MOE Research Advisory Committee (grant ER2030 to Jowett), Waterloo Centre for Groundwater Research, and its successor, Crestech, are appreciated. For their support, the authors are indebted to Walter Cook, Dale Henry and Gilles Castonguay of MOE.

REFERENCES

Crites, R., and G. Tchobanoglous 1998. *Small and Decentralized Wastewater Management Systems*. McGraw-Hill.

Ford, F. C. 1998. "Leachate Biofilters—Pilot And Full-Scale Examples", *Innovative Landfill Leachate Management Systems*, EPIC Conference Proceedings, Toronto, 41 pp.

Jowett, E. C. 1997. "Sewage and Leachate Wastewater Treatment Using the Absorbent Waterloo Biofilter", In M. S. Bedinger, A. I. Johnson, and J. S. Fleming (Eds.), *ASTM STP 1324,Site Characterization and Design of On-Site Septic Systems*, pp.261–282. American Society for Testing and Materials. West Conshohoken PA.

Jowett, E. C., D. A. Manz, B. Kim, S. Kim, and F. C. Ford. 1998. "Communal-Size Sewage and Leachate Treatment Using Waterloo Biofilters in Treatment Trains", *Environmental Science and Engineering*. Jan.98: 36–37.

2,4-DICHLOROPHENOXYACETIC ACID (2,4-D) DEGRADATION IN A MEMBRANE BIOREACTOR

J.F. Buenrostro-Zagal; A. Ramírez-Oliva,; B. Schettino-Bermúdez; S. Caffarel-Méndez (TESE, México);
and ***H.M. Poggi-Varaldo*** (CINVESTAV del IPN, México)

ABSTRACT: This research aimed at acclimating a 2,4-D degrading culture from wild inocula, and implementing and determining the performance of a membrane bioreactor for 2,4-D removal from an industrial wastewater. A mixed, aerobic microbial culture was acclimated to 2,4-D in submerged culture. The acclimated culture was able to use 2,4-D as the sole source of carbon and energy with an optimum at pH between 7 to 8. Gram negative microorganisms predominated in the consortium. The culture growth in 2,4-D followed an Andrews-like substrate inhibition kinetic model with μ_{max} = 0.06 1/h, K_s = 17 mg/L, K_i = 128 mg/L. The partition coefficient of the selective silicone tubing membrane in the reactor was 131 L_{aq}/L_{membr}, which suggested an important adsorption effect/solution of the membrane. The permeability coefficient determined at pH 0, 1 and 4 was practically independent of pH with an average value of 3.2×10^{-8} m/s. In continuous reactor experiments, unit reductions in the range of 110 to 120 mg 2,4D/L_{lumen}.h (32 to 34 mg/h.m^2) at lumen retention times of 4.1 to 15 min were achieved. Overall, the membrane bioreactor seems to be an effective configuration for removing 2,4-D while at the same time avoiding the active biomass exposure to the low pH and high salinity concentration of the influent.

INTRODUCTION

The 2,4-dichlorophenoxyacetic acid (2,4-D) is widely used for commercial herbicides formulation. It is toxic to plants, microflora and human beings (Papanastasiou and Maier, 1982; Dreisbach, 1984). In Ecatepec, Mexico, there is a big industry which manufactures 2,4-D and other herbicides. It discharges a low pH, high salinity and 2,4-D rich effluent; a very aggressive and toxic wastewater. The characteristics of this effluent makes difficult its treatment by conventional biological methods, since the biomass cannot withstand the extreme conditions of pH and salinity.

Recent developments in membrane separation techniques have shown the effectiveness of membrane bioreactors (MBR) for treatment of hostile effluents containing toxic and recalcitrant organic compounds (Brindle & Stephenson, 1996). Particularly, the MBRs based on selective extractive membranes seem a promising alternative (Livingston, 1993). This reactor configuration allows for a physical separation between the aggressive influent and the degradative biomass. The membrane performs as a selective barrier that allows the passage of the organic compounds to be degraded, while retaining the inorganic ions. Thus, this work aimed at acclimating a 2,4-D degrading culture from a wild aerobic inocula, and implementing and determining the performance of a MBR for 2,4-D removal from a toxic effluent discharged by a herbicide manufacturing plant.

MATERIALS AND METHODS

Mixed culture experiments. An aerobic mixed culture was developed from an agricultural soil (University of Chapingo, México) previously treated with 2,4-D and propagated/acclimated as described elsewhere (Sandoval, 1997). Culture medium contained (in g/L): K_2HPO_4, 1.5; $(NH_4)_2SO_4$, 0.4; $MgSO_4$, 1.5; trace elements; and 2,4-D, 0.25. The 2,4-D was the only source of carbon and energy. Medium and inoculum were fed to a Bioflo III, 3 L capacity bioreactor and incubated at 25°C, 300 rpm and 0.6 vvm air flowrate. The 2,4-D concentration was determined by HPLC, biomass was quantified as protein by Lowry's method. Broth pH was controlled at 6.0, 7.0 and 8.0 either with NaOH or HCl 0.1 N. Reactor content was sampled at different times during each run. Results were fitted to a substrate-inhibition kinetic model (Haldane-like or Andrews) as outlined in Papanastasiou & Maier (1982) and Bayley & Ollis (1986).

Bioreactor experiments. In the second phase, a membrane bioreactor was implemented. The reactor was a 3 L lab-scale fermenter (Bioflo III, New Brunswick) fitted with a hydrophobic membrane (a silicone rubber tubing 5.73 m long x 1.4 mm inside diameter, thickness 0.5 mm). Membrane was characterized in terms of its partition and diffusive transport coefficients (Livingston, 1993). The biomass was grown as a biofilm outside the membrane, while the wastewater (low pH and high salinity) was continuously circulated inside the membrane lumen. Three different influent flowrates were tested (retention times of 15, 8.15 and 4.1 min), two replicate runs each. The biomass was in contact with a neutral, aerated biomedium that was removed and supplied in batch mode. The feed wastewater had a pH=1 and contained 300 mg/L 2,4-D and 5000 mg/L NaCl. The biomedium composition was (in g/L): K_2HPO_4, 1.5; $MgSO_4$, 1.5; $(NH_4)_2SO_4$, 0.375; $CaCl_2$, 0.3; $ZnCl_2$, 0.3; $FeCl_3$, 0.3; $MnSO_4$, 0.3. Biomedium was mixed at 100 rpm and aerated at 100 vvm, its pH and temperature were kept at pH=7 and 25°C respectively. The 2,4-D concentration was determined by HPLC. Ammonia-nitrogen and chloride anion were determined by nesslerization and Mohr's method respectively in the biomedium.

RESULTS AND DISCUSSION

Mixed culture experiments. The 2,4-D removal was more rapid at pH 7 and 8 than at pH 6 (Table 1). Also, it can be seen the inhibitory effect of 2,4-D in the uptake rate, since the removal of the first 50% of 2,4-D takes 18 h ($t_{50\%}$), while the removal of the last 50% (at diluted concentrations) only lasts 9 h at pH=7 ($t_{100\%}$-$t_{50\%}$). Mixed culture growth results from the batch run at pH=7 were fitted to a substrate-inhibition model (Fig. 1 and Table 2). The agreement between the experimental points and the model was good. The higher 2,4-D uptake at higher pH might partially be explained by the acid-base equilibrium of the 2,4-D. At higher pH, the 2,4-D is mainly present as an anion, more soluble and perhaps toxic than the unionized 2,4-D. Papanastasiou & Maier (1982) also found a good fitting between 2,4-D growth specific rate and the substrate-inhibition kinetic model, when 2,4-D was the only source of carbon and energy at initial concentration of 50 mg/L. Sinton *et al.* (1986) argued that inhibition was probably due to intermediate metabolites such as the di-chlorophenol and quoted the similarity between the kinetic coefficient values found by Tyler & Finn (1974) for the latter compound and those reported by Papanastasiou & Maier (1982) for 2,4-D. Comparison of our results show that our mixed culture grew somewhat slower (a 60% lower μ_{max}) but with more affinity (lower K_s value) and with less inhibitory cffcct (highcr K_i) than thc mixed culture studied by Papanastasiou & Maier (1982), see Table 2.

TABLE 1. pH effects on 2,4 D removal by the aerobic mixed culture.

pH	$t_{50\%}$ (h)	$t_{100\%}$ (h)
6.0	32	45
7.0	18	27
8.0	16	25

Bioreactor experiments. The diffusive transport coefficient k_o was independent of the pH in the acid range representative of the pH conditions of the influent, Table 3. The partition coefficient value was 131 L_{aq}/L_{membr}, which suggested an important selective dissolution of 2,4-D in the membrane material. At the

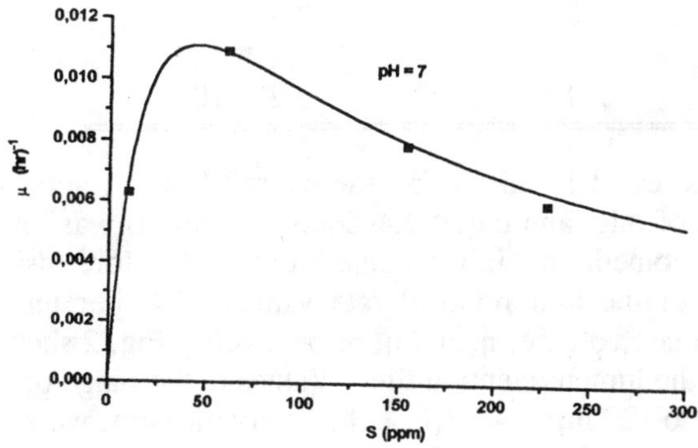

FIGURE 1. Model fitting of substrate-inhibition kinetics at pH = 7 (symbols, experimental results; line, model prediction).

TABLE 2. Kinetic coefficients of 2,4 D-degrading mixed cultures.

Source of inoculum	Culture conditions	μ_{max} (1/h)	K_s (mg/L)	K_i (mg/L)	Reference
activated sludge	acclimation, 20°C, batch, aerobic,lab 50-100 mg/L 2,4-D	0.15	40	31	Papanastasiou & Maier (1982)
soil, activated sludge	acclimation, aerobic, 2,4 dichlorophenol	0.23	11.7	35.7	Tyler and Finn (1974)
Agricultural soil exposed to 2,4-D	acclimation, 25°C, batch, aerobic, lab 250 mg/L 2,4-D	0.06	17.8	129	this work

TABLE 3. Diffusive transport coefficient of the silicone membrane.

pH	k_o (m/s)
0	$3.2 * 10^{-8}$
1	$3.1 * 10^{-8}$
4	$3.2 * 10^{-8}$

short retention times tested in this work, the overall 2,4-D removal efficiency determined in terms of inlet and outlet 2,4-D concentrations was low (no 2,4-D was detected in the biomedium), in the range 10 to 30%. More insightful could be the examination of the unit removal rate values (UR, per unit membrane surface or per unit lumen volume, η_s and η_v respectively). Fig. 2 shows the 2,4-D UR as a function of the lumen retention time. Values in the range of 32 to 34 mg 2,4-D/(h.m^2) or 110 to 120 mg 2,4-D/(L_{lumen}.h), fairly uniform, were obtained for the retention times tested. Short retention times are associated to high flowrates, and to relatively high interphase mass transfer coefficient values between lumen liquid phase and the membrane. Thus, the 2,4-D rate is probably limited by the membrane diffusive transport coefficient and the concentration driving force between lumen and outside 2,4-D concentrations, which are approximately constant and uniform. The URs obtained in this work were of the same order of magnitude than results from membrane bioreactors degrading phenol and other toxic compounds (Livingston, 1993). Circumstantial evidence of 2,4-D degradation and biomass growth is presented in Fig. 3. Chloride ion concentration in the biomedium increases with time at all retention times tested (Fig. 3A), suggesting a release of Cl^- via dechlorination of 2,4-D. On the other hand, ammonia-nitrogen concentration in the biomedium decreased with time, probably related to 2,4-D uptake by the bacteria and biomass synthesis (Fig. 3B).

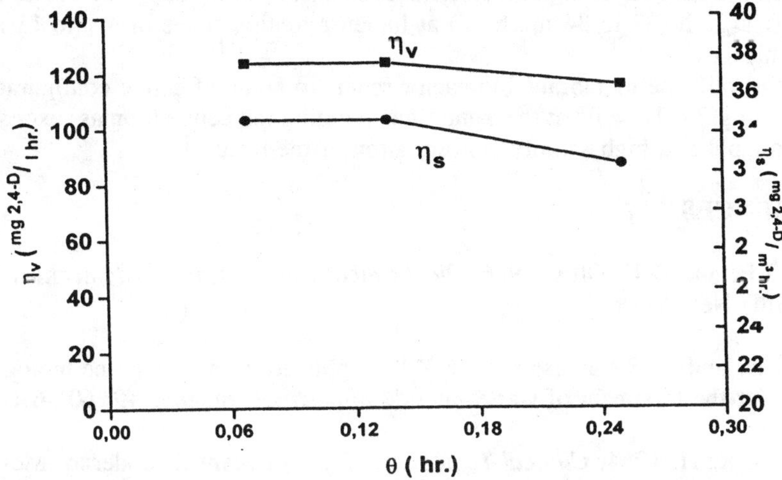

FIGURE 2. Unit removal rates of 2,4-D at different retention times

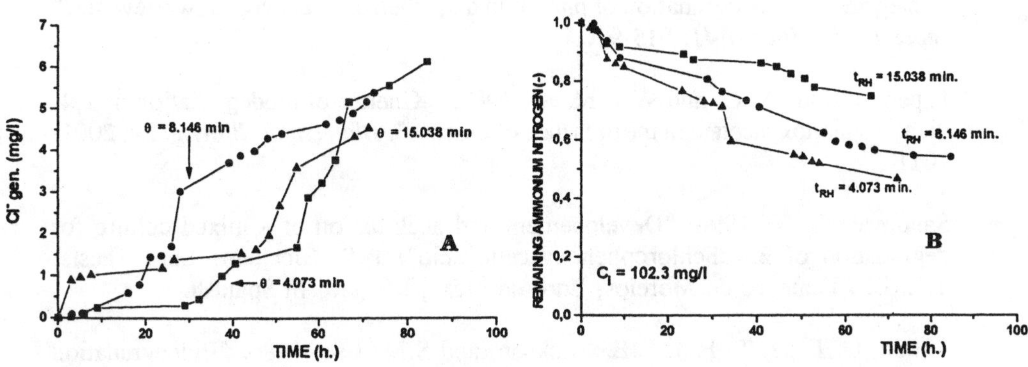

FIGURE 3. (A) Chloride ion release, (B) Ammonia-nitrogen decrease in the biomedium.

CONCLUSIONS

■ A mixed, aerobic microbial culture was acclimated to 2,4-D in submerged culture. The acclimated culture was able to use 2,4-D as the sole source of carbon and energy with an optimum between pH 7 to 8. Gram negative microorganisms predominated in the consortium. The culture growth in 2,4-D followed an Andrews-like substrate inhibition kinetic model with μ_{max} = 0.06 1/h, K_s = 17 mg/L, K_i = 128 mg/L.

- The partition coefficient of the selective silicone tubing membrane in the reactor was 131 L_{aq}/L_{membr}. The permeability coefficient determined at pH 0, 1 and 4 was practically independent of pH with an average value of $3.2*10^{-8}$ m/s.
- In continuous reactor experiments, unit reductions in the range of 110 to 120 mg 2,4D/L_{lumen}.h (32 to 34 mg/h.m^2) at lumen retention times of 4.1 to 15 min were achieved.

Overall, the membrane bioreactor seems to be an effective configuration for removing 2,4-D while at the same time avoiding the active biomass exposure to the low pH and high salinity concentration of the influent.

REFERENCES

Bailey, J.E. and D.F. Ollis. 1986. *Biochemical Engineering Fundamentals*. Mc Graw-Hill, New York, USA.

Brindle, K. and T. Stephenson. 1996. "The application of membrane biological reactors for the treatment of wastewaters". *Biotechnol. Bioeng*. *49*: 601-610.

Dreisbach, R. H. 1984. *Clinical Toxicology*. Edit. El Manual Moderno. México D.F., México. Spanish version.

Livingston, A. G. 1993. "A novel membrane bioreactor for detoxifying industrial wastewater 1. Biodegradation of phenol in a synthetically concocted wastewater." *Biotechnol. Bioeng. 41*: 915-926.

Papanastasiou, A. C. and W.J. Maier. 1982. "Kinetics of biodegradation of 2,4-dichlorophenoxyacetate in the presence of glucose". *Biotechnol. Bioeng. 24*: 2001-2011.

Sandoval, G. V. 1996. "Developement and acclimation of a mixed culture for degradation of 2,4-dichlorophenoxyacetic acid". B.S. Biochem. Eng. Thesis. T.E.S.E. Ecatepec de Morelos, Edo. de Méx., México. In Spanish.

Sinton, G. L., L.T. Fan, L.E. Erickson, and S.M. Lee. 1986. "Biodegradation of 2,4-D and related xenobiotics compounds." *Enzyme Microb. Technol. 8*: 395-403.

Tyler, J.E. and R.K. Finn. 1974. "Growth rates of a Pseudomonad on 2,4 dichlorophenoxyacetic acid and 2,4 di-chlorophenol." *Appl. Microbiol. 28*: 181-184.

DEGRADATION OF SYNTHETIC AND NATURAL DYES BY WHITE ROT FUNGI

J.G. Belote, E.P. Chagas, and ***L.R. Durrant***
DCA/FEA – UNICAMP (Campinas State University), Campinas-Sp, Brazil.

ABSTRACT: Here we describe the decolorization and degradation of the azo dyes orange G, new cocine and tartrazine and of the natural dyes cochineal (carminic acid) and annatto (norbixin) by *Phanerochaete chrysosporium* and *Pleurotus sajor caju*, following their growth or eight to twelve days in each of the dyes or in a combination of synthetic dyes or of natural dyes. Effluent from a natural dye plant was also studied. Both fungi showed similar decolorization of new coccine, orange G and cochineal as determined by ultraviolet-visible absorption spectroscopy. *P. chrysosporium* was able to efficiently decolorize the natural dyes, individually or in combination, and also the natural dyes containing industrial effluent. Degradation of the dyes, as measured by reversed-phase HPLC on a C_{18} column, was observed for decolorized samples and varied with each strain as did the type and activities of the ligninolytic enzymes present in the culture supernatants.

INTRODUCTION

Synthetic dyes have been increasingly used in the textile, paper, cosmetics, pharmaceutical and food industries because of their ease of use, cost effectiveness in synthesis, firmness and variety of color compared with natural dyes [1]. Azo dyes, the largest class of synthetic dyes, are characterized by the presence of one or more azo bonds (-N=N-) in association with one or more aromatic groups. Many studies indicate that these dyes are toxic or carcinogenic. The discharge of dye-containing effluent to main water bodies and wastewater treatment systems is currently causing significant health concerns to environmental regulatory agencies [2]. While coloured organic compounds generally impart only a minor fraction of the organic load to wastewater their color renders them aesthetically unacceptable. Color removal, therefore has recently also become of major scientific interest.

The ability of microorganisms to carry out decolorization has received much attention. Microbial decolorization and degradation of dyes is seen as a cost-effective method for removing these pollutants from the environment.

The white rot fungus *Phanerochaete chrysosporium* can decolorize, depolimerize, or mineralize high molecular mass chemically irregular aromatic polymers including lignin, pulp mill effluents and highly colored by-products and wastewaters from various industries.

The ability of *P. chrysoporium* and other white rot fungi to degrade, decolorize or mineralize these compounds is due to the action of the highly oxidative and relatively nonspecific ligninolytic system of these microorganisms. Extracellular lignin peroxidases (LiP), manganese-dependent peroxidases (MnP), laccases and hydrogen peroxide-generating enzymes, such as veratryl alcohol oxidases (VAO) are essential components of the ligninolytic enzyme system of various white rot fungi. These fungi have great potential for biotechnological applications in pulping, bleaching and environmental remediation processes.

In this study, we show the ability of two white rot fungi to decolorize and degrade natural and azo dyes used by food industries. The enzyme system produced during growth of the fungi in medium containing these dyes or the effluent of a natural dye producing plant was also determined.

MATERIALS AND METHODS

The fungal strains were *Phanerochaete chrysosporium* ATCC 24725 and *Pleurotus sajor caju* 020 obtained from the culture collection of the Systematic and Microbial Physiology Laboratory of the Food Engineering Faculty of Unicamp, Campinas-SP.

The dyes used were new coccine (NC), orange G (OG), tartrazine (T), cochineal (carminic acid) (C), annatto (norbixin) (NA), a combination of the synthetic dyes (CS), a combination of the natural dyes (CN) and also the effluent of an industry producing cochineal and annatto (EF).

Fungi were inoculated in malt agar (30.0 g/L); 5.0 g/L peptone and 15.0 g/L agar) and incubated (*P. chrysosporium*_- 37°C, *P. sajor caju* – 30°C) until extensive mycelial growth occurred. They were then divided into pieces of 1.0 cm^2 and for each 10.0 mL liquid culture media, one square was added into the flasks. The liquid media contained 0.5 % malt extract plus the dyes as follows: NC, OG, CS, C, NA and CN were used at the final concentration of 100.0 mg/L and 10.0 mg/L for tartrazine. The natural dye-contaning effluent used without sterillization was added asseptically to previously sterilized flasks. Following inoculation the flasks (50.0 mL medium/ 125 mL Erlenmeyer) were incubated for a maximum of 12 days at 100 rpm. Controls consisting of uninoculated flasks were also prepared. Supernatants samples were collected at each 48 h interval and were used for enzyme activities, decolorization and degradation determinations.

The enzymatic activities of lignin peroxidase (LiP) [3], laccase, using *o* -dianisidine (D) or siringaldazine (S), as substrates [4] and manganese-peroxidase (MnP) [5] were determined. Veratryl alcohol oxidase (VAO) was determined as LiP, but H_2O_2 was replaced by H_2O. One unit LiP, VAO, laccase, MnP was defined as 1.0 umol of substrate oxidized per minute per liter.

The decolorization of the dyes was followed by monitoring changes in their absorption spectra (200 to 800 nm) using a spectrophotometer Shimadzu 1201 and comparing to the respective control.

High performance liquid chromatography (HPLC) was used to determine the degradation of the dyes in the supernatants samples. These analyses were performed with a Zorbax ODS (0.46 x 15.0 cm) C_{18} reverse-phase column (SUPELCO Chromatography Products). Separation was achieved according to [6] and the retention time was determined using a UV detector (307 nm).

Chemical oxygen demand determination (COD) measurements were made using a Portable Datalogging Spectrophotometer DR/2010 HACH and was carried out to the catalogue instructions.

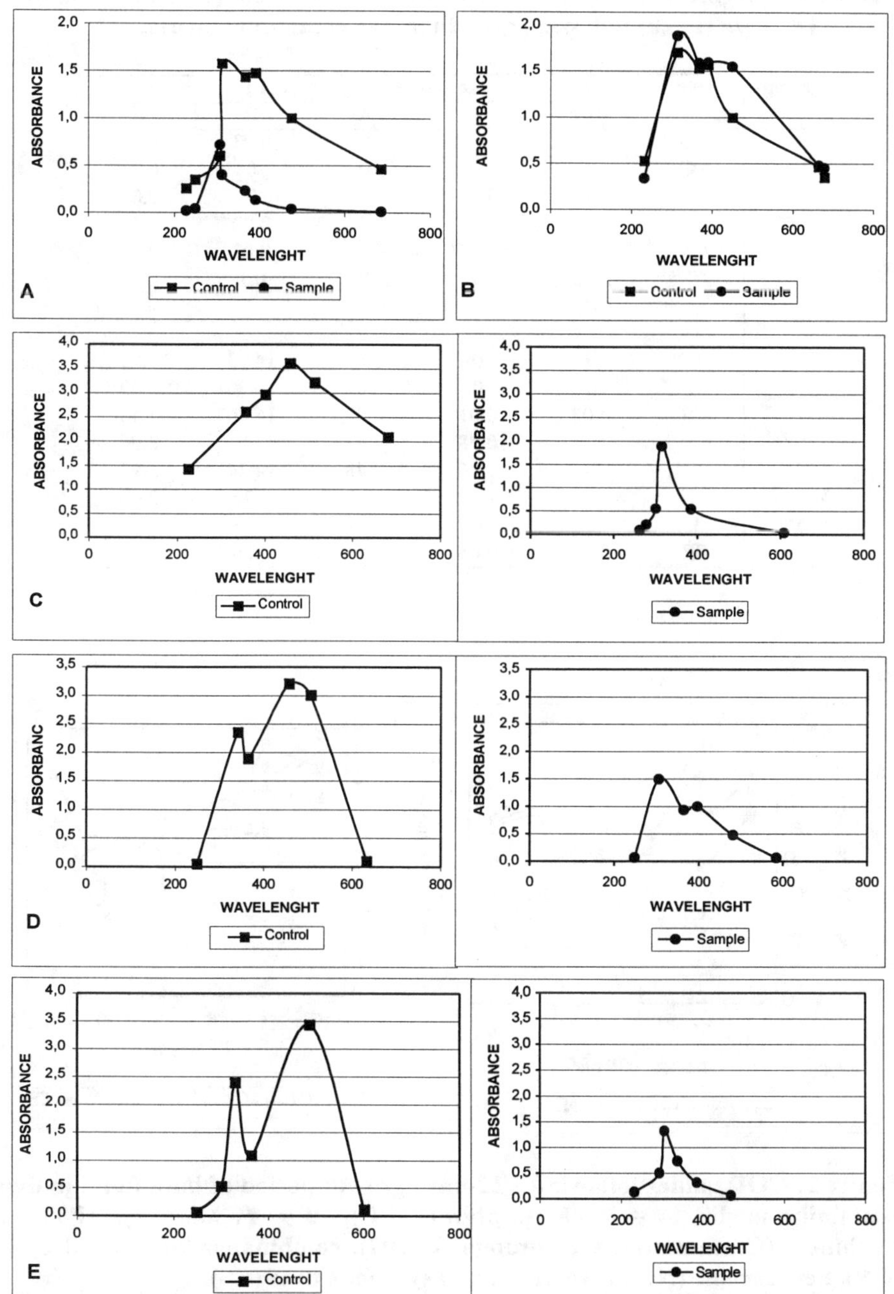

Figure 1: Absorption spectra of supernatants controls and samples following a 8-day incubation period A = *P. sajor caju* (020) in effluent; B = *P. chrysosporium* in effluent; C = *P. sajor caju* (020) in combination of natural dyes, D = *P. sajor caju* (020) in combination of synthetic dyes; E = *P. chrysosporium* in combination of synthetic dyes.

Table 1: Enzyme activities produced by *P. chrysoporium* (PC) and *P. sajor caju* (P020) following growth in dye containing media.

Fungi	Dye	Activities Enzyme (U/L)				
		MnP	LiP	VAO	D	S
PC	OG	0	0	0	0.03	0.09
	NC	0	0.05	0	0	0.21
	CS	1.16	5.42	0	0	0.07
	C	0	0.16	0	0.02	0.15
	AN	2.41	13.29	0.17	0.07	0.10
	T	23.3	0	0	0.1	1.1
P020	OG	0	0	0.23	14.03	4.58
	NC	0	0	0.07	20.46	2.09
	CS	0.08	0.01	0.08	16.66	5.44
	CN	0.13	0.98	0.93	14.01	2.06
	C	0	7.46	0.48	14.24	3.97
	AN	0	0	0.08	9.77	2.09
	T	0	0	0	0.23	0.5
	EF	0.13	0.08	0	0.26	0.14

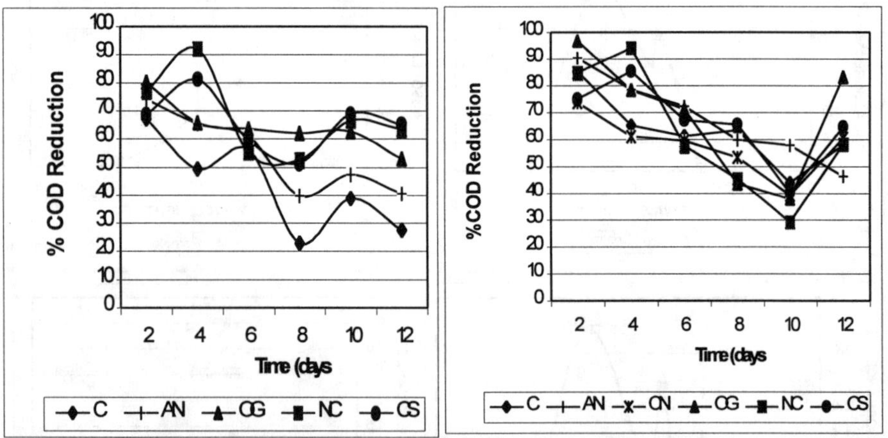

Figure 2: COD values following a 12-days growth period of both fungi in dye containing media A = *P. chrysosporium* (PC), B = *P. sajor caju* (P020) . Cochineal (C), annatto (AN), orange G (OG), combination of natural dyes (CN), new coccine (NC), combination of synthetic dyes (CS).

DISCUSSION

Differences between the two fungal strains regarding their ability to decolorize the dyes used in this work have been observed. *P. chrysosporium* was able to efficiently decolorize new coccine and orange G but only 60 % of the tartrazine's color. *P sajo caju* removed efficiently the colors of new coccine but only partially orange G (decolorization < 50 %) and tartrazine was very poorly affected. Tartrazine was the dye less decolorized by both fungi.

The chromatograms obtained following the HPLC analyses of the supernatants and controls showed that the azo dyes were also degraded indicating that the decolorization was not only a result of changes in the electron distribuition or of shitfts in the chromophore groups but possibly due to the breaking of the dye molecules.

Both fungi were able to decolorize the water-extracted annatto and cochineal, but only *P. sajor caju* decolorized the industrial effluent. *P. chrysosporium* was not able to grow in the oil-extracted annatto (results not shown), and the presence of this substance in the dye-containing effluent was problably the reason for the lack of growth and degradation.

Lowest COD values were observed following growth of *P. chrysosporium* for 8 days in all the dyes, whereas for *P. sajor caju* the lowest values for COD were observed following a 10 day growth period. Interestingly, an increase in the COD values was observed for both fungi after they reached the lowest COD, suggesting that the formation of a more complex molecule may have occurred possibly through a repolymerization mechanism.

Table 2: % Decolorization (absorption spectra) and % degradation (HPLC chromatograms) following growth of both fungi in azo dyes for eight days.

Fungi	Dye	% Decolorization	% Degradation
PC	T	60.0	100.0
	NC	95.0	100.0
	OG	96.8	100.0
P020	T	8.5	60.0
	NC	96.5	100.0
	OG	47.5	80.0

Both fungi exhibited ligninolytic activities which varied with the dye used for growth, but in general *P. sajor caju* presented higher activities of VAO and laccases in both substrates (*o*-dianisidine and syringaldazine) than *P. chrysosporium* which in turn produced generally higher LiP and MnP activities than *P. sajor caju*.

ACKNOWLEDGMENTS

J.G. Belote is grateful to CAPES and E.P. Chagas to FAPESP for financial support.

REFERENCES

[1] Marmion, D. M. *Handbook of U.S. colorants*. John Wiley & Sons, Inc., New York. Third ed. 573 p. 1991.

[2] Capalash, N.; Sharma, P. Biodegradation of textile azo-dyes by *Phanerochaete chrysosporium. World J. Microbiol. Biotechnol.* 8: 309-312. 1992.

[3] Tien, M.; Kirk, T.K. Lignin peroxidase of *Phanerochaete chrysosporium. Meth. Enzymol.* 161:238-249. 1988.

[4] Herrera, A.E.M. Produção e caracterização parcial de um composto de baixa massa molecular com atividade fenoloxidásica de *Thermoascus aurantiacus*. Campinas, SP, Brazil, 1995 (Ph.D. Thesis, Universidade Estadual de Campinas).

[5] Kuwahara, M.; Glen, J.K.; Morgan, M.A.; Gold, M.H. Separation and characterization of two extracellular H_2O_2 dependent oxidase from ligninolytic cultures of *Phanerochaete chrysosporium. FEBS Lett* 169: 247-250, 1984.

[6] Zghal, H.; Gueitin, C.; Halaqui, E. and Czchoc, M. Determination of sunset yellow FCF and ponceau 4R in sweets by high performance liquid chromatography. *Science des Aliments*. 15: 491-496, 1995.

AUTOTHERMAL THERMOPHILIC AERATED BIOREACTOR IMPACTS ON MICROORGANISMS. A PILOT STUDY

Michael Linich (The University of Newcastle, NSW Australia.)

ABSTRACT: An Autothermal Thermophilic Aerated Bioreactor (ATAB) pilot plant operating on food processing wastes is being studied to identify the microorganisms present. Incubations of the wastewater stream from the wastewater holding tanks to sludge (biomass) from the Dissolved Air Floatation (DAF) plant and the ATAB pilot plant were carried out at both 30^{0}C and 60^{0}C to estimate the impact of the process on culturable microbes. It was found that resident culturable microflora in the wastewater stream did not survive the ATAB process. The bioreactors performance appears to be related to the presence of thermophilic organisms not previously detected in the wastewater feed. Molecular methods using Amplified Ribosomal DNA Restriction Analysis (ARDRA) are now being applied to analyse the thermophilic microbial communities diversity and to identify important members. It can be concluded that the high autothermal operating temperatures create a pasteurisation effect on resident microbial species, some of which are pathogens, in the wastewater stream. The heat produced in this process is potentially recoverable for reuse. The stabilised and pasteurised sludge may be useful for value adding products.

INTRODUCTION AND BACKGROUND

Conventional wastewater treatment methods for the stabilisation and disinfection of bio-solids are generally expensive in both operator costs and the cost of chemical dosing. The resultant product is often unsuitable for sustainable environmental uses because of chemical contamination.

Food processing wastewater and their derived sludges, referred hereafter as 'biomass' typically exhibit inherent high energy value in the solid particulate fraction. The use of thermophilic processes for treatment prior to reuse is an emerging sustainable technology..

This study is focused on a poultry processing plant with a wastewater stream of approximately 30,000 litres/hr over an eight hour day. The wastewater is derived from a number of sources which includes drainage from marinade tanks, wash down from the factory floor (feathers, meat, fat) and the hose down from the trucks and crates (faeces, soil etc). In keeping with good environmental practice a proportion of this water is recycled before being collected into a 250,000 litre holding tank.

Dissolved Air Flotation (DAF) technology is currently used to treat the wastewater to remove approximately 80% of the Biochemical Oxygen Demand (BOD) and produces 8000 litres of liquid biomass per day. The treated water is subsequently discharged to sewer under a trade waste licence. The biomass is stored in a 35,000 litres holding tank which is periodically drained by a licensed contractor for disposal. The biomass normally produced in the DAF plant is highly putrifiable and attracts complaints from local residents during the summer months. Biomass viscosity and composition is variable on a daily basis but contains 5-8% solids normally. The mean cell resident time (MCRT) in the bioreactor is approximately 5 days.

The ATAB process introduces air into the reactor for biological stabilisation of the organic substances. The biology of the ATAB process resembles that of a high speed liquid composting system in that energy is released in the form of heat whilst at the same time organic solids are degraded. This generation of heat is carried out by aerobic thermotolerant or thermophilic organisms. Such processes normally operate between 50-60^0 C (Hamer 95).

The plant continues to operate autothermally at between 60-70^0 C and usually performs at 70^0 C. At this temperature, genuine thermophilic process microbes as opposed to the thermotolerant mesophiles are able to develop and assert their potential (Mason et. al., 87).

This study examined the microbial consortia present in the wastewater stream, the DAF treatment plant and the ATAB pilot plant. It examined the changes taking place in the culturable microbial population throughout the collection, treatment and digestion process. Currently molecular methods are being employed to analyse the ATAB biomass.

AIM

To examine the relative abundance of culturable fungi, actinomycetes and bacterial populations in the process wastewater stream through an analysis of liquid in:

A. Holding Tank Wastewater (DAF influent)
B. DAF Biomass (sludge feed for ATAB)
C. ATAB Biomass effluent (treated sludge) at both 30^{0}C and 60^{0}C.

METHOD

To support the growth of the three main groups of microorganisms (actinomycetes, fungi and bacteria) three different types of media were used. Sabourard Dextrose Agar (BBL) for the isolation and growth of fungi, Dermasel Selective Agar (Oxoid) for the isolation and growth of actinomycetes and Tryptic Soy Agar (BBL) for the isolation and growth of bacteria other than actinomycetes.

To quantify the numbers of microbes (actinomycetes, fungi and bacteria) in the wastewater of the influent of the DAF plant, the sludge produced from the

DAF plant and the effluent from the ATAB process, the following methodology was used.

1. 1ml of waste fluid from each source was diluted into 99 ml of sterile distilled water. Wastewater was collected from the holding tank, DAF sludge was collected from the front of the DAF tank and the ATAB effluent was collected directly from the pilot plant using 100ml sterile plastic containers.
2. Mixing was uniform for each step and comprised of a 3 minute agitation.
3. A further two 1/100 serial dilutions were done using the same protocol as described in the previous step. This resulted in a 10^{-2} , 10^{-4} and 10^{-6} dilutions.
4. Petri plates were labelled and dilutions were dispensed in such a way (either 1.0 ml or 0.1 ml) to give the following concentrations for each of the microbial groups as follows:-
   - Actinomycetes (10^{-3}, 10^{-4}, 10^{-5} and 10^{-6})
   - Fungi (10^{-2}, 10^{-3} , 10^{-4} and 10^{-5})
   - Bacteria (10^{-4}, 10^{-5}, 10^{-6} and 10^{-7})
5. Each plate was prepared in triplicate.
6. Using a micropipette, sterile disposable microtips and aseptic technique the dilute solutions were distributed to each of the Petri plates in accordance with the amount (either 1ml or 0.1ml) to arrive at the appropriate dilution.
7. 15ml of the Tryptic Soy Agar, at 50 ^{0}C, was poured onto the solution in the Petri plate and the plate was swirled gently on the bench to mix the contents and then left for the agar to harden (ie, pour plate technique).
8. The same was done for the Sabourard Dextrose Agar and the Dermasel Selective agars.
9. The Petri dishes were sealed with paraffin film and taped together.
10. The Petri plates were incubated in an inverted position. Two plates from each set were incubated at 60^{0}C and the third plate from each set incubated at 30^{0}C. The 60^{0}C plates were incubated in a humidified environment to prevent desiccation of the agar base for 48 hours.

RESULTS

Table 1 gives the absolute numbers of organisms found growing on each of the plates. Where colonies exceeded 250 the plate was labelled as TNTC (Too Numerous To Count).

DISCUSSION

Population dynamics. There are limitations to the methods used in this experiment in that only culturable organisms were studied. However significant changes in microbial communities are evident. Throughout the ATAB process we have observed a change in the microbial consortia present with the loss of major taxa. Bull and Slater (1982) state that changing end conditions induce changes in the microbial population composition. These observations are consistent with this generalisation and the ecological observations of populations in extreme environments made by Kahn (1998) and more recently by Hugenholtz, Goebel and Pace (1997).

TABLE 1. **Organisms culturable from Wastewater, DAF, and ATAB at 30°C and 60°C.**

	Dilution	Wastewater		DAF sludge		ATAB sludge	
		60°C	30°C	60°C	30°C	60°C	30°C
Actinomycetes	$10.^{-3}$	0	8	0	0	0	0
	$10.^{-4}$	0	1	0	0	0	0
	$10.^{-5}$	0	0	0	0	0	0
	$10.^{-6}$	0	0	0	0	0	0
Fungi	$10.^{-2}$	0	TNTC	0	TNTC	0	0
	$10.^{-3}$	0	TNTC	0	TNTC	0	0
	$10.^{-4}$	0	19	0	43	0	0
	$10.^{-5}$	0	0	0	4	0	0
Bacteria	$10.^{-4}$	0	59	0	TNTC	TNTC	2
	$10.^{-5}$	0	12	1	TNTC	19	0
	$10.^{-6}$	0	1	0	43	0.5	0
	$10.^{-7}$	0	0	0	0	0	0

Few, if any, of the culturable microorganisms which are present in high numbers in the wastewater stream are able to survive the treatment process. The benefits of a thermophilic process is that at the right temperature it will both effectively pasteurise pathogens and stabilise the sludge. Biomass applicable for this process would include any unstable liquid organic waste, of high energy (kilojoules) content, derived from food processing or human sewage. There is the added potential of combining sludges or liquid wastes to enhance the process.

The DAF process. The DAF process serves to concentrate suspended solids, on which microbes adhere, into a sludge which then served as the source of the sample for this analysis. Thus one would expect that the total number of microorganisms would be greater. Table one indicates that this is the case for both fungi and bacteria at 30°C but not for actinomycetes. This was confirmed by a second series of replicate experiments aimed directly at this hypothesis. Overall the DAF sludge concentrated the microbial population by a factor of some thirty six (36) times at 30°C. This is in direct contrast to an observed thousand (1000) fold decrease in the microbial population when the DAF sludge was incubated at 60° C.

ATAB. Observations of plate growth of the bacteria found growing in the ATAB biomass show growth is restricted to the surface of the agar despite the 'pour plate' technique being used. The surface growth suggests obligate aerobic metabolism. This was confirmed by growth of the bacteria on a stab subculture. These observations are consistent with the high rate of aeration in the bioreactor.

Bacterial numbers. At 30^{0}C the total number of culturable bacteria growing in the ATAB process is approximately 1000 times less than the total number of bacteria found in the DAF biomass. However at 60^{0}C this trend is reversed. There are about 10 times more bacteria present in ATAB biomass at 60^{0}C then found growing in DAF at 60^{0}C.

With growth of bacteria at 60^{0}C at 10^{-5} in the DAF samples it could be argued that this organism was either a contaminant or a spore of a thermophile which is present in the wastewater in very low concentrations. If the observed Bacillus populations are derived from the germination of spores then these spores are entering the ATAB process either via the air being sparged into the sludge, as the air is not filtered in any way, or they are present as ubiquitous spores in the wastewater stream. As only small volumes were analysed there may be sampling errors. If the latter is the case then the observations of the lack of growth of Bacillus at 60 ^{0}C in the wastewater and DAF fluids suggests that they are present only in very small numbers.

The role of microbes in ATAB is commonly thought of as dependent upon a consortia of thermophilic organisms acting together to produce a desirable end product. Studies on single substrate ATAB processes identifies two types of heterotrophic microbial consortia. One type contains more than 90% of single primary substrate utiliser (Hamer, 1990). The observations tend to support this type of consortia. However it is likely the molecular analysis will certainly give a more complex picture. The ATAB process observed here is dominated by only a few culturable organisms. Whilst it is fairly obvious that the ATAB influent is not a single substrate, its exact chemical composition is not known.

Bacillus species have been identified as the main organisms which grow in aerobic thermophilic processes (Sonnleiter and Fiechter, 1983) and furthermore it is members of this genera which produce thermotolerant protease exoenzymes (Mason et.al, 1992). The role of microbial exoenzymes in colloidal and particulate hydrolysis remains a matter for debate. But it appears to be a significant factor in such a process.

The fact that the microbes observed growing at 60^{0}C appear to be obligate aerobes is significant in that at 60^{0}C one would expect that availability of oxygen to the organisms would be a limiting nutrient. Whilst the operating temperatures may affect the solubility of oxygen the much higher oxygen diffusivity at this elevated temperature may balance out oxygen mass transfer into the system (Hamer, 1995). The aeration device used in this ATAB plant has enhanced the mass transfer of oxygen into solution in this treatment process.

It is important to note that the autothermal processes created temperatures which are equal to or greater than that at which domestic hot water is supplied. As the pilot plant was constructed from stainless steel it would be possible to further construct a heat exchange jacket to capture the heat generated by the process. High calorific sludges could be put to use in the development of sustainable technologies for resource recovery.

REFERENCES

Bull A.T., and Slater J.H. 1982. "Microbial interaction and community structure" in *Microbial interactions and communities* Vol 1, pp 13-44. Academic Press, London.

Hamer G. 1995. "Thermophilic Aerobic Processes for waste sewage sludge treatment" in *Environmental Biotechnology :Principles and Applications,* pp 206-220, Kluwer Academic Publications.

Hamer G. 1990. "Aerobic Biotreatment : The performance limits of microbes and the potential for exploitation" , *Trans. I Chem E.*, **68,** Pent B. May.

Hamer G. and Bryers J. 1985. "Aerobic thermophilic sludge Treatment : Some Biotechnologies Concepts" , *Conservation and Recycling* 8 (1) : 267-284.

Hugenholtz P., Goebel B. and Pace N. 1998 " Impact of culture -independent Studies on the emerging phylogenetic view of bacterial diversity" , *Journal of Bacteriology* :4765-4774.

Kahn L. B. 1998. "Microbial diversity in a geothermal hot pool" Master of Science thesis, University of Aukland., Aukland, NZ.

Mason C.A. et. al., 1992. "Aerobic thermophilic waste sludge treatment", *Wat.Sci Tech,* 25 (1).

Mason C.A. et. al., 1987. "Aerobic thermophilic biodegradation of microbial cells- Some effects of dissolved oxygen and temperature", *Appl. Microbiol. Biotechnol.* 25: 568-576.

Sonnleiter, B. and Fietcher A. 1983. "Bacterial diversity in thermophilic aerobic sewage sludge"- II. Types of organism and their capacities. *Eur. J. Appl. Microbiol. Biotechnol.*, 18, 174-180.

SUSTAINABLE REMEDIATION OF LAND CONTAMINATION

Theresa Kearney and Ian Martin (Environment Agency,
Solihull B92 7HX, United Kingdom)
Sue Herbert, (Environment Agency, Bristol BS12 4UD, United Kingdom)

ABSTRACT: The wider environmental effect of remediating land contamination is an important component of any decision about how to bring land back into beneficial use. In the UK, the Environment Agency is developing guidance to assist those selecting different remedial options to include wider environmental consequences in their risk management decisions. The developing framework will be based around a tiered approach moving from the initial identification of potential environmental effects through to a qualitative and/or quantitative appraisal of the relative differences between two or more remedial options. This paper describes the interim findings of this work and discusses the definition of the evaluation criteria and their operational boundaries within the context of the framework.

INTRODUCTION

The concept of sustainable development was first considered at the United Nation's Earth Summit conference in Rio de Janeiro in 1992 and was later followed by a paper entitled Sustainable Development: The UK Strategy in 1994. A number of definitions for sustainable development have been proposed, although the most commonly used definition in the UK is; ".... development that meets the needs of the present without compromising the ability of future generations to meet their own needs" (Brundtland 1987). At a strategic level, the remediation of contaminated sites supports the goal of sustainable development by helping to conserve land as a resource, preventing the spread of pollution to air and water, and reducing the pressure for development on green field sites (Environment Agency, 1996).

Decisions about which risk management option(s) are most appropriate for a particular site need to consider in an holistic manner the effectiveness of the remedial technique in dealing with the identified risks, the wider environmental effects of the remedial operation(s), and broader considerations of the relevant economic, social and political values that apply to the circumstances. Interpreting sustainable development in the context of land remediation is therefore a complex issue and requires guidance on specific components of the decision process, such as the environmental effect of different types of remedial options as well as overall guidance on the whole risk management process.

An underlying principle of the UK approach to identifying, assessing, and dealing with land contamination is that it should be carried out within a risk based framework. A risk can only be present if there is a plausible relationship between a source (e.g. oil spill), a pathway (e.g. groundwater) and a receptor (e.g. children, animals) on a site. In those situations where this is the case, the objective of the remedial works is to mitigate any potential or actual risks by controlling or breaking the source-pathway-receptor chain. Remediation of land contamination is therefore intended to reduce risks to human health, surface and groundwaters, ecosystems and to enable redevelopment of such sites.

Remedial objectives are fundamental to the process of selecting a suitable remedial strategy since they consider both the risks themselves and the manner in which the risks can be managed on a site-specific basis. They are developed via a staged approach in which the final stage provides performance criteria against which the success of a remedial strategy for a site can be assessed. Remedial objectives may take many forms, and in particular will include essential, or core environmental benefits such as reducing risks to required levels, and possibly desirable benefits, which are not essential but advantageous. In considering the wider environmental effects of two or more remedial options, it is important that different options can be compared taking into account the remedial objectives of a site. It is clear that selecting suitable remedial option(s) which are capable of achieving the remedial objectives would lead to a core environmental benefit, whilst, improving the site's biodiversity may be considered as a desirable or non-core benefit. In contrast, negative environmental effects due to remedial activities would include waste generation and gaseous emissions. It is important to ensure that in achieving the remedial objectives for a site, remediation processes do not lead to unacceptable harm, nor make excessive demands on resources or have deleterious environmental, social or economic consequences if it is to contribute effectively to the goal of sustainable redevelopment.

The aim of this paper is to describe a framework that is being developed for comparing the environmental effects between two or more potential remedial options at the stage of selecting a remedial strategy for a site.

MANAGEMENT OF CONTAMINATED LAND

The UK framework for managing contaminated land as detailed in the Handbook of model procedures for the management of contaminated land incorporates good practice procedures, including the use of risk assessment and risk management techniques, into a systematic process for identifying, making decisions, and taking appropriate action to deal with contamination (Figure 1) (DETR and Environment Agency, *in preparation)*.

It is important that the selection of the preferred remedial scheme examines the various risk management options being considered to meet the remedial objectives for a site in a consistent and transparent manner. The evaluation and selection primary model procedure provides such a framework (DETR and Environment Agency, *in preparation*). In using this procedure, it is assumed that unacceptable risks due to land contamination have been identified and appropriate remedial action needs to be taken. This procedure has two principle objectives;

(i) to identify those remedial options that are potentially applicable to the contaminants present and the site conditions, and
(ii) to compare potentially suitable methods against a set of screening parameters that are appropriate to managing the site.

These screening parameters can be both technical (e.g. suitability to treat contaminant types, site conditions) and non-technical (e.g. wider environmental effects and costs and benefits) against which the evaluation and selection of remedial options for a site is considered. Environmental effects of remedial operations is one of a number of screening parameters that can be used in the process of evaluating and selecting a suitable remedial scheme for a site. The remainder of this paper will focus on the approach being taken for assessing the wider environmental effects of two or more remedial options.

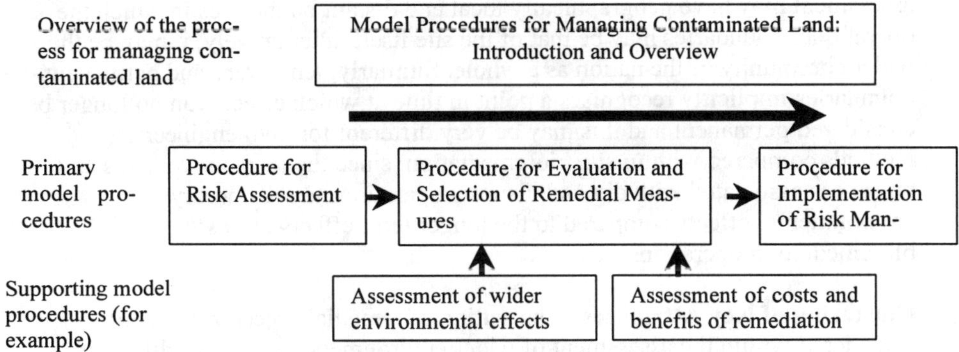

FIGURE 1. Framework of model procedures for managing contaminated land.

ENVIRONMENTAL EFFECTS

Environmental effects consider the wider environmental consequences of a particular remedial option on a site-specific basis. Such effects incurred during treatment of contaminated media will vary according to the type of contaminant(s) present and the treatment employed such that they may be temporary effects (e.g. lorry movements) or permanent (e.g. loss of soil function). The effects may also be positive (e.g. improving soil function or the ecosystem) or negative (e.g. loss of soil structure and integrity, release of volatile emissions). The importance of the environmental effects for each option considered will be dependant on the site itself, for example, nuisance issues (e.g. odours, dust, noise) associated with remedial options for a remote site may be less important compared with one in a city centre. In addition, the significance of such effects will vary at a local, regional and / or national level.

The framework being developed for assessing the environmental effects of two or more remedial options for a site is described in this paper, firstly by considering the operational boundaries of an assessment within which key attributes can be defined. Attributes are criteria (e.g. energy demand, chemical inputs, process emissions) that identify individual environmental effects. Measurement of the attributes within the developing framework will be based on a tiered approach that initially considers identifying potential environmental effects, and then moves into an optional qualitative and / or quantitative appraisal of the environmental differences between two or more remedial options for a site.

Operational Boundaries. Comparisons of the environmental effects of remediation will require an initial agreement of the operational boundaries for the remedial works on the site (e.g. resource inputs, generation of waste streams, gaseous emissions). The guidance aims to provide a framework for defining suitable operational boundaries for comparing two or more remedial options on a site depending upon available information and expertise. There is no defined set of generic operational boundaries applicable to all sites. Definition of boundaries therefore needs to be set on a site-specific basis and it is possible that any one assessment may have a number of boundaries associated with it. For example, an assessment may have geographically local and distant boundaries in which the operational boundaries may be that of the site itself, alternatively, it may be the wider community or the nation as a whole. Similarly, temporary and permanent boundaries implicitly recognises a point in time at which effects can no longer be considered permanent and this may be very different for civil engineering methods compared with *in situ* bioremediation, since the lorry movements and nuisances associated with dig and disposal operations are temporary environmental effects compared to the longer term effects of *in situ* bioremediation operations.

Clustering of Key Attributes. The setting of remedial objectives are not considered within the assessment of wider environmental effects of different remedial options. Remedial objectives are site specific and reflect the manner by which risks are to be managed. These objectives are considered within a tiered approach and become more detailed in moving from one tier to the next. In comparing the environmental effects of two or more remedial options, consideration needs to be given to identifying key attributes within the operational boundaries so that different remedial options can be compared for a given site within a similar risk management context .

The guidance considers the need to identify suitable clusters and attributes in assessing the environmental consequences of two or more options for a site. Clusters collate attributes on a site-specific basis under a number of broad headings such as core and non-core environmental benefits, and core and non-core environmental effects (Table 1). Clustering of attributes helps to simplify their assessment and identify any duplication, gaps, or redundancy. Core benefit clusters would include attributes such as the ability of the remedial options to treat the soil to a standard that meets the remedial objectives, whilst non-core environmental benefits may consider for example, the environmental effects in continuing the remedial works to remove further contamination after the remedial objectives for the site have been met. Alternative clustering of attributes may be based on (i) human health and safety, (ii) environment, (iii) land use (iv) process emissions or alternatively, (v) intrinsic site value, (vi) resources required for operation and (vii) externalities. Intrinsic site value considers attributes such as the site value post treatment. Resources includes the primary inputs such as raw materials and energy required during operation of the remedial option(s) on site but not those actually required to generate the primary inputs. Externalities include product and process waste streams generated during treatment.

TABLE 1. Clustering of key attributes to measure environmental effects between remedial options.

Examples of Clusters	Examples of Attributes
Core Environmental Benefits	Achieving the remedial objectives
Non-Core Environmental Benefits	Achieving remediation to a standard more than the remedial objectives Improving soil function Improving the ecosystem
Core Environmental Consequences	Energy use Resource inputs Generation of volatile and gaseous emissions
Non-core Environmental Consequences	Loss of soil function

Measurement of Environmental Effects. The guidance being developed is intended to provide a framework for assessing the relative environmental differences between two or more remedial options. The framework is likely to be based on a tiered approach (Figure 2). The initial tier involves identifying the characteristics of the site, the contamination and the environmental consequences of operating the selected options on the site. The second tier involves qualitatively observing the potential differences in environmental effects without the need to actually estimate their significance. For those sites where the choice of remedial option(s) is clear, this qualitative approach may be sufficient to aid decisions made with regard to the environmental effects of different remedial options. However, in those circumstances where this is not the case, a more detailed quantitative assessment may need to be considered prior to selecting the most suitable remedial options. The quantitative method is not intended to provide an absolute measurement but to clarify the relative differences between options through a detailed consideration of the attributes. The requirement to consider a more detailed assessment may result from;

(i) the need to ensure that the significance of any attributes have been considered in sufficient detail; and
(ii) the need to undertake a sufficiently robust assessment to allay the concerns of other interested parties.

The outcome of the assessment for evaluating the environmental effects of a number of remedial options for a site will enable the user to rank the options according to their environmental effects or consequences. This information can subsequently be considered in conjunction with other criteria (e.g. technical effectiveness, cost and benefits) in selecting the preferred remedial strategy for the site.

It is intended that the level of assessment builds in sophistication in moving from one tier to the next, with assessments carried out for previous steps feeding into the work required for subsequent steps and thereby reducing duplication of effort. It is important therefore that the available information for the potential remedial options at a site is compatible with the information requirements for the relevant tier within the framework.

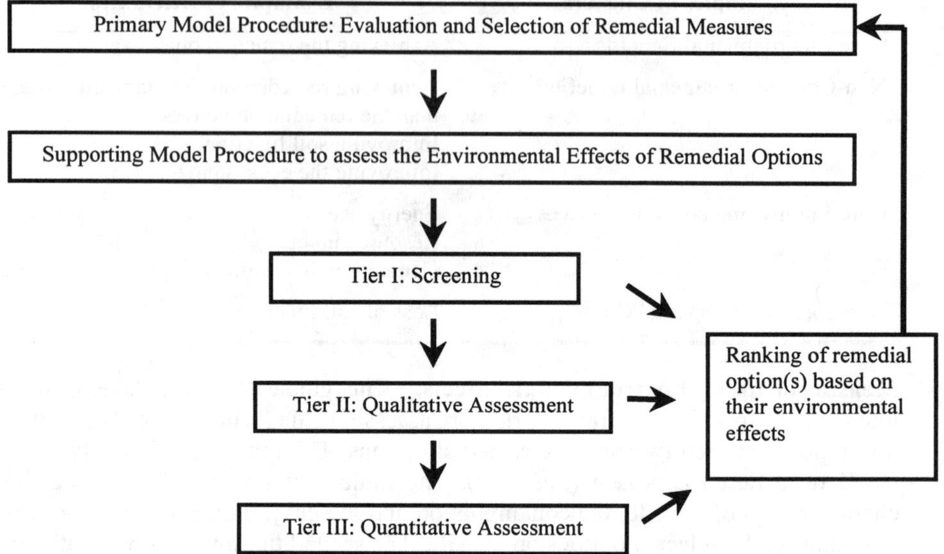

FIGURE 2. Schematic of the framework being developed for measuring the environmental effects of two or more remedial options for a site.

CONCLUSIONS

This paper discusses some of the issues in developing a framework for comparing the wider environmental effects of two or more remedial options that are being considered for a site. It is important in the UK situation that any framework being developed for considering the environmental effects of remediation, links with other selection parameters identified within the model procedures such as costs and benefits at the stage of evaluating and selecting suitable remedial option(s) on a site specific basis.

REFERENCES

Department of the Environment, Transport and Regions / Environment Agency. (in preparation). *Handbook of Model Procedures for the Management of Contaminated Land. CLR 11.* Department of the Environment, Transport and Regions, London.

Department of Environment. 1994. *Sustainable Development: the UK Strategy. CM2426.* HMSO, London.

Environment Agency (in preparation). *Sustainable Remediation: Environmental Merit. R&D Technical Report No. P238.*

G.H. Brundtland. 1987. *Our Common Future*. World Commission on Environment and Development.

ECONOMIC ANALYSIS TO SUPPORT REMEDIAL DECISION MAKING AND TECHNOLOGY SELECTION

Paul E. Hardisty (Komex Europe, Bristol, UK)
Anthony Brown (Komex H2O Science, Huntington Beach, Ca., USA)

ABSTRACT: Economic analysis can be used to account for the real overall cost of pollution, including private costs (to the problem holder) and external costs (to society). Overall decisions on the required level of remediation can thus be made with an understanding of the full economic implications. In many cases, remedial objectives are set with only a narrow (mostly private) view of the costs and benefits involved - the larger implications for the environment and society as a whole are rarely considered. However, it has become increasingly apparent that the external costs and benefits of remediation can be substantial, and their absence from analysis may result in remedial programs which are unsustainable and non-optimal. Selection of remedial objectives can be guided by a preliminary benefits analysis, based on a range of likely remedial objectives: "remediation to this level will be worth at least this much to society as a whole...". The benefits analysis should include net private benefits, and if possible wider external benefits, often described as the value of damage averted by taking action. Valuation of external benefits of remediation is not straightforward, and can include the value of damaged resources, option values, and intrinsic worth (existence and bequest values). Once the remedial objective has been set, and a threshold cost identified (equivalent to the total benefits of remediation), the question becomes how best to achieve the goal. It is safe to say that cost is one of the single most important factors in site clean-up decision making. A basic rule of remediation is often taken to be the maximization of contaminant mass removed for the money spent. However, remediation may also be governed by other objectives and constraints. In some situations, minimization of time, rather than cost, could be the constraint. Evaluation of the economics of a clean-up project is directly linked to the objectives of the site owner, the needs of society as represented by the regulators, and the constraints within which the remediation is to be performed. Once objectives and constraints have been clearly identified, a range of possible remedial approaches and technologies can be developed to achieve the set goal, and each option evaluated on a comparative basis. This can be achieved through comparison of life-cycle time-cost curves for each remedial option. This approach combines evaluation of technical feasibility and effect, with consideration of capital, operation and maintenance costs, and the dis-benefits of remediation, over a selected time horizon. By constraining remedial alternatives within cost and time boundaries, possible solutions can be evaluated with respect to specific criteria. In many cases, the hidden costs of certain remedial approaches are revealed.

INTRODUCTION

Most activities associated with production of goods and services in modern society are linked directly or indirectly to the production of wastes and pollution (World Bank, 1992). Benefits to society are accrued through this production, measured as GNP (Gross National Product). However, with the benefits also come costs associated with environmental damage. These may include degraded air quality, polluted rivers and coastlines, damaged aquifers and soil, and habitat destruction. Until recently, however, these costs have not been explicitly accounted for in the balance of accounts. If environmental costs are included in calculation of GNP, a more realistic picture of society's progress and well being is provided. Some activities which appear on first analysis to benefit society, may in fact be net detriments when environmental impacts are taken into consideration. This approach can be extended to the subject of site remediation. The full value of environmental damage caused by a contaminated site includes losses suffered by the problem holder (internal or private), and external costs (society suffers a loss). Typically, external costs are not accounted for in the problem holder's financial statements. However, under the polluter-pays principle, which has been adopted by many industrialized nations, private firms are now economically responsible for at least a portion of the damage. Thus, a more accurate view of a firm's balance of accounts would involve discounting corporate profits based on the costs of environmental damage which occurred during the production of those profits. Similarly, the benefits of expenditures on remediation can be seen as the wider (private and social) cost of the damage averted by undertaking the remediation.

Typically, one of the major impediments to site remediation is the perception that money spent provides no return to the corporation. In some cases this may indeed be true. However, there clearly exist situations where a certain level of expenditure for site clean-up is both warranted and beneficial to the problem holder and to society.

REMEDIAL DECISION MAKING

Many different factors go into the decision to remediate a site. These include the level and type of contamination, the nature of off-site impacts, the risks involved, the regulatory and political climate, corporate policy, the availability of suitable technology, and the financial resources available.

Economic Framework. The decision to remediate a contaminated site can be examined in an economic context. Cost-benefit analysis can be used to account for the overall cost of damages associated with pollution, and thus make overall decisions on the benefits of different levels of clean-up. This process can be used to assist firms in setting economically realistic and justifiable remedial goals for individual sites, groups of sites, or as corporate policy. In addition, regulatory bodies may use the analysis to require that at least some of the wider external benefits of remediation are included in the analysis (with the result that higher levels of expenditure on remediation may be found to be warranted). The process can work both ways, however: firms may also be able to argue, in certain cases, that lower levels of expenditure on remediation are warranted, because the net benefits of remediation are small.

A major obstacle to applying quantitative cost-benefit analysis for setting remedial objectives is the difficulty in monetising many environmental benefits. The value of an environmental resource includes it use-value (economic value accrued through direct of indirect use of the resource), option value (value in retaining the option to use the resource in the future, regardless of whether it is being used now), and non-use value (existence and bequest values). Use-values are relatively well understood, and in many cases can be estimated based on market values for land, or groundwater, for instance. However, estimation of non-use and option values is not straightforward, although most workers now feel that the non-use value of groundwater is considerable (Raucher, 1983). In many cases, non-use values can only be described qualitatively ("society values the ability to pass on an undamaged aquifer to future generations"), and incorporated in a semi-quantitative way. Benefit analysis can, however, be a powerful tool for setting a threshold cost constraint for the project ("achieving this remedial objective is worth at least this much"...).

Once the remedial objective has been set, and a threshold cost identified (equivalent to the total benefits of remediation), the question becomes how best to achieve the goal. Which of the myriad of available remedial approaches and technologies is best able to reach the intended goals. Cost-effectiveness or least-cost analysis can be used to provide efficient allocative decisions for reaching those overall goals.

The Decision Making Process. Finding the optimal remedial goal, and then choosing the best method of achieving that goal, requires a rational approach, based on good data. The following general steps should be considered: 1) *Understand the problems at the site*. The value of good site characterisation and good cost and benefit data can never be under-estimated. 2) *Assess the risks posed by the problem.* Using the tools of risk assessment, the implications of the problem are determined. Many different types of risk exist (human health, ecological, economic, public relations, personal and corporate liability), and one or more may be important at the site. These risks can be economically valued. 3) *Set remedial goals and constraints for the site.* Select a remedial goal based on corporate and regulatory requirements, risk and a preliminary objective-cost-benefit sensitivity analysis. 4) *Identify constraints and determine an optimal remedial approach.* Constraints set the limits of what is possible. The approach best able to reach the set goal within the applied constraints, is selected. 5) *Implement the optimal remediation;* 6) *Monitor Results.* Assess remedial progress through careful monitoring, and modify as necessary for efficient improvements. Monitor economic performance also.

Cost Benefit Analysis. Determination of the cost constraint is best accomplished using cost-benefit analysis. For a given project, the decision to proceed is taken when the net benefits of the programme exceed the costs, as:

$$\sum_{t=1}^{t=n}(B_t - C_t)\frac{1}{(1+i)^t} > 0 \tag{1}$$

where B_t are the benefits accrued by the project over time, C_t are the costs of the programme at time t, i is the discount rate, and n is the total period of time the programme is expected to run. Unfortunately, for environmental remediation projects, it is usually the costs which are easiest to quantify, and receive most of the attention from corporate decision makers. Financial risk derives from several sources, including uncertainty in the future discount rate, programme completion time, and the inherent difficulties associated with determining remedial costs for subsurface contamination. The environmental benefits of remediation are harder to quantify in economic terms, but are no less important.

Benefits of remediation can be expressed as:

$$\sum_{t=1}^{t=n} B_t = \pi_{nh} + \sum_{t=1}^{t=n}\left\{B_{ne} + B_{np}\right\} \tag{2}$$

where π_{nh} is that portion of the historical profits generated by the operation which can be attributed to the fact that spending did not occur on mitigative measures which would have prevented the present contamination. This distinction explicitly accounts for the fact money spent in the past on measures such as liners for ponds, double walled USTs, proper landfilling practices and spill control could have prevented the need for present day remediation. In addition, the fact that these funds were not spent would have shown up as profit in previous years. For this reason, this type of benefit is often referred to as a shifted benefit (Pearce and Warford, 1993). It is important to stress that the method for accounting for historical profits presented in no way suggests that operators should have employed present-day mitigative techniques in the past. Clearly many of these techniques were not available twenty or even ten years ago. Certainly in the case of gasworks sites in the U.K., for example, many of which were operational from the late 1890's to the middle of this century, it is difficult to assign a shifted benefit to a present day analysis, so this term should be included only with caution. However, this does provide a methodology for assigning a reasonable historical benefit to the firm. Clearly, the inclusion of this term would depend on the perspective taken: optimisation of societal benefit might call for some accounting for historic profits realised as a direct of contaminant generation; a private commercial interest would likely resist its inclusion in many cases. However, it is useful to consider this concept in a balanced and dispassionate discussion.

The term B_{ne} refers to the net environmental benefit accrued to society as a result of the remediation going ahead, and is an external benefit (from the firm's viewpoint). This could include the value to the public of access to a reclaimed coastline or wetland area, or the value of the groundwater resource being protected or remediated. This term does not however, explicitly account for "shifted irreversibilities", defined as costs shifted to future generations by non-reversible environmental impacts. Shifted irreversibilities would include species extinction, permanent loss of sensitive habitat, and the like. These societal benefits are more difficult to quantify for use in a cost benefit analysis. The net private benefits (B_{np}) must also be accounted for. These would include the value of fines, control orders or lost production not incurred because the remediation went ahead, and the public relations value of the clean-up. Increased land value, and reduced corporate liability, would also benefit the firm directly.

For example, the benefits of groundwater remediation can be defined by the change in expected damage (E(D)), expressed by Raucher (1983) as:

$$E(D) = p\left[qC_r + (1-q)C_u\right] \quad (3)$$

where p is the probability, in the absence of remediation, that damage will occur, q is the probability that groundwater contamination would be detected before tainted water was used, C_r is the cost of the most economically efficient response to the problem, and Cu is the cost incurred if contaminated water continued to be used in the same way as prior to the incident. This highlights the links between sound technical understanding of the problem, and the economic analysis: both p and q are highly dependent on a sound understanding of the hydrogeological regime and the behaviour of the contaminants within the groundwater system. The inclusion of a probabilistic component to the framework explicitly recognises the role of uncertainty in the decision making process.

Remedial Costs. The costs of a remediation project are much easier to determine. Using equation 2, the maximum remediation cost which would be contemplated would be equal to B_t. Note that the lower the risks posed by the problem, and the more flexible the regulatory environment, the lower the sum which would be expected to be devoted to clean-up. Thus, this type of analysis provides a powerful tool with which decision makers can identify a realistic and economically justifiable cost constraint for remediation. In addition, by adopting a common economic framework of this type, negotiation between responsible parties and regulators could be simplified.

Optimization of the Level of Clean-Up. Another way of considering the remedial goal is to consider the level of clean-up (or more generally the level of remediation) to be achieved, and the marginal cost and benefits of clean-up. In this case, marginal cost refers to the additional cost required to achieve an increment of clean-up. The marginal cost of clean-up typically drops as clean-up begins, due to economies of scale and apportionment of fixed costs, reaches a minimum, and then rises steeply after remediation of the most concentrated and easy-to-access contamination has occurred. Often, remediation of the last 10 % of the contamination can cost as much as clean-up of the first 90 % (NRC, 1994). Conversely, marginal benefit is high initially, and decreases as more and more of the contamination is removed. The optimal level of clean-up would occur when marginal costs are equal to marginal benefits.

REMEDIAL TECHNOLOGY SELECTION

Once the goals and constraints for the remediation have been determined, and the decision to remediate made, attention turns to the most effective technical means of achieving the objectives. A wide variety of remedial technologies have been developed over the past fifteen years for both soil and groundwater. Different methods are suitable for different situations, and involve different levels of cost.

Technology Life-Cycle Cost Curves. It is useful when discussing remedial technologies to consider the life-cycle cost curve of each method. A typical cost curve includes the initial capital cost, ongoing operation and maintenance (O&M) and monitoring costs, and capital improvement costs. To provide a consistent basis for comparison of costs in time, and recognizing the time value of money, costs are expressed as net present value (NPV) costs. Capital intensive methods (such as excavation and thermal treatment of contaminated soil, for instance) are characterized by cost curves which are initially very steep, but which reach the remedial objectives quickly. In contrast, bio-remediation and in-situ methods may have smaller initial capital costs, but significant annual O&M requirements, and may require considerable periods of time to reach specified goals. In this way, life-cycle cost curves can be generated for a number of remedial technologies. By applying the cost and time constraints for remediation which optimize net societal benefit, the most economically efficient solution can be identified.

CONCLUSIONS

Economic analysis can assist in rational decision making regarding site remediation. The fundamental questions of whether to remediate or not, and if so to what level, can be examined in economic terms. Establishing a maximum acceptable cost for remediation puts an important constraint on remedial goal selection. The cost of remediation ideally should be balanced against the benefits accrued by site clean-up. While it is clear that individual corporate interests are not always linked directly to social optima, there are clearly strong overlaps between what is good for a corporation, and what is good for the society within which it exists and operates.

REFERENCES

National Research Council. 1994. *Alternatives for Groundwater Clean-Up.* National Academy Press, Washington,. D.C.

Pearce, D.W., and Warford, J.J. 1993. *World Without End*, Published for the World Bank, Oxford University Press, Oxford.

Raucher, R.L. 1983. "A Conceptual Framework for Measuring the Benefits of Groundwater Protection." *Water Res. Research*, vol. 19, No 2. Pp 320 - 326.

World Bank, 1992. *World Development Report 1992: Development and the Environment.* Oxford University Press.

DUTCH RESEARCH PROGRAMME ON IN-SITU BIOREMEDIATION (NOBIS): A MARKET-ORIENTED R&D NETWORK— BRINGING KNOWLEDGE TOGETHER FOR SOLUTIONS

Jos H.A.M. Verheul and Harry J. Vermeulen
(NOBIS, Gouda, The Netherlands)

INTRODUCTION

Local cases of contaminated sites have often hampered the planning and the economic development of: urban areas (for example textile cleaning companies and former gas works sites), industrial areas (often large-scale, the sites are being sold, the user is not the owner) and natural areas (dump sites in the countryside, filling in waterways).

Meanwhile, the 'classical' ***uniform*** risk assessment of soil contamination based on concentrations has been abandoned in the Netherlands. This was necessary in order to achieve the desired cost reduction and to improve the functioning of the market. Presently the costs of soil remediation in the Netherlands are estimated at about $25 billion (formerly about $50 billion).

Apart from abandoning the uniform risk assessment as mentioned previously, cost reduction is also related to the *available* means (not only money, but also time and space) and cheaper research methods and remediation techniques. NOBIS has focused on the latter aspects and used the actual risks in location-specific circumstances as the basis for technology development.

THE NOBIS PROGRAMME

The Dutch Research programme In-situ Bioremediation (NOBIS) started in 1995. NOBIS is a collaborative venture between public and private parties. The present extent of the programme can best be described by the following key data:

- *number of projects*: about 100, subdivided into about 25 research projects, 50 feasibility projects and 25 implementation projects;
- *number of participating parties*: about 100 (industry, authorities, consultants, contractors, technological institutions and universities);
- *financing*: about 20 million dollars for projects, half of which is financed by the government and the other half by private parties;
- *turnover*: about 4 million dollars annually and a programme duration of 5 years.

A market study has shown that in-situ bioremediation may be an important *part* of the solution in 70% of the cases of soil remediation. In 15% of the cases, in-situ bioremediation offers the *total* solution. This implies that combinations of techniques are usually required. The in-situ technology mainly focuses on mobile contaminants.

Finally, new techniques for biological remediation of contaminated dredging sludge have also been developed within the NOBIS framework.

A METHOD FOR SOIL REMEDIATION

NOBIS is expected to supply the technical and scientific products that facilitate a problem solving method to tackle soil contamination. The products have to meet the knowledge requirements of the new government policy, as well as help get around the obstructions all other parties involved are confronted with in practical situations. Concentrations of contaminants are no longer the central issue, but rather the *actual* risks of the contaminants in *location-specific* circumstances.

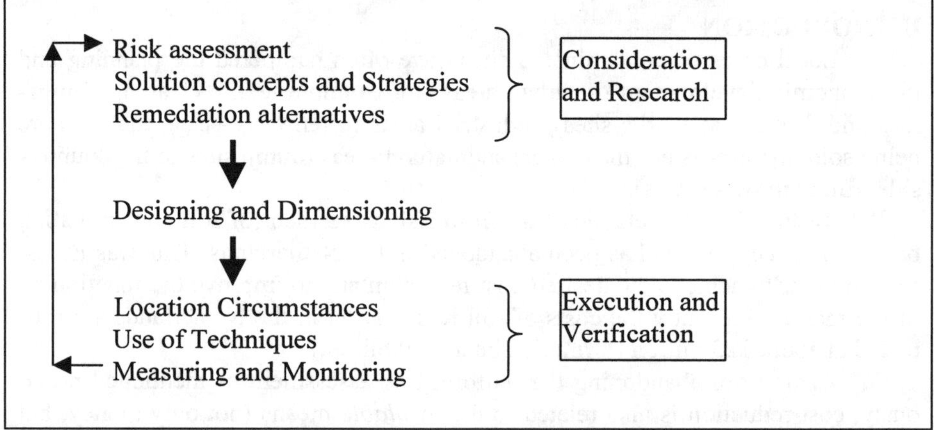

FIGURE 1. The soil remediation cycle. The basis for soil remediation is the reduction of risks. This is what the results of the efforts are tested against.

The developments in the field of technology will be discussed briefly on the basis of the cycle shown in figure 1. According to this schematic representation, basically any case of soil contamination can be tackled step by step. This cycle was not a basic principle for the research programme, but it has been derived from the methods used in the various projects. This paper will focus on the first three items mentioned in figure 1.

Risk assessment. The actual risk assessment as well as the establishment of priorities regarding the remediation are and will remain policy matters. However, the existing frameworks for weighing the pros and cons had little support from the soil remediation market. There appeared to be a need for *new* weighing methods based on technological principles that: (A) expressed the balance between the risk reduction, the environmental merit and the costs of the remediation alternatives [1] and (B) included a strategy for handling potential risks [2].

Moreover, (C) soil remediation often leads to lengthy groundwater purification projects that hardly ever achieve the intended remediation objectives. For these so-called stagnating groundwater purification projects, a method of weighing has been developed that allows for an interim assessment of whether continuation is still worthwhile [3].

So, the contamination of the soil is no longer assessed solely on the basis of concentrations. Risks have been linked to the intended use of the soil. This has created more room for new concepts and strategies and allows time and space to be more efficiently used.

Concepts and strategies. New concepts and strategies for remediation integrate planning development, water systems and intended function. The development of a new concept only results from good communication between all parties involved. Examples of such concepts are:

1. The remediation of dredging sludge in combination with the cultivation of willows as biomass. The advantages of this concept are: shared use of the land, significant reduction of the costs of dredging sludge remediation, proceeds in the form of biomass [4]. At a test farm in the north of the Netherlands this concept is being put into practice. It involves PAH contaminated dredging sludge. The sludge is spread among the willow fields. Critical factors are the possible spread of the contaminants to ground and surface water and ecotoxicological risks. So far, the monitoring data assembled over a period of about one and a half years show no signs of any spread. Bioassays with earthworms show no negative consequences for the worm population. Current plans are to apply the concept at a regional scale.
2. Application of natural degradation as a remediation alternative for (former) textile cleaning company sites. The remediation of the contamination of the soil at textile cleaning companies stagnated because of the high costs of the traditional remediation alternatives. Many of these companies are situated in inner city areas. Moreover, they are frequently small family businesses. Obviously, the financial resources of these companies are insufficient to cover the costs of soil remediation. Therefore, a simple and fast strategy has been developed to assess natural degradation as a remediation alternative. This has created new possibilities for the realisation of an approach geared to a specific branch [5].
3. As a final example, the management of former dumpsites can be mentioned. In the Netherlands are about 3000 closed dumpsites. New monitoring methods using sensor technology for the analysis of macro parameters and contaminants coupled with telemetric data transmission offer possibilities for a substantial reduction of the costs [6].

Other concepts involve the remediation of former gas works sites in combination with urban development [7]; the development of natural areas in the countryside at and around former dump sites in combination with the processing of dredging sludge from the same area [8]; incorporation of closed oil extraction in the countryside in combination with less intensive land farming [9].

The weighing of remediation alternatives. A decision support system has been developed for weighing the various remediation alternatives. This system enables objective mutual comparison and optimisation of the alternatives. These are assessed with regard to their contribution to Risk reduction, Environmental merit and Costs [1]. The costs and benefits for the environment are weighed, as well. For mobile contaminants a system has been developed, that determines the cost effectiveness of a measure [10].

A remediation alternative can be chosen using the following strategy. Apart from the primary risk assessment, the following points are involved:

1. *Take the time.* Natural processes in the soil are slow. This implies that the time factor is of vital importance for biological remediation options. Therefore, it is important to take the time in order to be able to use the advantages of in-situ degradation. Taking the time means attacking a contaminated situation at an early stage, so that, when another application for the soil arises some years later, the location is suitable for that new purpose. Taking the time also implies that the in-situ degradation process is in progress while the (new) company activities continue as usual.
2. *Use the 'self cleaning capacity' of the soil.* All sorts of biological degradation processes take place in the soil. Generally, it is worthwhile to investigate whether the natural degradation process *as such* will be sufficient to obtain an acceptable risk reduction over time [11]. A number of factors play a role in this assessment, including the natural 'rate of disappearance' of the contaminants; the geochemical characterisation in relation to the degradation products; the capacity (potential) of the soil to remove a certain quantity of contaminants over a period of time; the 'global' modelling of the geohydrology and the compound behaviour to obtain an overall impression of the location and to determine the dominant processes [11].
3. *Stimulate natural processes.* A high return on investments can be obtained by stimulating the biological degradation processes that are already present. Limited injection of nutrients, oxygen, other electron acceptors or donors and a subtle control of the groundwater flow may be sufficient to stimulate the degradation processes to such an extent that the requirements for acceptable risk reduction will be met. This less intensive remediation can sometimes be combined with control measures that have already been taken or will be taken anyway. As mentioned before, the time factor is very important in this approach. The term within which the risk reduction will have to be realised will usually have to be weighed against the technological interventions.
4. *Intensive in-situ remediation if necessary.* Despite all efforts, the previously described line of thought may lead to the conclusion that the stimulation of natural degradation will not sufficiently reduce the observed risks. In that case, a more intensive approach is required [12].

Apart from the previously mentioned REC method [1], the application of a method to quantify the financial risks of a remediation alternative is also being worked on [13]. Insurance companies are expected to play an increasingly important role in this matter, as well [14].

Research and characterisation. Remediation alternatives on the basis of natural processes require new methods and techniques for research and characterisation of the soil matrix. The concentrations of contaminants are not the central issue here, but rather the biological and chemical processes that are responsible for the degradation of contaminants. The risk reduction that has to be achieved in relation to the desired development of the location (concept) determines what remediation alternatives are possible. Thus the research and the characterisation are at the service of the intended solution. This is in contrast to the past, when soil research was organised around defining the problem. NOBIS has generated a wide range of new and/or improved research methods. Strategically these methods all aim at efficiently mapping the processes in the soil. For VOCl- compounds, this mainly involves concentration contour analysis of degradation products in relation to the geochemistry supported by geohydrological modelling. In addition to the previously mentioned approach in the textile-cleaning branch, this method was developed and tested at various locations [15,16,17]. Measuring hydrogen as the ultimate electron donor plays a central role here. [18]. Based on this research approach, various pilot tests have been started which will lead to a full-scale application [15,16]. For volatile aromatics (BTEX), a similar approach has been realised. However, it has to be noted that the intrinsic degradation of benzene is a significant bottleneck in the field characterisation [19]. This has still not been well demonstrated, although the degradation of benzene under denitrifying circumstances has been proven in the laboratory [20]

A major aspect of the research into biodegradation of mineral oil is a new method to characterise mineral oil. This involves the development of a methodology to distinguish in advance what part of the oil can be removed by biodegradation, stripping or flushing [21]. This method also fits in well with the risk assessment of mineral oil. On the basis of this method of analysis, it is also possible to state in advance what residual concentrations are feasible. Another important development is the direct link between analysis data and soil structure in the design and dimensioning of air-sparging systems [22]. Pilot tests may often play an important role in establishing the feasibility of remediation projects [23,24].

Location circumstances, Implementation and monitoring. In the implementation of a technique, measuring and monitoring is of crucial importance. The design has to be flexible to such an extent that improvement on the basis of monitoring data are allowed for. Monitoring on the basis of gauging observations is generally not very reliable. The use of sensortechnology simplifies the monitoring, leads to and more reliable results and is cost-effective. Sensors for the measurement of the pH, temperature, EC and oxygen values have been developed. Sensors for the measurement of BTEX and VOCl are still being developed. The first BTEX prototypes will enter the testingphase in the spring of 1999. All sensors can be assembled in to one instrument suitable fore monitoring in one-inch gauges [25].

RETROSPECTIVE.

Based on 3.5 years of NOBIS, the following conclusion can be drawn:
A programme in which research is executed on the basis of bottlenecks experienced in practice and in which all other parties involved collaborate on a solution,

offers excellent opportunities to efficiently and effectively develop methods and techniques for soil remediation.

The NOBIS approach will be continued by a new foundation: soil knowledge development and transfer (SKB). This programme started in January 1999 and has an intended duration of 10 years. The programme has been budgeted at $ 7,5 million a year.

REFERENCES.

[1]. Nijboer, M., et.al. 1998. "The REC decision support system for comparing soil remediation alternatives". *NOBIS report 95-1-03.*

[2]. Schurink, E., et.al. 1998. "ICM and aftercare by means of flexible emission control. *NOBIS report CO-379600/72* (In Dutch).

[3]. Hetterschijt, R., et.al. 1999. "Assessment of effectiveness of in situ remediation. Modelling fate and transport of contaminants. *NOBIS report 95-2-11, phase 2.*

[4]. Harmsen, J., et.al. 1999. "Growing of biomass to stimulate bioremediation in technical and economical perspective". In: *In Situ and On-Site Bioremediation , San Diego 1999.*

[5]. Van den Berg, J.H. et.al. 1998. "Intrinsic biodegradation and bioscreens for remediation of soil at laundry sites. *NOBIS report 96-2-01, phase 2 and 3* (In Dutch).

[6]. Van Breukelen, B., et.al. 1998. "Natural attenuation at the Coupépolder landfill? Hydrological, geochemical and biological characterisation". *NOBIS report 96-3-04, phase 2.*

[7].Heersche, J.A.N.M., et.al. 1999. Biotechnological remediation of former gaswork sites. *NOBIS report 96-3-04* (In Dutch).

[8]. Van den Akker, J.J.H., et.al. 1999. Ecological risk reductions of former dumpsites by means of isolation with dredging sludge. *NOBIS report 97-1-21* (In Dutch).

[9]. Brouwer, L., et.al. 1998. "Remediation Trial at NAM's Monitoring Station No. 1 in Schoonebeek". *NOBIS report 95-1-44* (In Dutch).

[10] Beinat, E., et.al. 1999. "Cost effective soil remediation. A decision support system for pollutants in the mobile regime". *NOBIS report 96021* (In Dutch)

[11]. Sinke, A., et.al. 1999. "Stimulation of the acceptance of natural attenuation in the Netherlands". In: *In Situ and On-Site bioremediation. San Diego 1999.*

[12]. De Kreuk, J., et.al. 1999. "Full-scale anaerobic biodegradation of per- and trichloroethylene". *NOBIS report 95-2-19* (In Dutch)

[13]. Hetterschijt, R., et.al. 1999. "Quantification of financial risks of remediation alternatives. *NOBIS report 98-1-10.* To be published in autumn 1999.

[14]. Van der Kooij, A., et.al. 1999. Insurance of in situ bioremediation projects. NOBIS project to be started in 1999. *NOBIS project number 98-1-07.*

[15]. Van Aalst-van Leeuwen, M.A., et.al. 1997. "Biodegradation of per- and trichloroethylene under influence of sequential reduction/oxidation conditions". *NOBIS report 95-1-41* (In Dutch).

[16]. Bosma, T.N.P., et.al. 1999. “Risk reduction on a industrial site by means of in situ measures. *NOBIS report 95-1-06* (In Dutch).
[17]. Doelman, P., et.al. 1998. “Combi-remediation: combined biodegradation of PER and BTEX by engineered mixing of groundwaterplumes. *NOBIS report 97-1-15, State of the art.*
[18] Rijnaarts, H.H.M., et.al. 1999. “Hydrogenconcentration in groundwater as parameter of the in-situ condition for the biodegradation of chlorinated hydrocarbons. *NOBIS report 96024, in press.*
[19]. Baten, H. et.al. 1997. “Biodegradation of BTEX under anaerobic conditions on the sites Slochteren and Schoonebeek 107”. *NOBIS report 95-1-43* (In Dutch).
[20]. Rijnaarts, H.H.M., et.al. 1999. “Intrinsic and enhanced biodegradation of benzene in strongly reduced aquifers”. In: *In Situ and On-Site Bioremediation. San Diego 1999.*
[21]. Weytingh, K.R., et.al. 1998. “Improvement of the in situ bioremediation alternative by adding the concept of imbibition and drainage to existing theory. Phase 2: oil characterisation”. *NOBIS report 95-2-02, phase 2.* (In Dutch).
[22]. Weytingh, K.R. et.al. 1999. “Improvement of the in situ bioremediation alternative by adding the concept of imbibition and drainage to existing theory. Phase 3: Contour analysis of mineral oil by soil parameters and chemical analysis. *NOBIS report 95-2-02, phase 3* (In Dutch).
[23]. Bonneur, A., et.al. 1998. “In situ remediation of layered and poorly permeable soils. *NOBIS report 95-1-16.* (In Dutch)
[24] Hennink, M., et.al. 1999. “In situ remediation of clay soils. *NOBIS report 95-1-05* (In Dutch).
[25] Heimovaara, T.J., 1999. “Development of an optical sensor for BTEX and chlorinated ethylenes. *NOBIS report 97-1-09, in press.*

AUTHOR INDEX

This index contains names, affiliations, and volume/page citations for all authors who contributed to the eight-volume proceedings of the Fifth International In Situ and On-Site Bioremediation Symposium (San Diego, California, April 19–22, 1999). Ordering information is provided on the back cover of this book.

The citations reference the eight volumes as follows:

5(1): Alleman, B.C., and A. Leeson (Eds.), *Natural Attenuation of Chlorinated Solvents, Petroleum Hydrocarbons, and Other Organic Compounds*. Battelle Press, Columbus, OH, 1999. 402 pp.

5(2): Leeson, A., and B.C. Alleman (Eds.), *Engineered Approaches for In Situ Bioremediation of Chlorinated Solvent Contamination*. Battelle Press, Columbus, OH, 1999. 336 pp.

5(3): Alleman, B.C., and A. Leeson (Eds.), *In Situ Bioremediation of Petroleum Hydrocarbon and Other Organic Compounds*. Battelle Press, Columbus, OH, 1999. 588 pp.

5(4): Leeson, A., and B.C. Alleman (Eds.), *Bioremediation of Metals and Inorganic Compounds*. Battelle Press, Columbus, OH, 1999. 190 pp.

5(5): Alleman, B.C., and A. Leeson (Eds.), *Bioreactor and Ex Situ Biological Treatment Technologies*. Battelle Press, Columbus, OH, 1999. 256 pp.

5(6): Leeson, A., and B.C. Alleman (Eds.), *Phytoremediation and Innovative Strategies for Specialized Remedial Applications*. Battelle Press, Columbus, OH, 1999. 340 pp.

5(7): Alleman, B.C., and A. Leeson (Eds.), *Bioremediation of Nitroaromatic and Haloaromatic Compounds*. Battelle Press, Columbus, OH, 1999. 302 pp.

5(8): Leeson, A., and B.C. Alleman (Eds.), *Bioremediation Technologies for Polycyclic Aromatic Hydrocarbon Compounds*. Battelle Press, Columbus, OH, 1999. 358 pp.

KEYWORD INDEX

This index contains keyword terms assigned to the articles in the eight-volume proceedings of the Fifth International In Situ and On-Site Bioremediation Symposium (San Diego, California, April 19-22, 1999). Ordering information is provided on the back cover of this book.

In assigning the terms that appear in this index, no attempt was made to reference all subjects addressed. Instead, terms were assigned to each article to reflect the primary topics covered by that article. Authors' suggestions were taken into consideration and expanded or revised as necessary. The citations reference the eight volumes as follows:

5(1): Alleman, B.C., and A. Leeson (Eds.), *Natural Attenuation of Chlorinated Solvents, Petroleum Hydrocarbons, and Other Organic Compounds*. Battelle Press, Columbus, OH, 1999. 402 pp.

5(2): Leeson, A., and B.C. Alleman (Eds.), *Engineered Approaches for In Situ Bioremediation of Chlorinated Solvent Contamination*. Battelle Press, Columbus, OH, 1999. 336 pp.

5(3): Alleman, B.C., and A. Leeson (Eds.), *In Situ Bioremediation of Petroleum Hydrocarbon and Other Organic Compounds*. Battelle Press, Columbus, OH, 1999. 588 pp.

5(4): Leeson, A., and B.C. Alleman (Eds.), *Bioremediation of Metals and Inorganic Compounds*. Battelle Press, Columbus, OH, 1999. 190 pp.

5(5): Alleman, B.C., and A. Leeson (Eds.), *Bioreactor and Ex Situ Biological Treatment Technologies*. Battelle Press, Columbus, OH, 1999. 256 pp.

5(6): Leeson, A., and B.C. Alleman (Eds.), *Phytoremediation and Innovative Strategies for Specialized Remedial Applications*. Battelle Press, Columbus, OH, 1999. 340 pp.

5(7): Alleman, B.C., and A. Leeson (Eds.), *Bioremediation of Nitroaromatic and Haloaromatic Compounds*. Battelle Press, Columbus, OH, 1999. 302 pp.

5(8): Leeson, A., and B.C. Alleman (Eds.), *Bioremediation Technologies for Polycyclic Aromatic Hydrocarbon Compounds*. Battelle Press, Columbus, OH, 1999. 358 pp.

A

B

C

D

E

J

K

L

M

N

O

P

Q

R

S

T

U

V

W

Z